Japanese Electronics Technology:
Enterprise and Innovation

Japanese Electronics Technology: Enterprise and Innovation

by Gene Gregory

Professor of International Business
Sophia University, Tokyo

JOHN WILEY & SONS
Chichester · New York · Brisbane · Toronto · Singapore

Copyright © 1985 and 1986 by Gene Gregory

Originally published by The Japan Times Ltd.

First edition: May 1985
Second edition: February 1986

Library of Congress Cataloguing in Publication Data:

Gregory, Gene.
  Japanese electronics technology, enterprise and
  innovation.

  Includes index.
  1. Electronic industries—Japan—Technological
  innovations.   I. Title.
HD9696.A3J334634 1986   338.4'7621381'0952   86-9189

ISBN 0 471 91038 4

British Library Cataloguing in Publication Data:

Gregory, Gene.
  Japanese electronics technology:
  enterprise and innovation.—2nd ed.
  1. Electronic industries—Japan
  I. Title
  338.4'7621381'0952   HD9696.A3J3

ISBN 0 471 91038 4

Printed and bound in Great Britain by
Anchor Brendon Limited, Tiptree, Essex

For Ann

Preface to the Second Edition

SOMETIME AROUND the end of 1985, NEC Corporation emerged as the world's largest producer of semiconductors. The leading Japanese microelectronics device maker moved up from the No. 3 position worldwide to the top rank, surpassing American firms that have held the lead since the early days of the semiconductor era. And not far behind, close on the heels of second and third ranking Motorola and Texas Instruments, were Hitachi and Toshiba. Three of the five largest semiconductor makers in the world are, at this juncture, Japanese, and these three are distinguished by their high level of integration of electronics production and the thrust of their research and development effort.

Indeed, this new configuration of the global microelectronics industry was to be further manifest at the 1986 international microelectronics devices conference in Los Angeles, where two Japanese firms were scheduled to introduce 4-megabit dynamic random access memory (DRAM) devices already test-manufactured using improved methods of cutting sub-micron geometries in silicon wafers. The race for the next generation of ulta-large scale integrated (ULSI) circuit markets is on and Japanese makers have grasped the technological lead.

If, as the indicators suggest, this transformation is permanent, or as permanent as anything can be in such a rapidly changing industry, it must be numbered among the most important industrial events of the past twenty-five years. It is the first time a non-Atlantic firm has taken the lead in a major high technology industry. It marks a further critical stage, following that of consumer products, in the epochal shift of the world electronics industry's epicenter to East Asia. And it sets the stage for yet another round of protectionism in the United States that could have far-reaching deleterious effects upon those sectors of high technology production that still enjoy a leading worldwide position. Should this happen, prospects are that production of computers and other advanced electronic equipment will follow the pattern of consumer electronics, moving to offshore locations where the supply of lower cost components can be assured.

This prospect makes it all the more important that the factors shaping the continuing Japanese electronics revolution be understood. The first edition of this book was admittedly a first and incomplete attempt to put the critical factors together in a single volume. It is, so far, the one attempt we have. And for this reason, it is even more timely, even more relevant, and even more needed than when it was first published last year.

To enhance its relevance and contribute to the understanding of this latest transformation in global electronics industry structures, a new chapter analyzing the impact of the VLSI project on the Japanese semiconductor manufacturing equipment industry has been included in this second edition. The processes by which original technologies were developed and rapidly diffused throughout the Japanese industry are described, with special focus on the dynamic institutional interactions that have driven technological advance of the industry and become a model of industrial cooperation.

As a result, industrial policymakers and corporate strategists in the electronics sector throughout the world now find it necessary to reassess their methods and manners to take into account fully the Japanese approach to innovation. To do this, it is first necessary to understand the essential features and achievements of this new force for change in electronics technology.

January 1986, Tokyo   G.G.

Preface

THE ASCENDANCY OF the Japanese electronics industry is cause for reflection about the accuracy of long-term technological forecasting. When, in 1957, I first began systematic in-depth observations of the Japanese electronics industry, no one expected that in less than three decades it would emerge at the forefront of advanced technologies then led by large-scale or science-based venture businesses in the United States. It was not even apparent in those early years that Japanese manufacturers would be able to gain significant shares of world markets for basic home electronic appliances such as radios and television receivers, which until then were the exclusive province of domestic manufacturers.

As inconceivable as it may now seem, prior to 1957 there had been very little international trade in electronic products. And, of course, except for the performance of camera makers in the early 1950s, there was little to suggest that Japanese industry had the capacity to manufacture high quality products embodying state-of-the-art technology and launch the kind of global effort required to alter this established order.

Yet portents of the future were present, had anyone fully understood their import. Indeed, the very reasons that brought me to Japan at that time were harbingers of things to come.

An early advocate of harnessing the forces of solid-state electronics for economic development in Southeast Asia, I had been commissioned to survey the global electronics industry to determine what aspects of this new technology might be useful to developing countries and what sources could be tapped for its transfer. The logic of this process seemed to be inherent to the technology. Transistors would make possible a radical change in consumer electronics equipment, freeing radios in particular from reliance on electric power sources not commonly available in rural Asian communities of the time. A potential revolution, therefore, loomed visibly on the horizon. And in that revolution were perceptible forces that could transform the entire process of industrial development.

This perception was far from universal, however. Not only had American industry not yet fully grasped the implications of solid-state electronics for consumer use, but the proposition that this technology had any important commercial or industrial implications for developing Asian markets was considered down-right absurd. At the time, no U.S. manufacturer was prepared to license upstart Southeast Asian entrepreneurs to assemble consumer electronic products even for local consumption, much less for export.

ix

Leading European electronic firms were then only beginning to con-
template seriously the potential of the transistor for consumer applications. Nor were they at all inclined to take seriously the notion that they
should be thinking in terms of licensing any part of this technology to the
former colonies. They had not the slightest inkling of the potential threat
solid-state electronics entailed for their traditional exclusive preserves in
theretofore dependent territories.

Significantly, not only had Japanese makers already advanced in the
development of transistorized radios, but they understood the logic of
licensing assembly of those products in Southeast Asian markets as the
most effective market entry strategy. It was Sony, then known as Tokyo
Tsushin Kogyo K.K., that licensed the first assembly of transistor radios
in Hong Kong, Manila and Saigon, achieving rapid inroads into local and
Commonwealth markets.

Two features of this seminal development were, in retrospect, to
become the critical ingredients of Japanese advance in electronics. the
first was an innovative application of new technologies to consumer
needs. And the second, equally important, has been the skillful use of
feedback information from the marketplace as the necessary ingredient of
effective corporate strategy. As a result, not only have products been
tailored to consumer needs, but product strategies have tended to be adjusted to shifting politico-economic realities.

The collection of articles presented in this volume, written over a
period of some seven years, chronicle and analyze developments in the
various major sectors of the Japanese electronics industry a quarter of a
century later, at the end of the 1970s and the first half of the 1980s, placing them in a global context. The various articles identify and examine
the forces at play at the time of writing, in the industry itself and in world
markets for its products — forces that have kinetically interacted to bring
about the transformation of global patterns of production and
technological advance.

Together this collection reflects a total reality that is greater than the
sum of its widely diverse parts described in the individual articles. Seen in
this holistic perspective, the development of the Japanese electronics industry appears as a dynamic interaction between the pull of global market
forces and the push of innovating enterprises invigorated by the exigencies of economic security and a compelling urge to excel.

The introduction describes this interaction in its global context before
moving to an analysis of technological change, competitive market forces
and the corporate strategies and structures that evolved to manage them.
The role of invention and innovation in the Japanese electronics industry
is examined in the second section of the book, in part to redress a com-

mon misunderstanding of the character and forces molding Japanese technological advance. It is curious that while innovation is commonly understood to be the vital ingredient of competitive advantage in high-technology industries, Japanese firms are assumed to have achieved their obvious prowess in world markets mainly through reliance on technology developed by the very companies they have out-performed.

Innovation in consumer electronics, semiconductors, computers, factory automation and telecommunications has been remakable for both the advances in basic technologies and the rapidity with which those advances are brought to market. The resultant growth in these sectors, described in separate sections, reflects major changes in electronic technology that mark a new era in which Japanese semiconductor manufacturers have moved ahead of the former American front-runners to take the leadership in world markets. At this critical turning point, the long supremacy of the West in high technology was eclipsed by the Risen Sun.

The final chapter of the book puts this development of Japanese industry in a broader context, tracing the influence of Japanese technology, trade and investments on the development of a regional industry. As I had foreseen in the 1950s, but to a far greater extent than I could have imagined, the electronics industries of East and Southeast Asia have set the pace of the most rapid economic development in history. Clearly, by the mid-1980s, East and Southeast Asia together constituted the epicenter of the global consumer electronics industry, backed by the world's most advanced microelectronics R&D and production facilities. South Korea and Taiwan are rapidly emerging as contributors to the innovative force of this regional complex, giving further impetus to Japanese creativity and providing an example which China clearly intends to emulate.

If, indeed, the 21st century turns out to be an "Asian Century," as some seers have predicted, it will very likely be the result in no small measure of the momentum of electronics technology and the peculiar capacity of East Asian production systems to manage that technology most efficiently.

It is my hope that the publication of this collection of writings, which reflect the time and circumstance of their original publication, will contribute to a better understanding of this phenomenon and lead to a broader pattern of cooperation between industries in Japan, East Asia and the rest of the world. That certainly is the promise of information technologies for which electronics provide the essential hardware.

If the cooperation I have received in the original work of launching modest electronic ventures in Southeast Asia, the continual inquiries that made these writings possible and finally their preparation for publication in this volume are any measure, that hope has a very good chance of fulfillment. I could not begin to acknowledge the debts incurred to a vast legion of executives in industry, engineers and scientists in research institutes and officials of government, not only in Japan, but throughout Asia as well as in Europe and the United States.

But I owe a particular expression of thanks to the Japan Foundation which awarded me a professional grant in 1977-1978, making much of the research for this writing possible. To editors who shepherded the manuscripts through the original publication stages, I am especially grateful. Special thanks are also due to the Institute of Comparative Culture of Sophia University, the *Far Eastern Economic Review, Euro-Asia Business Review, Scientific American, New Scientist, Management Today, Research Management, Communications International* and Electronics Association of Japan, all publishers who have supported me in my writing and graciously agreed to reprinting of the articles in this volume.

The final version of this book displays the editorial skills and patience with which my wife, Ann, read the earlier manuscripts and guided the arduous transmutation of articles that followed no pre-designed plan or order into the logical arrangement of a book. Without her understanding and careful reading of the text, its publication would not have been possible.

My acceptance of full responsibility for all conclusions, judgment and factual errors will absolve these good people of any fault whatsoever.

April 1985, Tokyo

Gene Gregory

Contents

INTRODUCTION.

Japan and the Global Electronics Revolution

INTRODUCTION

Japan and the Global Electronics Revolution

FOR THE PAST half century, the world electronics industry has been one of the fastest growing manufacturing sectors, with a compound rate of growth of over 10 percent per annum at constant prices since 1935. Propelled by new military applications during World War II, electronics technology has expanded and been rapidly diffused to account for a share of industrial output equal to that of automobiles in the most advanced countries. Microelectronics technology is rapidly overtaking steel as the foundation stone of industrial systems. As mechanical products and systems are replaced by electronics devices and a vast array of new products and processes is developed using microcomputers, electronic technology is radically transforming industrial activity in virtually all of its aspects.

During the remainder of this century — given foreseeable changes in data processing, telecommunications, industrial and office automation — electronic technology will be the main axis around which the industries of advanced and developing countries alike will be restructured.

Just as with development of the steam engine and steel manufacture, the rapid advances in electronic technology have produced swift changes in global patterns of productive activity and trade. In the early period of the electronic industry's development, the leading companies were British and German; in the United States, the largest manufacturer of radio equipment prior to World War I was a subsidiary of Marconi, the leading British firm in the industry which shared the world market with Telefunken. Both Marconi and Telefunken had established strong patent positions backed by well-organized research and development efforts which enabled them to stay in the forefront of technological advance and assimilate innovations in other countries. By contrast, the early American pioneer firms in the industry were generally poorly managed, resulting in frequent failures, and were restricted in their operations to the market in

Published originally as "The Global Electronics Revolution," Tokyo: *Sophia International Review*, Vol. 4, 1982.

3

the United States. Although major electrical equipment manufacturers such as General Electric and Westinghouse, and the leading telephone company, American Telephone and Telegraph, had developed firm positions in their home markets and all three maintained relatively strong research and development organizations, they played no significant role in the early international communications or radio equipment markets.

I.

With the development of public broadcasting in the 1920s, the center of gravity of the electronics industry began shifting perceptibly to the United States. As public broadcasting was developed by private enterprise with government encouragement, it expanded rapidly, creating a mass market for radios which enabled U.S. manufacturers to achieve scale advantages in manufacture and to take the lead in innovation, especially in the design and production of home radio receivers.

Likewise, in television, the invention and improvement of cathode-ray oscilloscopes took place mainly in Germany, England and Russia. But it was RCA which hurdled the major development problems, with heavy investments in R&D from 1924 to 1939, to launch the first commercial television broadcasts. As a result, RCA took the lead in television systems technologies, followed in the prewar period by Telefunken and EMI. This lead was widened during and after the war with RCA's introduction of color television systems fully 13 years ahead of European competitors. The struggle between rival French and German systems retarded development in European color television, strengthening the pre-eminence of the American industry.

But, despite its technological prowess, the American radio and television receiver industry remained largely domestic, exporting only a small percentage of total production. Excepting some relatively modest investments — mainly in European and Latin American markets — U.S. manufacturers were not inclined to overseas manufacturing of consumer electronic products. In fact, in the post-World War II period the only U.S. manufacturer of consumer electronics products with anything approaching global stratgegies and structures was I.T.T., which has been further distinguished by its notable absence from the U.S. marketplace. Until the end of the 1950s, European manufacturers were the leading exporters of radios, television sets and hi-fi equipment, and Philips of Holland was the world's sole electronics company with a global strategy of production and sales.

It was only with the development of the electronic computer that

American leadership of the electronics industry was firmly established. Although the first successful computer — the electromechanical Z3 and Z4 — had been built in Berlin between 1936 and 1941 by Konrad Zuse, official support for the development of an electronic computer was refused by the Third Reich government. With the end of the war, the lead in computer development passed to the United States, where Remington Rand emerged front-runner in the early 1950s, yielding to IBM with the latter's introduction of the first transistorized computer in 1955. IBM was, of course, already a very successful office-machine manufacturer with a strong worldwide sales organization and well-established production facilities in key European markets. Building on this base during the 1960s, IBM pursued a strategy of global rationalization of production, sales and services and was followed by over a dozen other American firms, making the United States the heartland of the world electronic capital equipment industry.

As electronic capital equipment manufacture consists largely of assembling components, and ultimately depends upon the ability to mobilize sizeable resources for R&D and market development, the ascendancy of the American computer industry was assured by two parellel developments: the invention of the transistor, and heavy U.S. government expenditures on military and space programs.

First generation computers had been built with electronic tubes (18,000 of them in the first ENIAC model completed in 1946), but here the U.S. industry had no particular advantage over European manufacturers. Philips, Telefunken, GEC and Marconi were as advanced as their American competitors in electronic tube technology and could very quickly imitate any important advances made in the United States. It was the invention of the transistor that changed the position of the U.S. industry completely. For a quarter of a century, after 1952, when the Bell basic patents in transistors were made available to the industry, almost all important inventions and innovations in semiconductor devices were made in the United States, with a lag of one to four years before manufacture began in Europe or Japan.

Although Remington Rand led the commercialization of the electronic computer with the first installation of UNIVAC I in 1951, it was IBM that pioneered the transistorized computer in 1955 and went on to make industry history with its highly successful 1401 transistorized series introduced in 1960. By January 1961, IBM's share of the U.S. computer market had reached 71 percent and, with the exception of the British and Japanese markets, was roughly the same in world markets for the succeeding decade.

American military and space programs contributed to the strength of the computer industry in several important ways. In the 1950s, government R&D contracts accounted for approximately 60 percent of total expenditures at IBM, and government purchases of data processing equipment for military and space programes provided a ready market for successive generations of computers. At the same time, these programs provided direct and indirect stimulus to the semiconductor industry, increasing its competitive advantage in world markets. U.S. computer makers were usually in a better position to take advantage of advanced semiconductor technology than were smaller European and Japanese firms throughout the decade of the 1960s and well into the 1970s. Thus, in 1965, when IBM introduced the first third generation computer, using hybrid integrated circuits produced in its own plant, the world computer leader also became the world's largest manufacturer of advanced microelectronic devices.

II.

The U.S. ascendancy remained inviolate throughout the second and third generations of computer and semiconductor technology. But by 1974 a new challenger had emerged when the leading Japanese computer manufacturers unveiled more advanced general purpose computers than the IBM 370 models then available. Again, in 1978, Fujitsu and Hitachi introduced what was then heralded as the world's largest and fastest computer system, and a year later when IBM announced its new fourth generation computers, all four leading Japanese mainframe manufacturers responded with competitive models within the short period of three months. In 1978, before the announcement of the fourth generation computers, Japanese exports of EDP equipment more than doubled that of the previous year, suggesting to many observers that the Japanese computer industry might be on the verge of repeating the earlier feats of automakers in world markets.

Then, in 1979, for the first time since it surpassed UNIVAC in the 1950s and went on to take the lion's share of the world business, IBM ceded leadership in a major market, dropping behind Fujitsu in Japan. A year later, IBM's announcement of its top-of-the-line H series central processing units looked very much like a defensive move, with the new machine's rated performance barely matching Hitachi processors already available for delivery. And by the time 1980 drew to a close, Nippon Electric Company had made its own bid for the large-scale computer market with its System 1000 ultra-large processor, which reputedly had twice the memory capacity of the largest computer then on the market.

Although U.S. computer manufacturers still retained the largest share of world markets, with the advent of the 1980s a fundamental change had taken place in the global structure of the industry. Technological leadership was now shared between the Japanese and American industries, and the Japanese had definitely gained some important competitive advantages in world markets. The Pacific Basin had become the bipolar center of the world's electronic data processing industry, relegating to history the era of American hegemony.

This was only part, albeit a vital part, of a much broader trend, however. Since the late 1950s, there has been a remarkable shift of the center of gravity of the entire electronics industry to the Pacific. By 1970, East Asia had become the epicenter of the world consumer electronics industry, with Japan in undisputed leadership; virtually all the revolutionary innovations in consumer electronic products since the first transistor radio have come from Japanese industry. The electronic calculator was developed in Japan, where leading manufacturers still remain in the vanguard of world innovation and production. Between them, Japan and Hong Kong share the lead in electronic watch production. Japanese manufacturers developed and successfully commercialized consumer video taperecorders. And if Asian output of passive components far exceeds that of North America or Western Europe, it is largely due to Japanese leadership in miniaturization and high quality standards.

In still other sub-sectors, Japan has moved to the forefront in either technology or production. Fully 50 percent of the world's robots were employed in Japan in 1979, and most of these were special purpose machines, custom-designed by the users or their subcontractors. By 1981, Japan's share of the world robot park had risen to 70 percent. Fujitsu Fanuc set the pace for world numerical controls technology throughout the 1970s. Half of the world's ground stations for satellite communications were built by a single Japanese telecommunications equipment maker during that same period. And Japanese technology was the state-of-the-art in critical fields such as optical fibers, pulse code modulation and digital exchanges.

As the center of consumer and industrial electronic equipment manufacture shifted to the Pacific Basin, with each successive generation of semiconductors the Japanese industry gained in prowess. Applications of transistors and integrated circuits by Japanese industry itself were at once more varied and made at a remarkably faster pace than by other advanced countries throughout the 1970s. Between them, computer makers and audiovideo manufacturers, which are often the same, account for 50 percent of the Japanese semiconductor demand. Diffusion has been especially rapid in those sectors where Japanese manufacturers hold major

INTRODUCTION

positions in world markets: cameras, watches, calculators, copiers and motor vehicles. And the rapid pace of application has been sustained in telecommunications, medical electronics, home appliances and toys.

In 1979, Japan became a net exporter of integrated circuits, marking a significant strengthening of Japanese manufacturers' position in world markets, especially for advanced memory devices. Fujitsu's lead in fielding 64K random access memories (RAM), ushering in the era of very large-scale integration (VLSI), signaled Japanese technological advantage in the coming big-volume segment of the market. And in 1980, Nippon Telephone and Telegraph, Fujitsu and NEC succeeded collectively in developing the world's first 128K RAM which was shortly followed by the announcement of 256K bit VLSI chips by Japanese manufacturers. In the fifth generation of semiconductor devices, technological leadership has shifted from the United States to Japan[1].

It was not here, at the forefront of applied semiconductor technology that Japanese manufacturers were making their greatest advances in world markets, however. By 1978, it was estimated that Japanese producers were supplying 35 percent of U.S. demand for 16K RAM, the most rapidly growing segment of the market. Whether as a result of under-estimation of market growth potential or an inability to expand production facilities during the recession of 1975-76, or both, the U.S. semiconductor industry was incapable of meeting home demand in the years that followed. Major U.S. equipment manufacturers such as NCR reported depressed profits for 1978 as a result of the shortage of the 16K RAM integrated circuits. IBM went into the open market for the first time to purchase substantial quantities of the device. And a number of leading U.S. semiconductor makers themselves had to purchase Japanese 16K RAMs in order to satisfy their customers[2].

Moreover, Japanese manufacturers were supplying their customers with devices that exceeded the quality and reliability of U.S. competitors, with rejection rates one-half to one-third those of U.S. memories[3]. Rather than meet Japanese competition for 16K RAM, U.S. producers began rapidly to evacuate the market, shifting production to 64K RAM, with higher density and higher value-added, thus leaving the market open to Japanese imports. It was more important for the U.S. industry to assure early market share advantage in the new 64K RAM, than wage a

(1) For further details see the author's article "Industries of the Future: Semi-conductors — Another Leap into the 1980s", Far Eastern Economic Review, 5 December 1980.
(2) Trade and Technology, Hearing before the Sub-Committee on International Finance, Committee on Banking, Housing and Urban Affairs, U.S. Senate, 96th Congress, Second Session, Part III, 15 January 1980.
(3) Business Week, 3 December 1979.

defensive effort to shore up positions in the market for devices that would soon enter the declining phase of the product life cycle.

But the contest over world markets for the new generation of integrated circuits was over almost before it had begun. By mid-1981, Japanese semiconductor manufacturers had won 70 percent of the market for 64K RAM microchips.

Significantly, the success of the Japanese makers in taking such a commanding lead is attributable mainly to their early mastery of complex 64K RAM production technology. Entrepreneurial companies of Silicon Valley, in their haste to execute complicated new designs while introducing the new manufacturing techniques required to produce the new generation of devices, failed to understand the complexity of the new technology. Firms which owed their success to the brilliance of individual designers, notable for their readiness to move from one maker to another, found themselves unsuited for the exigencies of advanced LSI and VLSI production.

The higher level of integration of the new devices made long-range planning, experience obtained through sustained team effort and a much higher level of manufacturing discipline of highest priority. In production which requires, above all, a stable work force, the high mobility of engineers and designers, for long considered vital to the technical leadership of American semiconductor makers, has become a major handicap.

This remarkable lead of Japanese industry in the forefront of advanced microelectronics technology strengthens its position in the increasingly fierce competition for world markets among the electronics industries of East and Southeast Asia. Far from being a simple extension of the Japanese industry, as is often asserted, the various Asian electronics industries have been increasing their shares of world markets for consumer electronic products at the expense of Japanese makers.

III.

From the beginning, the Hong Kong industry, which was the first to develop a capability for competing with Japan in the late 1950s, had a life of its own. Chinese entrepreneurs, including among them trained engineers who had come to Hong Kong from Shanghai as refugees after 1948, were quick to see the opportunities which electronics manufacture afforded outward-looking Hong Kong enterprise. Although Tokyo Tsushin Kogyo K.K. (renamed the Sony Corporation) did give the Hong Kong electronics industry an important early boost with technical assistance tied to subcontracting of production for the Commonwealth markets, the rapid growth of electronics production in Hong Kong was basically the achievement of a working alliance which developed between local entrepreneurs and large mass merchandisers in major North American and European markets.

It was this successful experience of the Hong Kong industry, as much as that of the Japanese, which provided the model for the development of electronic manufacture in Taiwan, South Korea, Singapore and Malaysia. And indeed, Hong Kong entepreneurs played a direct role in this development in Taiwan, Singapore and Malaysia before Japanese electronic firms became important investors in those countries.

The locational dynamics at work in the East Asian electronics industry are not widely understood and therefore deserve special attention. A good deal of this misunderstanding is due to the eschatology which serves to explain the demise of Western electronics industries faced with Asian competition. The conventional paradigm is that cheap Asian electronic manufactures produced by low cost labor and marketed by aggressive Japanese traders invade Western markets, competing unfairly with the higher cost home production. According to this view, the electronics industry has developed in those East Asian countries with an abundance of cheap labor, and largely because Japanese firms have been quick to transfer their labor-intensive production. Products of these industries in turn simply flow through Japanese trading channels to the markets of North America and Europe with all the advantages of preferential treatment accorded imports from developing countries.

There is, of course, as with all conventional wisdom, an element of truth in this paradigm. But the reality is at once more complex and fundamentally different. There are countries in Asia where labor is cheaper than in Korea, Taiwan, Hong Kong and Singapore, yet the electronics industry has not developed appreciably in those countries. Nor is that lack of development due to an absence of Japanese investments. The Japanese have invested in the electronics industries of Indonesia, the Philippines and Thailand, but those industries have yet to reach the take-off stage of development. The critical determinants of growth must therefore be factors other than low labor costs.

Essentially, the dynamic processes which have shaped the development of the Asian electronics industry involve all the various aspects of comparative advantage, but especially linkage with world markets and the ability to assimilate, adapt and manage technological change.

Contrary to common popular impression, in the early stages of this process the main foreign agents of change were not the large multinationals — American, Japanese or European. Nor were they Japanese trading companies, which played a secondary role as a channel for

Japanese consumer electronic products to foreign markets in the 1950s. Rather, the catalytic agent in this process has been the mass merchandisers which control major distribution channels in principal markets.

It is a little-understood historical fact that the Japanese consumer goods industry itself developed largely under the impetus of first American and then European mass merchandisers who subcontracted their production, not necessarily to the lowest cost producer, but to the most efficient in terms of quality, manufacturing cost, transfer costs, and delivery. As late as 1960, the largest producer of transistor radios in Japan (Toshiba) did not market any of its products under its own brand name in foreign markets; in North America they were sold under the house labels of Sears Roebuck, J.C. Penny and other large chain stores, the largest of which maintain buying offices in Tokyo. And throughout the various stages of growth of the Japanese industry, this linkage between mass merchandisers and their subcontractors in Japan has been an important factor in the dynamics of change. Even the move of Japanese color television production to the United States, particularly Sanyo's takeover of Warwick, was directly influenced by Sears, a long-time customer of both Sanyo and Warwick, as well as a major shareholder in the latter company.

But already in the late 1950s, department stores, large drug chains and supermarkets were shifting their procurement of the most price-competitive consumer electronic goods to Hong Kong, where the industry grew under the aegis of local entrepreneurs almost as rapidly as textile workers could be retrained as assemblers of electronic products. During the first half of the 1960s, the growth of demand was so great that Hong Kong firms established branches in Taiwan, where trained labor supply was still more abundant, to meet incoming orders.

IV.

The main reaction of the Japanese industry to Hong Kong competition was to constantly shift production to higher value-added products, which took the form first of more sophisticated, multifunction radios and radio-recorders and then television receivers, hi-fi and video recorders. Rapid innovation and automation of production were recognized as the prime imperatives for successfully competing with the new challengers from rapidly industrializing countries in the region. Since this competition threatened Japanese markets abroad, protection was clearly not the appropriate remedy.

The secondary strategy of major Japanese electronics manufacturers, and the primary strategy of smaller, specialized radio makers, was to

11

move labor-intensive production offshore — mainly to Taiwan, after 1966, when the Kaohsiung Export Processing Zone was established. Between 1966 and 1970, about 40 Japanese electronic firms invested in Taiwan for the manufacture of radios, taperecorders and components[4]. But by that time the indigenous Taiwan industry had already reached the point of take-off.

American manufacturers, who considered radio receivers as marginal and the task of meeting Asian competition with offshore production not worth the trouble, were content to follow the mass merchandisers in subcontracting production with Japanese or Hong Kong manufacturers. In only one or two instances did U.S. multinationals take the acquisition route, taking over Hong Kong firms directly, and these eventually ended in failure. Hong Kong entrepreneurs were more efficient on their own home ground.

For their part, European manufacturers were so absorbed with the developing technological gap between them and the United States in semiconductors and electronic capital goods that they failed to articulate any strategic response at all to the emergence of Asian competition in consumer electronics and passive components. Throughout the 1960s and 1970s they continued to lose world market share.

As a result, by 1978 the Hong Kong electronics industry had risen to share with Japan the leadership in world radio production. In that year, Hong Kong manufacturers produced twice as many radios as Japan, although the total value of Japanese exports of sophisticated multi-functional radios was twice that of Hong Kong's radio exports[5]. There had gradually emerged, since the inception of Hong Kong production in the late 1950s, an effective segmentation of the market and a division of labour in radio production between Japan and the developing countries of Asia; the upper segment of the market has been supplied from Japan, and the lower end from the developing countries, especially Hong Kong.

The dynamics of television production and sales have traced quite a different locational pattern from that of Asian radio production, however.

When the U.S. mass merchandisers began gaining market share with Japanese imports in the 1960s, American manufacturers were not prepared to abandon the field as they had in radio competition. Instead, leading U.S. manufacturers responded by relocating production offshore

---

(4) Yoshihara, Kunio. *Japanese Investments in Southeast Asia*. The University Press of Hawaii, Honolulu, 1978. p. 155.

(5) See *Yearbook of Industrial Statistics 1978*, United Nations, New York, 1979, p. 632; *Yearbook of International Trade Statistics 1978*, United Nations, New York, 1979, p. 447.

to wholly-owned subsidiaries established in Taiwan and then just over the border in Mexico. By 1973, virtually all the major U.S. television manufacturers were producing in Taiwan for the North American market.

Once again the Japanese industry responded with differentiated strategies. From 1971, after the first revaluation of the yen, to 1973, virtually all of Japanese black and white television production was moved to offshore Asian production sites. During these three years, Japanese manufacturers obtained approvals for 27 investments in Taiwan, 87 in South Korea, 14 in Malaysia and 13 in Singapore[6]. Color television production, with few exceptions, was retained in Japan, where heavy investments were made in automation and change to all-solid-state designs. Although some color television receivers were manufactured in other Asian countries, these were usually the lower-cost, small-screen models.

Significantly, the higher value-added obtainable in color television production through the introduction of the latest semiconductor technology was a decisive factor influencing Japanese locational strategy. By redesigning color television receivers using the most advanced integrated circuitry, eliminating as many components as possible and introducing automatic insertion of the remaining components on printed circuit boards, Japanese manufacturers were able to reduce assembly costs below those of Taiwan and Korea where labor-intensive processes were the mode and production scales did not allow heavy investments in continuing improvements in automation.

V.

As semiconductor prices declined, monochrome television receivers underwent the same design changes, and new manufacturing systems made it possible for production to be gradually returned to Japan from offshore sites. But these radical technological changes in television production opened another locational option which the Japanese industry was forced, by the rising protectionist tide and increased transport costs following higher fuel prices, to consider. Changes in marine transport economics rendered disadvantageous the cost of shipping bulky complete television receivers long distances by sea; all other things being equal, it became less costly to export modules to be assembled in cabinets and with cathode ray receiving tubes purchased in the marketplace. And for any firms which were tempted to offset this advantage with greater efficiency

---

(6) Yoshihara, *op. cit.*, p. 155.

of production in Japan, the threat of import controls and the actual imposition of orderly marketing arrangements (OMA) by the United States was sufficient inducement to opt for relocation of color television production, at least, to sites in major markets of North America and Western Europe. Manufacture in the United States was not only a means of offsetting restrictive measures on imports from Japan, but it effectively checked attempts of U.S. manufacturers to regain market share with products of offshore facilities, especially those located in Taiwan and Mexico.

Sony led the way, establishing successful production facilities in San Diego in the first half of the 1970s, and was followed by six other leading Japanese consumer electronics manufacturers in the last half of the decade. As a result, Japanese makers were, by 1978, in a position to benefit from highly automated Japanese production close to the customer and at the same time join forces with U.S. labor in requesting restrictions on imports from Taiwan, Korea and Mexico. Thus, some U.S. firms in Taiwan began phasing out production and, to protect and expand market share, at least two Taiwan firms established manufacturing facilities in the United States, even though exports to the American market continued to increase somewhat in 1979. Korean shipments of color television sets to the American market came under increasing pressure, and have been restricted, as have those of Japan and Taiwan, by a separate orderly marketing agreement with the United States. Following the pattern of Japanese and Taiwanese electronic equipment manufacturing, leading Korean electronic firms were reportedly among the bidders for Decca in 1979 and had by then begun exploring other acquisition possibilities in major markets. In 1981, Gold Star began construction of a color television assembly plant, with an initial capacity of 120,000 sets per year to be upgraded to 400,000 sets annually, in Huntsville, Alabama, in the United States.

Investments by Asian manufacturers in color television assembly facilities located in principal markets did not spell the end of production in Asia, however. Quite to the contrary. They have, in fact, served to spur the continued growth of that production in Japan, Taiwan and Korea where production records were still being broken in 1980. At the same time, during a period when labor-intensive production was losing its competitive cutting edge, transport costs were rising and protectionist forces were growing, Singapore emerged as one of the world's ten major exporters of television receivers. In just three years, 1975-1978, Singapore shipments of television receivers increased almost fourfold.

Meanwhile, the American television industry was undergoing a major transformation. Of the seventeen manufacturers which constituted the industry in 1970, only five were still producing and under American

14

ownership in 1981. Long-established consumer electronics manufacturers such as Emerson, Arvin, Setchel-Carlson, Packard Bell, and Admiral had abandoned the field entirely, while Motorola and Whirlpool (a major shareholder in Warwick) sold their television interest to leading Japanese manufacturers, and both Magnavox and Sylvania-Philco were acquired by Philips of Holland. In addition to these acquisitions by Japanese and European multinationals, five more major Japanese television manufacturers established production facilities in the United States, four of these as a direct response to the OMA adopted as a result of American pressures on the Japanese government and industry.[7]

VI.

If the locational patterns in East Asian radio and television production have been quite different, largely because of differences in technology, the semiconductor industry has followed yet another, quite distinct, development path. Although Japanese firms got an early start in semiconductor manufacture in the mid-1950s, and Korea, Taiwan, Hong Kong and Singapore were the major markets for Japanese IC and discrete semiconductor exports, they did not play the leading role in the early diffusion of semiconductor technology in the region.

The diffusion of semiconductor production technology was largely independent of the commercial flow of components to the East Asian consumer goods manufacturers. Japanese investments in semiconductor manufacture in the region — three in Korea, two in Taiwan, and two in Malaysia — were established mainly for re-export, especially to Japan.[8] Offshore semiconductor production was begun by Japanese firms in the 1960s as a response to cost advantages obtained by U.S. makers through assembly in the area.

From the outset, the diffusion of semiconductor technology in East Asia had been largely effected by American firms which were competing ferociously for market share at home and abroad. Fairchild, a latecomer to the U.S. industry which had spun off from Schockley Laboratories in 1957 to introduce planar transistor technology in 1960, integrated circuits in 1961, and the MOS transistor in 1962, pioneered overseas assembly as part of its strategy to gain market share as rapidly as possible for maximum advantage from these new technologies. Hard-pressed for resources

---

(7) *Colour Television Receivers and Subassemblies Thereof,* United States International Trade Commission, Washington, D.C., May 1980. p. A-17.
(8) *The Japanese Semiconductor Industry 1980,*BA Asia Ltd., Hong Kong, 1980. p. 158.

to exploit all three major innovations simultaneously, Fairchild concentrated available resources on mesa and planar transistors and integrated circuits. Assembly in Asia was seen as a means of gaining maximum learning curve and scale economies advantages, as well as the advantages of lower cost and efficient labor in those stages where this factor was a most critical one. By 1966, Fairchild, the then-leader in advanced semiconductor technology was thus able to capture 13 percent of the total U.S. semiconductor market and, together with Texas Instruments, Motorola, Signetics and Westinghouse, to control 80 per cent or more of the integrated circuit market.[9]

Like the consumer electronic industry, Asian semiconductor production outside Japan was begun in Hong Kong. In 1962, Fairchild established a plant to assemble semiconductors, taking advantage of provisions of the Tariff Schedule of the United States, items 806.30 and 807.00, under which imports of assembled articles are dutiable only to the extent of the value-added abroad. Demonstrating that it was as innovative in production strategies as in product development, the U.S. industry's newcomer shipped semiconductor components from California to its Hong Kong plant for assembly and re-export to the United States. And since savings on assembly costs were substantially greater than the cost of transportation in both directions, semiconductors could be priced more competitively for world markets.

Following its success in Hong Kong, Fairchild installed another facility in Korea several years later, paving the way for similar facilities of Signetics, Motorola, Control Data, Applied Magnetics and International Micro Electronic Company. Philco-Ford then set up operations in Taiwan, Sprague Electric in Hong Kong, and National Semiconductor and Texas Instruments in Singapore. Other U.S. semiconductor firms followed by establishing production facilities of this nature in Malaysia, the Philippines and Taiwan, as well as Brazil and Argentina.

The magnitude of U.S. investments in the East Asian region and their effects on locational patterns of world semiconductor production is of a major order. In 1978, Singapore, South Korea, Taiwan, Malaysia and Hong Kong were numbered among the world's top ten exporters of semiconductors. In that year, U.S. imports from offshore semiconductor assembly plants amounted to US$883 million, most of which was shipped from East Asia. Imports by U.S. semiconductor makers from their offshore facilities accounted for fully 76 percent of total U.S. semiconductor imports during the same year.[10] Total value-added in overseas assembly

(9) Tilton, John E. *International Diffusion of Technology: A Case of Semiconductors*. The Brookings Institution, Washington, D.C., 1971. Chapters 3 and 4.

was only 39 percent, however, down from 57 percent in 1974 due to the increasing value of those parts of the complete IC device produced in the United States.

But this is not the full measure of output and shipments from semiconductor manufacturing facilities in East Asia. Total exports from Singapore, Malaysia, Hong Kong and South Korea in 1978 amounted to almost US$2,000 million, and if Taiwan, the Philippines and Thailand are added, the amount exceeded the total semiconductor exports of the United States. Or, put in global perspective, the semiconductor manufacturing complex in the developing countries of East Asia, taken together, was the leading supplier to world markets.[11]

VII.

Having attained this position largely by encouraging foreign investment in assembly operations, East Asian governments then began planning for the next stage of growth. Faced with rising production costs and the uncertainties inherent in substantial reliance on multinational enterprises for supplies of components, semi-finished materials, technology and market, industrial policymakers in East Asia adopted programs for developing integrated national semiconductor industries as the foundation for the rapid promotion of electronic capital goods manufacture and the consolidation of consumer goods production.

Abandoning its role as an offshore assembly area, South Korea set out in 1979 to become the next major electronics capital goods manufacturing country — to produce both advanced integrated circuits and computing systems. Gold Star and Taihan Electric Wire, both broad-based conglomerates with global sales networks, set the course with two new semiconductor plants at a new 2,000-acre electronics industrial park in Gumi, 284 km southeast of Seoul. To stimulate and guide the Korean electronics industry in its new directions, the government created a novel catalyst in the form of the Korean Institute of Electronics Technology (KIET), with initial funding of approximately US$60 million in seed money from the World Bank and the Korean government. The success of this plan, a U.S. semiconductor maker predicted, would mean that the

---

(10)   U.S. Census Bureau, Foreign Trade Data Printouts, cited in *Trade and Technology, Op. cit.*, p. 85.

(11)   *Yearbook of International Trade Statistics 1978, op. cit.*, p. 447; *Competitive Factors Influencing World Trade in Integrated Circuits*, U.S. International Trade Commission, Washington, D.C., 16 November 1979, p. 14.

combined output of integrated circuits at Gumi could eventually be as large as that of the world industry's leader, Texas Instruments.[12]

Similar moves were taken in Taiwan, where 2,000-hectare science-based Hsinchu Industrial Park was established in 1980 and a series of measures was adopted to spur the rapid development of information industries. Among the initial investments at Hsinchu were two for the production of semiconductor materials and components, by Sino-American Silicon Products, Inc. and Advanced Device Technology, Inc. Unlike past export-oriented investments in semiconductor assembly, these new ventures were intended to buttress the growth of manufacturers of computer and communications systems and take advantage of local demand of consumer goods makers.

As the East Asian electronics industry moves into more technologically concentrated product areas such as semiconductors, computers, digital and optical communications, and other industrial electronic products, major structural and locational changes have followed in a dynamic process of interaction. Labor-intensive manufacturing not susceptible to automation has shifted to new locations in Malaysia, Thailand, the Philippines and Indonesia. But the transformation from consumer electronic goods sub-contracting and semiconductor assembly operations to an inegrated electronics industry entails new resource requirements as well as global planning, manufacturing, sourcing of materials and marketing. New support industries, including intermediate materials and software, are necessary. Moreover, since domestic markets are characteristically limited in size, to attain the necessary level of production for scale and learning economies and for sustaining essential research and development, multinational strategies and structures are imperative.

Only a few electronic firms in Hong Kong, Taiwan and Korea, such as Atlas Electronics, Tatung, and Gold Star, have had much experience to date in multinational operations, and this experience has been limited to a few countries. But the preceptive examples of the Hong Kong shipping industry and Korean construction firms indicate just how deftly entrepreneurs in these countries are able to adapt to the competitive environment of multinational enterprise. As in the first stage of the development of the East Asian electronics industry, the prevailing appetite for change and competition which remains unsatisfied and undaunted throughout East Asia will be a decisive factor in the transition to manufacture and marketing of advanced electronic products.

---

(12) "South Korea Starts Own Silicon Valley", *Electronics*, 10 May 1979.

VIII.

As the first century of the electronic industry's development has shown, the continuing revolution in technology has been attended by an expanding and increasingly rapid diffusion of electronic production, a gradual but perceptible translocation of the manufacturing epicenter, and an accompanying shift in innovative activity. As electronic technology has spread, the potential for innovation has also increased and the tempo of innovation has accelerated.

At work in the locational dynamics of the electronic revolution has been a compound push-pull effect. The pull of gravitational forces attracts technology to the point of optimal productivity of both labor and capital, where the resistance to change is lowest, and required human and material resources for change are most abundant. At the same time, there is the push of inefficiency, institutional rigidity, resistance to change, and lack of resources or the ability to mobilize them for appropriate goals in those countries with the oldest electronics industries.

Market size, industrial organizational structure, and financial systems have been critical and often determining factors in the calculus of dynamic locational change. Those firms which have adopted and successfully manage global marketing strategies tend to flourish, whether they are located in North America, Europe or Asia; those which opt for national strategies can survive only with major infusions of public funds. High and increasing R&D, capital and marketing costs require efficient, flexible and rational financial systems; capital availability at competitive costs is vital to success in all key sectors of the industry. Although physical plant and other capital requirements of the industry vary widely between sectors and between stages of production within sectors, the critical association of electronics technology and its diffusion is manifest by a closer correlation of industry growth rates with capital availability and costs than with relative cost of labor.

The capital intensivity of research and development, production and marketing have raised the threshold of survival in the mainstream of the industry to levels that can be attained only through global rationalization and integration of activities. As a result, economies of scale and learning tend to be critical for major electronics components, consumer goods and industrial equipment manufacture.

At the same time, it has become apparent that within this global edifice there are many niches for specialty manufacturers, operating on a smaller scale with much lower resource requirements.

In such a competitive and fluid environment, earlier national industrial structures are no longer sustainable. In such a global marketplace, old national oligopolies tend to become largely irrelevant. As electronics applications rapidly become more pervasive, the number of new entrants into the industry proliferate at an astonishing rate.

Firms and governments around the world, in advanced and developing countries alike, are attempting to emulate the success of U.S., Japanese and East Asian industry leaders. European schemes, for example, include relatively high protective tariffs, direct import controls, preferential procurement, government seed money to spur research and develop new markets, direct government investment in new enterprises established to capture a part of the market for advanced electronic products, joint ventures between domestic and overseas corporations, direct investments in home industries by foreign multinationals, acquisition of U.S. companies for instant access to new technology and markets, and investments in high-growth areas such as East Asia. To the extent that these European initiatives succeed, and the ambitions of countries such as Brazil and Mexico are realized, a new multipolar constellation of the global electronic industry will emerge.

The rapid growth and diffusion of the global electronics industry clearly has increased the cost of maintaining any status quo, old or new, and created an environment conducive to rapid technological and industrial structural change. The ascendancy of the East Asian electronics industry and the shift of the world industry's center of gravity to the Pacific Basin are important, not because these developments threaten Asian dominance of the world industry, but rather because they mean the end of a period when one country or group of countries is capable of such dominance. The rapid development of the Japanese electronics industry, plus the diffusion of electronic technology to Hong Kong, Taiwan, Korea, Singapore and Malaysia have forced firms elsewhere to develop new technology quickly to maintain markets at home or abroad, adopt global strategies which optimize returns on investments in R&D, readjust product lines and rationalize production systems in response to changing technology and comparative advantage, or accept the alternative of abandoning the field entirely.

The essential ingredient in this intensively competitive environment is global vision. The principal certainties are continuing technological change spurred by the emergence of new centers of invention and innovation, a continuing extension of the electronics industry to new countries whose industries must excel in world markets to succeed, and perforce, the imperative of global vision by both industrial strategists and public policymakers. It is no longer tenable to assume a static relationship between "advanced industrial countries", "new or semi-industrial countries", and "less developed countries." The arcane notion that technological genius and industrial prowess are the birthright of a select group of countries has little relevance to the real world of the electronics age. New entrants must be anticipated. National adversary attitudes which attempt to resist or slow the pace of change will inevitably prove counter-productive. Successful competition in this new era is not inherently a zero-sums game between winners and losers in a technological race, but is essentially a cooperative game in which nations and firms eventually benefit from all new technological progress. Moreover, those nations and firms who play the game cooperatively are most likely to reap the rewards of potential synergy which derive therefrom.

PART I.

The Japanese Electronics Industry

CHAPTER 1

The Making of a Revolution

THE ELECTRONICS REVOLUTION, like the industrial revolution which was its precursor, is working an epochal transformation in economic structures that entails fundamental revisions in behavioural patterns of enterprises, nation states and international institutions.

The industrial revolution was based on a profligate use of mechanical energy, while electronic technology, which represents one of the greatest intellectual achievements of mankind, requires remarkably little energy. This fact alone is pregnant with some obvious — some not so obvious — consequences for institutional arrangements.

At least until the oil crisis of the 1970s, the state was an adequate, if far from perfect, instrument for macro-economic management; most industrial technologies could be developed effectively and the wealth created be more or less rationally allocated under the aegis of national institutions. The industrial revolution, dependent on the availability of cheap energy and materials, will be inhibited and eventually slowed by lack of these essential ingredients. In its stead, the electronics revolution, propelled by intellectual achievements, is destined for continuing growth as its knowledge base increases and broadens.

Inherently dependent on human intellect for its development and on social institutions for its management, electronic technology is essentially global in all its critical aspects. Its basic ingredients are at once potentially more universal and in fact more mobile, less susceptible to monopoly control and hence much less pervious to political restriction. Its rapid diffusion throughout the world in recent years is dramatic manifestation of this universality.

Essentially a large-scale technology requiring the mobilization of massive resources for development, electronics requires institutional arrangements which defy the limits of national boundaries. Costs of innovation are increasing exponentially with each new generation of elec-

Published originally as "Electronics Revolution: Success in Innovation is the Main Problem Ahead," *Far Eastern Economic Review*, 14 December 1979.

tronic devices and have reached the point where no single nation or firm can long sustain the mounting charges. Likewise, at the stages of production and consumption, national circumscriptions give way ultimately to the imperatives of global rationalization; although the basic technology is large-scale, global rationalization of production and consumption allows the smallest polities — witness Hong Kong and Singapore — to participate fruitfully and meaningfully in these vital processes.

It follows, therefore, that the electronics revolution differs from its predecessor in still another critical aspect. The industrial revolution has been, from the outset, largely an Atlantic affair and the fact that it was born and bred in Europe and America had far-reaching institutional and political consequences. Although the electronics revolution has it roots deep in those same fertile grounds of technological innovation, it has obtained an added dimension from the role which Japanese and other Asian electronics industries have come to play. Indeed, the dynamic thrust of some sectors of the electronics industry has shifted from the Atlantic to the Pacific, with Japan as the lodestar of a new industrial galaxy.

Some institutional consequences of these epochal traits of the electronics revolution — and especially the role of Japanese industrial prowess in its development — are now assuming recognizable patterns warranting special attention.

As early as 1957, Japanese industrial policymakers identified the electronics industry as a priority sector for development. At the time, few Japanese firms could be classified as electronic manufacturers; not one among them had distinguished itself as an exporter of electronic products. Total production of the industry, such as it was, had reached barely ¥60 billion (US$167 million) at the end of fiscal 1955 and growth rates had not reached the point of take-off.

The record since then is now legendary. The Japanese electronics industry has successively emerged as the world leader in the production of a broad range of home entertainment products, electronic organs, electronic watches, electronic calculators, numerical controls for machine tools, robots and virtually every component used in the various types of equipment. By the mid-1970s, Japanese output of computers, communications equipment, office machines and semiconductors was second only to the United States. Electronic medical equipment was emerging as a new Japanese speciality and an increasing number of products in the electronics field were the creatures of a newly discovered — or rediscovered — Japanese innovative genius.

Significantly, in recent years, when other industries were languishing in the doldrums of cyclical recession or structural depression, total production of the Japanese electronics industry continued to develop at an astonishing pace. The total value of its output, second only to that of the United States reached ¥6,380 billion in 1978 — or more than 100 times that of 1955 — and the industry had by then become one of Japan's major industrial sectors with its share of total industrial production on a par with the motorcar and steel industries.

Present parity is preliminary to tomorrow's pre-eminence, however. As electronic technology permeates one sector of the industrial system after another, it will soon surpass all traditional industries in size and diversity. Thus, during the recession years 1975-78, output of electronic products increased 50 percent compared with a 33.5 percent rise in motorcar production. Significantly, the motor industry itself entered the electronic age during this period.

But the full importance of the Japanese electronics industry — its part in transforming productive structures within Japan and its role in the world economy — is not adequately reflected in these indices of growth and permeation, however spectacular they may be. Nor do the impressive innovations, production shares, employment figures and export records attained by Japanese electronic manufacturers accurately depict the role the industry has come to play.

In 1974, when the Industrial Structures Council published its *Long Range Vision* of Japanese industry, the electronics sector was clearly specified as a key industry around which virtually the entire future production structure was to be developed. Plans were outlined for accelerating the pace of innovation, especially in basic components and computers, as well as for the systematic diffusion of core innovations and their derivatives through the full range of Japanese manufactured products and services. Automation became the *mot d'ordre* in all branches of Japanese industry and, in the electronics industry itself, Japanese firms proceeded to automate production to an extent unparalleled by any other electronics industry in the world.

In part because of this coherent and comprehensive medium- and long-term program to develop the electronics industry as the key sector for the projected knowledge-intensive industrial complex, Japan is now well on the way to becoming the heartland of the world electronics industry.

It is true, of course, that the United States remains by far the largest producer of electronic products in terms of total output. In 1975, the United States accounted for 41.2 percent of total world electronics production, compared to 15.6 percent by Japan — a difference which is considerably reduced if output is calculated on a per capita basis.

The United States nonetheless remains the leader, again in terms of volume output, in most important computer, communications equipment, avionics and integrated circuit production. In these sub-sectors American industry will most likely continue to play a leading role for the foreseeable future, with the Japanese industry moving out in front in some special product categories in each field.

But the division of labor which has been the pattern in the world industry of the last two decades — with U.S. pre-eminence unchallenged in industrial electronics and Japanese supremacy clearly established in consumer electronics — is now over. Although the Japanese industry retains undisputed world leadership in consumer electronics, by 1978 its production of industrial electronic products had surpassed the output of household goods for the first time in postwar history.

As in consumer electronics, Japanese manufacturers of industrial products have some impressive export achievements to their credit. Japanese numerical controls, for instance, have established themselves in the forefront of the technology race and are leaders in the world market. More than 50 percent of the robots in service around the globe are of Japanese origin. Japanese communications equipment manufacturers have attained global superiority in microwave telecommunications and have equipped more than half the world's earth stations for satellite communications.

This is just the beginning of things to come. In its final report, appropriately entitled *Facing the Future; Mastering the Probable and Managing the Unpredictable,* the Organisation for Economic Cooperation and Development (OECD) Interfutures group underscored the role of the electronics industry as the expanding sector of industry for the remainder of this century and noted especially that Japan has the strongest and most consistent program for industrial electronics of all advanced industrial countries.

Not only does that program already count among its major accomplishments substantial advances in automation, computers and communications equipment, but it also has to its credit the successful and timely development of VLSI (very large scale integration) technology required for future industrial electronics equipment generations. As the Interfutures report stresses: "The strength and consistency of the Japanese program lies essentially in the fact that work upstream in the components field produces very rapid results at the level of automated capital goods, in particular by the intensive use of numerical controls in machine tools and, more recently, the large-scale development of micro-processors (90 percent of the new models of numerically controlled machines are equipped with them). Japan's present leadership in the robot industry is a particularly significant illustration of this aspect."

The significance of this decisive quality of the Japanese program to achieve rapid diffusion of basic innovations and their derivatives is not confined to industrial electronics, however. While it is true that this phenomenon will inevitably produce, and indeed already has brought about, qualitative and quantitative increases in the productivity of all key Japanese industries, there are other features of the Japanese program for developing the electronics industry which give it even more far-reaching effects.

In the first place, the Japanese plan for the development of the electronics industry does not owe its distinctive strength and consistency only — or mainly — to the importance it gives to industrial electronics: rather its unique characteristic and principal source of prowess is its comprehensiveness, embracing the full gamut of electronic products and their wide-ranging applications. The Japanese industry is at once the most diversified and the most highly integrated electronics industry in the world and from these two features the industry has achieved a synergistic force that distinguishes it from all others.

What this means in practice is readily illustrated by a few examples. Without the signal success of the industry in developing hand-held and desktop electronic calculators, Japanese semiconductor makers, who supply LSIs for these calculators, would not have achieved the economies of scale and learning which constitute the firm foundation of their competitive strength today. Without that foundation, it would have been more difficult, if not impossible, to undertake the ambitious research and development (R&D) effort that has yielded timely advances in VLSI technology: without those advances, success in computer development as well as in other fields of industrial electronics would have been very unlikely.

Similarly, strength in TV production is yielding a wide range of advantages for the production of visual display terminals used in data processing and telecommunications. Integrated circuit applications in photography have strengthened the lead of Japanese camera manufacturers who have in turn diversified into the production of photocopy machines and facsimile equipment, which are important users of electronic devices.

Synergy in product development, production and marketing have found their organizational expression in highly integrated electrical/electronic equipment manufacturers.

As the Interfutures report correctly emphasises, competitive strength in the mainstream of the electronics industry is increasingly dependent upon the degree of integration of production within individual enterprises. The process of innovation is becoming more and more costly as the sustained basic research effort, necessary to stay abreast of technological development in the industry, steadily becomes both more intensive and extensive. Also, given the rapid pace of technological change in the industry, it is more essential than ever that the development of basic component technology be tied to the development of final products in which the components are ultimately used. These imperatives give added impetus to the trend in the industry towards large-scale organizations and vertical integration.

Furthermore, since much proprietary hardware and related software technology is contained in integrated circuits, electronic equipment manufacturers are now finding it imperative to integrate upstream into component manufacture. Production of integrated circuits for watches by Seiko set a pattern of integration that has since been followed in other sectors by Canon, Ricoh, Yamaha, Pioneer and Nippon Denso of the Toyota group.

Thus leading audio equipment manufacturers, optical equipment firms and watchmakers have developed their own in-house IC production capability. As a result, these sectors are marked by increasing concentration of production in larger firms. Small firms in these sectors can no longer keep pace with the rapid advance of capital-intensive technology, nor do they have the required resources to achieve the necessary degree of vertical integration to stay in the race.

To a very large degree, of course, this trend towards upstream integration and concentration among specialized equipment manufacturers has been an indispensable defensive measure against the already highly integrated electrical/electronic equipment giants: Hitachi, Toshiba, Matsushita, Sanyo and Sharp. Clearly, the main strength of the Japanese electronics industry is derived from the success with which these leading diversified makers have grasped the lead in semiconductor development and production. To assure a market for semiconductor devices, as color TV reached the limits of growth, these major appliance manufacturers have been further diversifying into the production of audio equipment, office machinery, watches and other products formerly the domain of smaller specialized firms.

This diversification contributes significantly to the speed of technological diffusion, economies of scale and learning and the reduction of risks, thus sustaining a substantially higher level of investment in R&D and automation than would otherwise be possible.

One further effect of this process has been the erosion of clear distinctions, born of electro-mechanical technology, between consumer and industrial products. The personal computer will before long be a common household appliance. Facsimile machines will be as ordinary to the household as the ubiquitous black telephone, and visual telecom-

munications terminals as universal as TV receivers: microprocessor applications are likely to be as numerous in home appliances as in factory
equipment.

As the distinction between industrial and consumer electronic products disappears, specialization in industrial and consumer electronic
manufacturing is no longer determined mainly by production technology
but to an increasing extent by marketing structures and strategies.
Whether or not an electronics equipment manufacturer produces a particular product will depend very largely upon its access to markets. Thus
the speed of diversification of consumer electronic manufacturers into
the field of office and industrial equipment will be governed by the
possibilities, costs and time requirements for developing new (global)
marketing structures.

Here, Japanese electronics firms in general enjoy an important advantage over those of most European companies, as well as all but a few
American giants. Almost all Japanese electronics companies have, as a
sine qua non of survival, adopted, and to a large extent mastered, global
strategies and structures.

In an age of multinational enterprises, this may not seem so surprising. Yet it is still true that in Europe there are only one or two integrated
electronics manufacturers with a global reach. Apart from Philips of
Holland and Siemens of Germany, other diversified and integrated electronic equipment manufacturers are essentially national, or at most European, in their marketing and production strategies and structures. In addition to these two giants, only a few specialized equipment manufacturers such as L.M. Ericsson of Sweden and Racal of Britain, have
developed strong positions in the world markets.

U.S. computer, telecommunications equipment and semiconductor
manufacturers are notable for their global strategies of course. But, quite
to the contrary, U.S. consumer electronics and passive component makers
have remained national in their scope of operations, manufacturing
abroad notably in a few low-wage countries to remain competitive in the
U.S. market where they are faced with severe competition of Japanese
and other Asian products.

However, as the Japanese industry has clearly demonstrated and
other Asian industries have learned, the consumer electronics sector is
essentially as global in its basic economic rationale as are the computer,
telecommunications and semiconductor sectors. Consumer electronics
manufacturers that have remained confined within their limited national
or regional markets have been unable to compete with more efficient
firms which rationalize production and markets globally. Most of these
national firms have floundered on the brink of bankruptcy for years,

vainly seeking their salvation in costly neo-protectionist measures, only to accept ultimately take-over bids by stronger foreign competitors organized for global operations.

In the United States, Magnavox was thus acquired by Philips, Motorola's TV television division by Matsushita, and Warwick and Fischer by Sanyo. Similarly in Germany, almost all consumer electronic manufacturers have been acquired by major Dutch, Japanese or French firms in quest of global structures. Now Britain's fractured industry is undergoing the same transformation. Having failed to develop their own global operations, they have no practical alternative to amalgamation into other worldwide networks.

Common wisdom regarding these developments is regrettably, if understandably, fashioned by national sentiments rather than by any profound understanding of the nature of the technological forces which make them necessary, and on the whole desirable. The advent of solid state microelectronics as the key technology of industrial systems has fundamentally transformed the nature of economic activity, revolutionizing production processes and patterns of consumption. Remarkably, even though aggregate economic growth has perceptively slowed in advanced countries, the pace of change in the ever-expanding electronics complex of industries remains more or less rapid. Those countries and those firms which are capable of adapting to this momentous change and of managing it will emerge as the leaders in the new electronics era.

What is important about the new structures fashioned by these forces is not the national origins of the new technology or firms that succeed in managing its development. Rather the critical factor is the extent to which firms and national economies participate in the application and diffusion of this technology. During the first two decades of systematic development of Japanese electronics, Japanese firms in the industry were almost totally dependent on outside sources of supply of basic technology. Yet they managed admirably to excel in world markets. Indeed, technological dependence was a major contributing factor in the formula of success.

Even more striking examples are found in Korea, Taiwan, Hong Kong and Singapore where local electronic firms have developed spectacularly without their own technology and even without their own marketing organizations. Yet their development has been based almost entirely on their participation in world markets.

The key to success in each of these cases was a global outlook, supplemented by a commitment to adapt both strategies and structures to the realities of world markets in order to optimize available resources. As markets changed, technology changed, new resources were accumulated

32

and experience gained, and the initial strategies and structures were altered accordingly to optimize new realities.

Optimal advantages in the context of this industrial system are not to be obtained through national technological independence or superiority, but through the development of global objectives pursued through what Dr. Michiyuki Uenohara, associate senior vice-president and general manager of Central Research Laboratories at Nippon Electric, likes to call "symbiotic cooperation."

Within an increasingly interdependent system, firms and nations are challenged to find and develop new modes of cooperation which will permit full exploitation of economies of scale and learning inherent to advanced technologies. Practically, this means an increasing internationalization of production processes based not only on product specialization but also on a fragmentation of processes which makes it possible for parts of the total process to be performed in those countries where they can be done most advantageously, by those firms which can do them best.

This phenomenon, already manifest in some sectors of the electronics industry, will tend to be generalized. Thus, more firms will follow the practice of American semiconductor manufacturers who produce a single component combining capital-intensive processes in advanced countries with labor-intensive processes in developing countries to obtain competitive prices for the end product sold in world markets. For the same reason, Japanese electronic equipment manufacturers combine production capabilities of their Japanese, offshore Asian and U.S. or European facilities in the production of a single product.

Ultimately, of course, the main stimulus for this internationalization of production is competition. Competition and symbiotic cooperation, rather than being opposites, are very much like two sides of the same coin.

Undoubtedly, as the Interfutures group predicts and recent experience clearly suggests, we are likely to see more intense competition between firms and countries to control the trend of the new international divisions of labor and thereby maximize the advantages of new technologies for their own productive systems. Of course, this competition will not be confined to the advanced industrial countries or the enterprises based there. Developing countries, some of which have already demonstrated impressive ability in integrating their production forces into global economic activities and structures, will play an increasing role in this competition. But the nature of electronics technology is such that this very competition must inevitably beget new patterns of cooperation.

Structure of Demand in the Electronics
Industry Worldwide, 1970–1985 (%)

		1970	1975	1985
Total demand	(1)	(56.8)	(91.3)	(204.8)
Consumption		20.7	19.0	17.4
Industry and private services		30.5	39.2	45.4
Administration	(2)	48.8	41.7	37.2
Total		100.0	100.0	100.0

(1) In billions current US dollars.
(2) Including military demand and communications systems.

Especially in consumer electronics, Japanese manufacturers have in the past played a key role in setting the mode of both competition and cooperation, and they are likely to continue to do so for the foreseeable future. But as they advance technologically and their weight in the global electronics industrial system increases, Japanese manufacturers will be called upon to devise new organizational approaches which, while assuring the benefits of competition, will achieve a higher level of cooperation in world markets, including, of course, in the Japanese market itself.

This will become steadily more important since, for a number of reasons in addition to those microeconomic advantages inherent in Japanese industrial organization, the Japanese electronics industry will continue to advance more rapidly than most others in advanced countries. Social institutions, cultural values, a pervasive recognition of economic dependence and geopolitical imperatives all conspire to assure a remarkably high structural adaptability of Japanese industries to the opportunities of technological change and contingent changes in comparative advantage in the marketplace.

Since industrial structural change is likely to be more rapid in Japan than in other advanced countries, where ossification of structures is induced by social trends towards oligarchies and the reduced capability of governments to manage economic and social policies, it follows that the focal point of innovation will shift gradually to Japan — especially in key technologies such as electronics.

There are clear indications that this shift is currently underway. While R&D investment in electronics in advanced industrial countries generally has been declining in recent years, it has been rising steadily in Japan. More important still, during this period an increasing share of R&D expenditures by other industries in Japan has been devoted to applications of electronics and electronic capital equipment in their respective sectors.

Although much remains to be done in the domain of basic research, the consensus necessary for innovation and technological progress is present in Japan to an extent it no longer exists in other advanced industrial countries. It is also noteworthy that the development of a broadening range of new electronic products as well as the achievements of key research efforts such as the joint industry-government VLSI project, belie the popular notion that the Japanese are natural imitators and not innovators. Of course, that theory never did suit the other commonly accepted typecasting which emphasizes the unique qualities of Japanese cultural values, art forms and social institutions — precisely those aspects of civilization in which pre-industrial innovative talents were most likely to find their pristine expression.

In all probability, the main problems which will confront the Japanese electronics industry in the future will derive from the success of its innovation rather than from any lack thereof.

Since the Japanese response to pressures for industrial structural adjustment generally will differ appreciably from that of other advanced industrial countries, applications of electronics technology throughout the industrial system — and hence the electronics industry in particular — are likely to progress faster in Japan than elsewhere. But other advanced countries, caught in the squeeze between these pressures and the rigidies of their industrial structures, will very likely respond to Japan's success, as they are at present, with new and improved forms of protectionism. Experience suggests that this response will be counterproductive, resulting in further divergence among advanced industrial countries, with consequences which could possibly destablize the world politico-economic order.

Japanese success in key industrial sectors such as electronics, which are inherently global in their organization and pervasive in their effects upon economic activity, will require new approaches to cooperation between developed countries and between them and developing countries not only to avoid the worst of all worlds, but to assure the best of them. Although this may seem to be primarily a Japanese concern, it is a concern that must be universal if suitable cooperative arrangements are to be found for the management of increasing global interdependence in the electronics industry and for the economy in all its aspects to assure the full benefits of the incipient electronics revolution to all mankind.

CHAPTER 2

Strategies and Structures

As LONG AGO as the mid-1950s, Japan's industrial policymakers targeted the 1980s as the decade when the information age would emerge in full bloom, with electronics as the basic technology that would power the economy into the 21st century. Right on schedule, in 1980 the Japanese electronics industry soared into orbit as the leading growth sector in a distinctly new stage of Japan's industrial revolution.

As became increasingly apparent during the 1970s, and was fully demonstrated by 1980 performance, the electronics industry is not only destined to play a leading role in sustaining Japan's continued economic growth but has also emerged as an important participant in the revitalization of economic structures around the globe. Overall, the Japanese electronics industry grew by 22.9 percent in 1980 as output reached ¥8.7 trillion (US$37 billion), twice the value of total production of the industry in 1975.

Despite world recessions that followed the oil crises of 1974 and 1978-79, Japan's output of electronic products continued to expand in all sectors at double-digit rates, demonstrating a remarkable resistance to downswings in the business cycle. This growth was sustained in the face of creeping protectionist measures in industrialized countries that forced major changes in the industry's export patterns. Moreover, increased home output reflects only indirectly the rapid extension of manufacturing to major overseas markets for consumer products and components, a development that over the past five years has entailed global reorganization of production and reallocation of resources on a massive scale.

In home production alone, however, throughout this period the Japanese electronics industry outpaced the American industry and in 1980 grew twice as fast in all sectors except computers, miscellaneous semiconductor devices and military electronics.

Japanese integrated circuit output rose 48.9 percent in 1980, more

First published as "Technology: Buzz, Tweet, Zap! Japan Wins," Far Eastern Economic Review, 21 August 1981.

than double the growth in United States production. Admittedly, the Japanese growth was from a smaller base, and more important, exports were concentrated largely in the high growth segments of the markets. But it was nonetheless remarkable that in 1980 Japanese semiconductor makers succeeded in capturing 40 percent of the world market for 16K random access memory (RAM) devices, and approximately 70 percent of the market for 64K RAM chips, using the latest very large scale integrated circuit (VLSI) technology.

In Japan itself, integrated circuit consumption was spurred by sharp increases in output of products using large quantities of advanced semiconductor devices. Video taperecorder (VTR) production rose 125 percent, robot output by 85 percent and medical electronic equipment by 22.2 percent, while mechanical systems were increasingly replaced by microcomputers in electrical appliances, cameras and vehicles. Contrary to predictions that consumer electronic production in Japan had peaked, total output of home entertainment equipment rose 28.1 percent in 1980. The takeoff of VTR sales was, to be sure, a major factor propelling the continued high growth of this sector. During 1980, 11 Japanese VTR producers turned out a total of 4.4 million units, more than twice as many as in 1979. Demand, which continued to grow at a high rate, is expected to reach 8.4 million units in 1981, making VTRs a ¥1 trillion business. Not only the major consumer electronic manufacturers but also camera and sewing machine makers have entered the VTR market or have announced their intention of doing so.

With this succession of new entrants, Japan's monthly production of VTRs will be increased to a million units by the end of 1981. At the same time, to avoid possible trade friction while expanding overseas sales, leading Japanese makers are planning joint production in several European countries. Japan Victor plans to join with AEG-Telefunken of West Germany, Thomson-Brandt of France and Thorn-EMI of Britain in a regional production scheme. Matsushita Electric Industrial has been negotiating with Robert Bosch to manufacture VTRs jointly in West Germany.

These moves to invest in overseas manufacturing early in the product lifecycle mark a radical new departure from past practices in the industry. During the 1960s and 1970s, Japanese consumer electronic equipment makers shifted production overseas only after governments had imposed restrictions on imports or production in Japan had lost its competitive advantage owing to aging technology and rising labor costs. In the case of VTRs, however, rapid growth of demand abroad makes it difficult to supply markets exclusively from Japan-based output and political realities preclude such a strategy.

Fig. 1   Labor Productivity in Japan's Industries

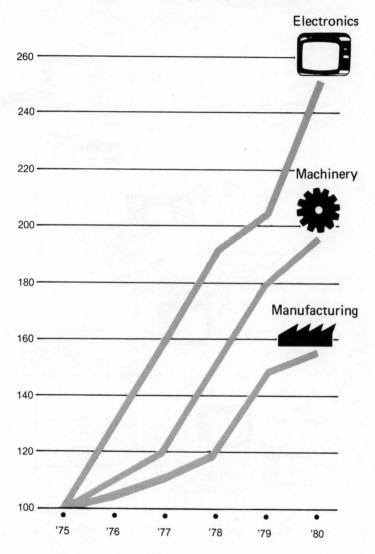

Fig. 2 How Increasing Original Technology Affects Japan's Market Share

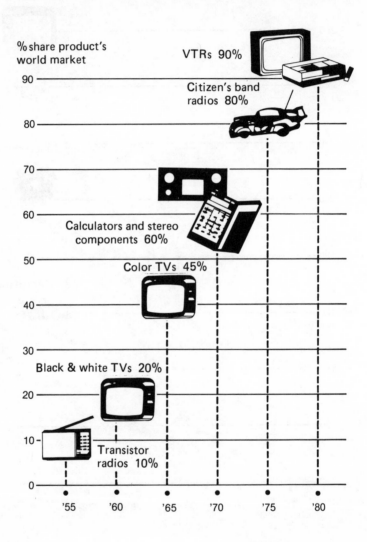

% share product's world market

VTRs 90%

Citizen's band radios 80%

Calculators and stereo components 60%

Color TVs 45%

Black & white TVs 20%

Transistor radios 10%

90 — 80 — 70 — 60 — 50 — 40 — 30 — 20 — 10 — 0

'55 '60 '65 '70 '75 '80

Thus, overseas production is not likely to impinge on domestic output for the foreseeable future. Rather, since VTRs use 3-5,000 electronic parts, many of which are not made abroad, increased production abroad will be an additional boon to semiconductor and other component producers in Japan.

But VTR production was by no means the sole stimulus to the consumer electronics sector in 1980. Taperecorder production rose 31.5 percent, in terms of value, over the previous year's level, due to new stereo recording and playback features, and the application of high value-added components such as microcomputers. For similar reasons, output of component stereo sets also rose 25.8 percent during 1980.

Even production of color TV sets rose by 18.1 percent in volume to a record-breaking 11.7 million units, despite protectionist measures in some key markets and apparent saturation of demand at home. In addition, output rose abroad — the industry now operates 39 overseas manufacturing or assembly facilities. In the United States alone, subsidiaries of Japanese manufacturers accounted for 30 percent of total color TV production, or roughly 3 million sets, bringing the total worldwide output of the industry to well over the 15-million mark. The lesson seems to be that protectionism abroad is good for the Japanese electronics industry. Indeed, the main effect of protectionist measures in the United States has been to give American consumers the privilege of paying more for TV receivers, to force the Japanese industry to be more efficient through extensive automation and to hasten the pace of rationalizing production on a global scale by major Japanese makers.

Protectionism has also served to speed the processes of technological change, giving to the Japanese consumer electronics sector a renewed vitality and catapulting production into a new stage of high growth. In the past, leading Japanese consumer electronic products — particularly transistor radios and TVs — were manufactured with technology originally imported from abroad and sold in world markets in direct competition with the original U.S. or European makers. Now the leading electronic products of Japanese industry are original, home-grown products of Japanese technology.

The immediate effect has been to give Japanese manufacturers a major lead in world markets at the outset, when new products are introduced, with competition confined largely among Japanese makers themselves. In the case of VTRs, Japanese makers currently account for 90-95 percent of total world output. So far there are few signs of anxiety abroad simply because there is hardly any production to be protected.

But an increased tempo of diversification and innovation has not been the only response of Japanese manufacturers of consumer electronic products to rising protectionism. A concerted effort has also been made to diversify markets, with resultant rapid increases in shipments to Asia, Latin America and Europe. By 1980, Asia was the largest market for Japanese color TVs, accounting for 33.2 percent of total exports, or 1.5 million units.

In parallel with the diversification of export markets, production underwent a major relocation. Manufacture of products losing international price competitiveness was moved to developing countries, from which both the Japanese and world markets are increasingly supplied. Beginning in 1974, manufacture of products that have caused trade disputes has been shifted to import-substitution production in the United States, regional-supply production in Western Europe and export-oriented production in developing countries.

The success of these multi-directional strategies is the stuff of which epochal industrial change is made. Japan has entered the information age as the epicenter of the world's consumer electronics industry and is better equipped than any other country to play a leading role in the diffusion of sophisticated information technologies from the factory to the office and home.

The timing could hardly be better. Since the development of VLSI technology, virtually every major Japanese consumer electronic equipment and computer manufacturer has introduced personal computers to the market with an explosive burst of manufacturing and marketing zeal reminiscent of the advent of electronic calculators a decade earlier.

NEC, the Japanese leader in microcomputers, expects to sell 120,000 personal computers in 1981, ten times its 1979 sales volume. Not all of these sales will be made in the domestic market, of course. NEC already exported 5,000 of its best selling PC8001 model to the U.S. market in the first half of 1981 and expected to sell 15,000 in that market before the end of that year.

Following rapidly on the heels of NEC, five more leading manufacturers — Canon, Hitachi, Oki, Sharp and Toshiba — entered the U.S. personal computer market in 1981. Industry sources estimate that the U.S. market for personal computers has already surpassed the US$2 billion level, with annual sales of 500,000 units. And by 1982, according to U.S. Department of Commerce predictions, Japanese manufacturers will account for 30-40 percent of that market. If this forecast proves accurate, sales by Japanese makers in the U.S. market during 1982 will skyrocket seven times over the expected 30,000 units to be shipped in 1981. And this is but a foretaste of expected expansion of this market for the remainder of the decade.

Nothing better illustrates the magnitude of the Japanese electronics industry's prowess. In 1979, Japanese personal computer makers were barely able to hold 20 percent of their own domestic market in the face of U.S. competition. By 1980, Japanese personal computer sales accounted for 80 percent of the home market. This year, 1981, Japanese manufacturers entered the U.S. market, and within a year's time they are expected to capture a major share of that market in the face of fierce competition from established suppliers.

The explanation for this dramatic shift in market share is clear. The potential market for personal computers is great and global, and Japanese electronic equipment makers are better positioned than others to take advantage of it. Not only is the mass production technology required for production of personal computers the strong suit of the Japanese electronics industry, but the extensive global marketing networks developed for the sale of calculators and other office equipment are tailor-made for a rapid take-off in the race for worldwide market share.

The industry's strength is not confined to mass production, however. In fact, since 1978 production of industrial electronic equipment has surpassed output of mass-produced consumer electronic products, reflecting the steadily growing demand at home for automation and rationalization of factory and office operations.

Since the oil crisis, the use of computers in manufacturing has increased sharply, resulting in a rapid proliferation of robots, automated production systems and unmanned warehouses. In 1980, which was designated by the Japan Industrial Robot Association (JIRA) as "Robot Diffusion Year One," production of robots increased 80 percent in value over 1979, adding 19,900 new units to Japan's burgeoning robot park.

JIRA and the Nomura Research Institute predict that Japan's robot market will continue to grow at least 20 percent annually, which will mean installations of 55,000 new units worth ¥290 billion in 1985, and as many as 94,000 units valued at ¥600 billion in 1990. This would mean that by 1990 Japanese industry would be installing each year a number of robots equal to that of the total world robot park in 1980.

Employing the latest developments in microcomputers, lower cost robots appeared on the Japanese market in 1981, opening a vast new potential for their use by middle-sized and small manufacturing companies. The cheapest robot available on the Japanese market previously sold for about ¥15 million, which is beyond the reach of most firms.

Now Sankyo Seiko Manufacturing, an electronics parts maker, has developed a compact assembly robot priced at ¥3.6 million, and Okamoto Seisakusho, a steel furniture maker, is offering a model selling at ¥3 million for use in loading metal sheets on press machines.

Although robots are the fastest growing segment of industrial electronics production, they still represent a small proportion of total output. Computers, of which robors are the progeny, accounted for 42.1 percent of 1980's production of industrial electronic equipment and 14.9 percent of the overall electronics industry.

Despite substantial growth in computer output in 1980, the 15.2 percent growth rate was considerably lower than that of the larger U.S. industry. Japanese computer output is still directed almost entirely towards the domestic market, which remains relatively dependent on imports. In 1980, computer imports rose to ¥16 billion, exceeding exports by 25 percent.

Indications are that this picture will change radically in the 1980s; however, as Japan's five major computer makers have all overtaken IBM in basic computer technology and have been steadily expanding their share of overseas markets with increasing investments and new links with foreign partners. As a result, exports of computers rose 31.1 percent in 1980 and are expected to grow at higher rates in subsequent years.

The competitive strength of Japan's computer industry is expected to increase further with the lead it has taken in the development of VLSI technology. During the 1960s and 1970s, the initial phase of Japan's information revolution, computers were used mainly in the industrial sector, confined largely to manufacturing purposes and operated mainly by experts. But this pattern is now undergoing a radical change. New generations of computers, using advanced semiconductor technology already available, will be cheaper, faster, smaller, easier-to-use machines operated by voice instructions. Improved versatility and greater accessibility will accelerate diffusion of computer use to broad new segments of Japanese society.

In June 1981, the information industry committee of the Industrial Structures Council described in considerable detail the forthcoming explosion of computer use in offices, education, medical care and homes. Anticipating this new demand, which is expected to rise at an annual rate of 40-60 percent in coming years to become a multitrillion yen business in 1990, computer builders have developed a variety of new office automation equipment including personal computers, small business computers and word processors.

Toshiba increased its sales of computers by 59.3 percent in 1980 by concentrating mainly on office computers. It expects to increase deliveries to the market by another 43 percent in 1981. Mitsubishi Electric also hopes to boost sales of office computers by 30 percent in 1981, and both NEC and Hitachi have introduced new low-priced desk top business computers in what promises to be the fastest growing segment of the business.

Japanese-language word processors are likely to be another product leader in the office automation market. In 1981 Toshiba plans to sell 2,100 word processors in the home market, up 102 percent from 1980. A rash of new products have been introduced by Canon, Hitachi, Matsushita, NEC, Nippon Univac, Sharp and Ricoh.

As computer production moves from custom and batch systems suitable for a narrow industrial market at home to a mass market with global dimensions, the Japanese computer industry will probably close the growth-rate gap with the U.S. industry by the end of the decade, erasing the last major advantage left to American electronic equipment manufacturers.

Although the American electronics industry is considerably larger than the Japanese industry in total output, as much as 45 percent of its production is geared to military requirements. The Japanese industry, on the other hand, is totally market-oriented, offering a far wider range of products — ranging from electronic games to super computers and robots. This broad mix of consumer, office, industrial and medical electronic products provides the Japanese industry with extraordinary flexibility in the market place and enables the industry to sustain relatively high research and development expenditure.

Vertical integration from semiconductors to end products enables Japanese manufacturers to obtain much faster return on investment in basic integrated circuit research than is possible for specialized merchant houses. since it allows rapid application of new devices in all products of the company for which they are suitable. Since returns on investment in new technology are generally greater and faster, the vertically integrated, highly diversified Japanese electronic manufacturers are therefore able to allocate massive resources to investments in research as well as in new plant and equipment.

Mitsubishi Electric regularly spends 12 percent of integrated circuit sales on research. During the recent wave of new VLSI development, some makers spent as much as 21 percent of current sales of semiconductors on research, and another 22 percent on new plant and equipment. This kind of capital commitment, even in a financial system famous for its capacity to sustain long-term investment, can be made only by widely diversified and vertically integrated manufacturers able to assure rapid diffusion of new technology once it has become operational.

At Matsushita Electric, 10,000 researchers and engineers working in 23 research laboratories, beginning with the most fundamental level of materials development and embracing all aspects of production, provide technological backup for 110 operating divisions producing all sorts of home appliances, industrial equipment, components and basic materials.

As a result, Matsushita Electric is Japan's largest patent holder, with more than 98,745 industrial property rights at home and abroad. To sustain this level of activity, Matsushita's investment in research in 1980 exceeded ¥100 billion, adding 8,671 patent and other industrial rights in 65 countries.

Unlike the electronics industry in the United States, where global rationalization is limited mainly to semiconductor and computer manufacturers, diversified and integrated Japanese electronic firms have developed worldwide manufacturing and sales networks. It is this global organization of highly diversified and integrated manufacturing and marketing operations that give to Japanese electronics firms the optimum market size and the versatility for the development of coming generations of information technology. As computers change from industrial products mainly used by specialists to mass produced consumer products, the global structures and strategies of Japanese electronic companies will yield a huge advantage over more specialized American manufacturers.

PART II.

Technology

CHAPTER 3

The Propensity for Innovation

ON APRIL 1, 1976, after long months of negotiations and preliminary organizational arrangements, five major Japanese electronic companies joined together in a cooperative research program with specific objectives of developing computer-oriented very large scale integration (VLSI) technologies. Four years later, at the end of March 1980, the program was officially terminated and the laboratory itself was liquidated the following June. By any measure a remarkable success, the VLSI cooperative research program marked the beginning of a new era in Japan's industrial development. Not only had the Japanese electronics industry, through this undertaking, closed the last remaining technological gap with its U.S. counterpart, but by 1980 the aggregate level of Japanese technology had surpassed that of all other advanced industrial countries.

Although the general level of Japanese technology had actually overtaken that in the United States by 1973, Japanese leadership in sectors such as consumer electronics, numerical controls, robotronics and telecommunications was not matched by basic semiconductor technology, which continued to lag behind American firms.[1] With the VLSI project this structural weakness was corrected. The 100 researchers who pooled their efforts, working cooperatively but in separate company teams, produced some 700-odd patented inventions in the short span of four years, enabling Japanese semiconductor manufacturers to take the lead in the development of the first 256K bit VLSIs and to move forward with preparations for production of microchips with a memory capacity of 1 megabit.

Parallel to the computer-oriented cooperative project, which was completed four years earlier than the U.S. counterpart (Table 1), a special communications-oriented VLSI development project involving some of the same companies had already succeeded in developing the world's first 128K read-only memory devices after two years of joint research under

First published as *The Japanese Propensity for Innovation: Electronics.* Tokyo: Sophia University, Institute of Comparative Culture, Business Series No. 86, 1982.

Table 1. National VLSI R&D Projects

Japan	Apr. 1976— Mar. 1980	¥30 bil.	VLSI Technology Research Association	Development of new generation of computers
United States	Sept. 1978— Aug. 1984	US$200 mil. (¥44 bil.)	Joint development by the Department of Defense and the defense industry	Development of VLSI for military use
Britain	1978—1983	£70 mil. (¥32 bil.)	INMOS (national project company), etc.	Development of 64K bit memory chips
France	1978—1982	Fr.F.600 mil. (¥30 bil.)	Thomson CSF group, etc.	n.a.
West Germany	1979—1981	DM300 mil. (¥34.5 bil.)	Joint research of government agencies, Siemens, Valvo, etc.	n.a.

Note: Conversion rates: US$1 = Japanese Yen 220; £1 = ¥457; Fr.F.1 + ¥50; DM1 = ¥115.

Source: VLSI Technology Research Association; Nikkei Sangyo Shimbun.

the auspices of the Musashino Electrical Communication Laboratory of the Nippon Telegraph and Telephone Public Corporation (NTT). For consumer electronics applications, moreover, leading equipment manufacturers such as Matsushita Electric, Sanyo and the Sony Corporation conducted independent research projects to develop superchips for their requirements.

I. Western Perception, Japanese Reality

The overtaking of U.S. semiconductor manufacturers in VLSI technology, more than any other single event, tended to focus attention on a fundamental change in technological leadership. Popular perceptions

---

(1) Especially informative are the findings reported by Dale W. Jorgenson and Mieko Nishimizu, "U.S. and Japanese Economic Growth, 1952-1974: An International Comparison," *The Economic Journal*, Volume 88, No. 352, December 1978; and Mieko Nishimizu, "Technological Superiority: A Milestone in the Postwar Japanese Growth," in Leon Hollerman, *Japan and the United States: Economic and Political Adversaries*. Boulder, Colorado: Westview Press, 1980.

abroad, which cast Japanese industry in the role of imitators, clearly were outdated. But industry leaders, economists and public-policy makers in the United States and Europe had only begun to appreciate the apparent fact that Japanese competitive power in sector after sector had been attained, not through low labor costs or unorthodox trade practices, but as a result of technological prowess.

If this appreciation was slow in coming it was at least in part because the available statistics seemed to buttress popular perceptions. In 1968, Japanese expenditures on research and development totaled at only 1.8 percent of GNP, slightly more than one-half the rate of expenditure in the United States and far below the absolute outlays of U.S. industry and government combined. Although the share of U.S. expenditures on R&D declined as a percentage of GNP throughout the first half of the 1970s, total outlays of more than US$24 billion were still almost several times that of Japan in 1975 (Figs. 1 and 2).

As a percentage of sales of manufactured products, R&D expenditures of Japanese industry as late as 1977 amounted to only 1.6 percent. When compared with the average 3.1 percent spent by U.S. industry, and 3.3 percent spent by West German manufacturers, Japanese outlays were indeed modest.

Nor was the performance of the Japanese government, in its share of overall R&D expenditures, any more impressive. Indeed, public outlays on research and development constituted a much lower share than in other industrial countries, seemingly compounding the lag represented by the lower share of total R&D expenditures as a percentage of GNP.[2] In 1978, government expenditures accounted for only 28 percent of total outlays for R&D in Japan, compared with 57.5 percent in France, 50.4 percent in the United States and 46.7 percent in West Germany. (Table 2)

Seen in normal ethnocentric perspective, superior genius for invention in Western industrial countries was apparently buttressed by superior resources.

But the advance of Japanese electronic manufacturers in VLSI technology shattered what remained of this time-worn illusion. Clearly, these measures of innovation are limited by the difficulty of assessing performance by inputs, as well as by the problem of deriving qualitative conclusions from quantitative averages. What counts in innovation, as in final production through-put, is obviously results obtained. And, as preponderant experience suggests, the results obtained will be largely in-

(2) Science and Technology Agency, *Indicators of Science and Technology: 1980,* Tokyo: June 1980.

Fig. 1   R&D Expenditure in Leading Countries

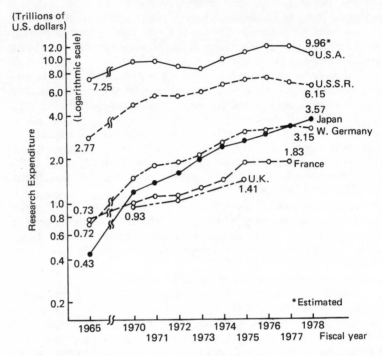

Source: Science and Technology Agency, 1979 White Paper, Tokyo: 1980.

fluenced by the allocation of resources among industries and between the public and private sectors.

Public Expenditure

In the first place, contrary to much conventional wisdom of the 1960s and 1970s, a large share of public expenditures in total outlays for R&D is not necessarily a sign of strength. In general, behavior clearly shows, this is an area where private firms are far more adept than government agencies. Efficiency in the use of available resources for R&D has undoubtedly been largely served in post-war Japan by the reluctance of the government to get involved in R&D — especially in the latter stages of development work. "As a matter of principle," the Economic Planning Agency pointedly noted in its annual survey for 1978-1979,

Fig. 2   Trends in Real Industry-financed Expenditure on R&D of Business in Selected OECD Countries

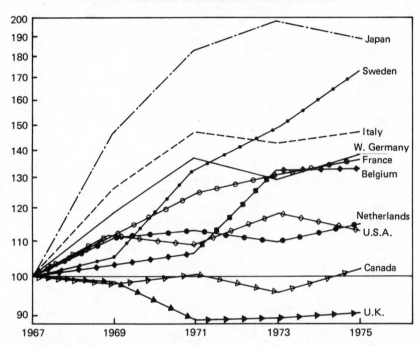

Source: OECD

"technological innovation should be the job of innovators — namely, business themselves."[3]

In keeping with this principle, rather than for a lack of funds, Japanese government support for R&D remained conspicuously low through the 1960s and 1970s. Significantly, if the proportion of total R&D expenditures provided by government in Japan was smaller than in any of the other advanced industrial countries, it was minute for the electronics and electrical industries.[4]

In 1975, the year before the VLSI project became operational, the Japanese government's share of R&D funding for the electronic and elec-

<hr>

(3)   Economic Planning Agency, *Economic Survey of Japan 1978-1979,* Tokyo: The Japan Times, 1980, p. 137.
(4)   Science and Technology Agency, *White Paper on Science and Technology,* Tokyo: 1975.

Table 2. R&D Expenses in Leading Countries
(Percent of Gross National Income)

	Japan	U.S.A.	U.K.	Germany	France	U.S.S.R.
(Total R&D Expenses)						
1968	1.77%	3.10%	2.49%	2.19%	2.36%	3.69%
1969	1.81	3.00	2.44	2.27	2.18	3.82
1970	1.96	2.91	(2.30)	2.42	2.17	4.04
1971	2.05	2.76	—	2.66	2.16	4.26
1972	2.06	2.67	2.27	2.60	2.09	4.59
1973	2.09	2.58	—	2.48	1.99	4.65
1974	2.17	2.57	—	2.55	2.05	4.66
1975	2.11	2.58	2.29	2.71	2.05	4.82
1976	2.13	2.54	—	2.61	2.05	4.59
1977	2.11	2.53	—	2.68	—	4.52
(National Defense R&D Expenses/Total R&D Expenses)						
1968	1.13%	33.08%	21.75%	9.31%	18.18%	
1969	0.99	30.66	21.10	8.74	15.88	
1970	0.93	29.20	22.42	7.78	22.80	
1971	0.91	28.90	—	6.55	23.24	
1972	0.89	29.12	24.98	5.29	21.27	
1973	0.79	28.01	—	6.61	21.94	
1974	0.67	27.36	—	6.28	20.21	
1975	0.65	26.11	25.70	5.68	19.29	
1976	0.64	24.03	—	5.72	18.81	
1977	0.67	23.72	—	5.82	18.21	
(Government's Share in R&D Expenses)						
1968	28.2%	60.7%	51.1%	47.0%	69.1%	
1969	26.3	58.1	50.4	46.3	66.7	
1970	25.2	56.6	(51.7)	46.6	63.1	
1971	27.4	56.0	—	48.3	63.3	
1972	27.2	55.4	55.1	49.9	62.5	
1973	26.4	53.3	—	50.6	61.4	
1974	26.5	51.2	—	51.3	61.4	
1975	27.5	51.6	(51.7)	49.0	60.1	
1976	27.2	50.6	—	48.4	57.9	
1977	27.4	50.5	—	48.5	56.7	

Source: Science and Technology Agency, *1979 White Paper*, Tokyo: 1980.

trical machinery sector was a trivial 2 percent, while government funding accounted for 44 percent of R&D expenditures for the industry in the United Kingdom, 38 percent in the United States, and 30 percent in France.[5] In that year, the electrical and electronics industry in OECD

(5)   OECD, *Technical Change and Economic Policy*, Paris: 1980.

Table 3. Japan: R&D Expenditure by Private Industry, 1970–1979

(billions of yen)

	1970	1975	1979
All industries	823	1,685	2,665
Manufacturing	761	1,537	2,447
Food	21	46	68
Textiles	14	23	33
Pulp & Paper	6	14	14
Publishing & Printing	3	6	5
Chemicals	175	322	490
Petroleum & Coal	10	17	26
Rubber products	10	29	46
Ceramics	18	42	73
Iron & Steel	36	89	120
Non-ferrous metals	18	26	40
Fabricated metal	14	29	55
Machinery	72	116	186
Electrical machinery	97	167	312
Communications & Electronics apparatus	130	289	383
Motor vehicles	79	196	373
Other transport equipment	16	94	72
Precision machinery	19	36	77
Other machinery	22	53	75
Transport, Communications & Public Utilities	39	81	144

Source: Science and Technology Agency, *1979 White Paper*, Tokyo: 1980.

countries spent about US$13 billion on R&D and employed 225 thousand researchers, representing approximately 30 percent of all industrial R&D resources in those countries. Yet, remarkably, with virtually no government support from 1967 to 1975, Japanese industry alone was responsible for 40 percent of the overall increase in electronics R&D in all OECD countries combined.[6]

Private Expenditure

This and other evidence not only clearly shows that, if Japanese industry is out-innovating electronics industries elsewhere, the reason is certainly not to be found in government funding; it also suggests that the explanation has much to do with specialization and industrial structure (Table 3) Although total R&D expenditures of all industries in Japan amounted to only 1.6 percent of sales of manufactured products in the

(6)   OECD, *Trends in Industrial R&D in Selected Member Countries, 1967-1975*, Paris: 1979, p. 31-35.

mid-1970s, outlays of electronics manufacturers were several times this level and the investment in R&D by Japanese semiconductor manufacturers was approximately 22 percent of sales. And, at the same time, investments of semiconductor manufacturers in plant and equipment were running at the rate of another 21 percent of sales.

Thus, although expenditures on R&D by Japanese industry as a whole increased by 300 percent during the 1970s, those by key industries rose much more rapidly. Further indication of this trend is provided by the top ten firms on the Tokyo Stock Exchange, ranked by total turnover. In Fiscal Year 1969 these manufacturers spent an average of 4.8 percent of sales on R&D. By 1973, expenditures had risen to 6.2 percent and in 1979 they reached 7.8 percent.[7]

But this evidence still does not tell us much about the actual innovative performance of Japanese industry. It suggests some reasons for higher efficiency in R&D, without providing an adequate measure. The fact that there has been a substantial increase in R&D expenditures, or that those expenditures have been more rationally allocated, does not prove that there has been an increase in the rate of innovation.

Patent Applications

If an accurate measure of innovative activity is woefully unavailable, patent application data provide a ready and useful index of technological output. Serious scrutiny of patent registries is recommended in this case, if for no other reason than that the innovators, the electronic manufacturers themselves, attach no small importance to their own patent positions as an element of competitive power and ability to control markets. But it is also true that in the electronics industry there is an apparent correlation between patent applications and other manifestations of change in relative innovative performance.

From 1957, when Japanese industrial policymakers first targeted the electronics industry for rapid development, to 1976, total annual patent applications filed with the Japanese Patent Office increased fivefold, from 33,000 to 161,000.[8] Already in 1962, Japanese patent applications exceeded those in West Germany, and in 1968 they surpassed those in the United States. By 1976, applications filed by Japanese industry with the Patent Office were running 60 percent higher than those in the United States and were some 46 percent above those of West Germany.[9]

---

(7) *Japanese Industries: The Technologies and Potential Growth Areas in the 1980s,* Tokyo: The Nikko Research Center, Ltd., October 1980, p. 3.

(8) This data does not include utility model applications which in Japan were relatively higher than patent applications for inventions.

Fig. 3   Number of Patent Applications by Country

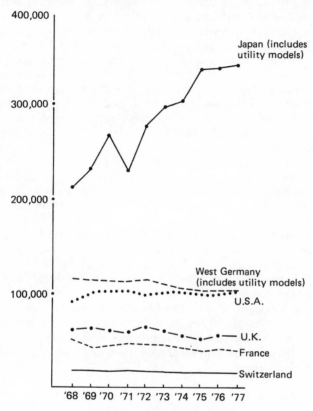

Source: Patent Agency, Annual Report 1979, Tokyo: 1980

During the 1970s, while patent applications in West Germany and the United States stagnated or declined, those in Japan rose by some 50 percent. (Fig. 3) And, in the United States itself, Japanese patent registrations increased nearly six times in the decade from 1966 to 1976, making Japanese nationals the largest single foreign national holders of U.S. patents. From 8,100 applicants in 1966 (less than one-half the number of

(9)   Ministère de l'Industrie, Institut National pour la Propriété Industrielle, *La Protection des Inventions en France et à l'Etranger,* Paris: 1964; see Graphiques I and 10 (sec). Japanese Patent Agency, *Annual Report: 1979,* Tokyo: 1980, reproduced in *Focus Japan,* July 1980, p. JS-A.

57

British applicants), the number of patents issued to Japanese applicants rose 300 percent to 25,100 in 1976 (more than twice the number of patents issued to British applicants).[10]

The resultant improvement in the propensity of Japanese corporations to develop new technologies is clearly reflected in the increasing number of patent rights owned by major Japanese companies.

The total number of patents registered by 566 leading Japanese manufacturing enterprises surveyed by the *Nihon Keizai Shimbun* in 1981 amounted to 392,000 at the end of fiscal 1980, up 35 percent from fiscal 1975. And, remarkably, the number of patent applications by the top ten of these companies in 1980 had risen to 167,000, or 30 percent more than just five years earlier. Still more significantly, six of the ten top patent-holders in 1980 were electronics equipment and component manufacturers, reflecting the increasing importance of this sector in Japan's industrial constellation and the leading role these Japanese firms are playing in the broadening global reservoir of electronics technology. (Table 4)

Note that while Britain boasted of ten times as many Nobel Prizes as Japan during this period, Japanese universities were graduating ten times as many engineers as those of the United Kingdom, and Japanese industry continued to increase outlays for both R&D and plant and equipment in stark contrast to the performance of British industry.

But patent statistics are at very best a crude measure of the rate of innovation. The average importance of the patents granted at one time and place may differ widely from the importance of those granted at another time and place. The proportion of total inventions that are patented may also vary significantly.

Trade in Technology

Direct evidence of a rise in the rate of innovation is better reflected in statistics of trade in technology. The reputation of Japanese industry for assimilation of foreign technology is, of course, reknown. Systematic technology transfer, if not invented by Japanese industry, was vastly improved upon by it. As a result, Japan was able to buy the bulk of the technology it needed to catch up with other advanced industry countries for approximately US$9 billion, or a fraction of the US$50 billion spent annually on R&D in the United States at the end of the 1970s.[11]

(10)   Science and Technology Agency, *Indicators of Science and Technology,* Tokyo: May 1981.
(11)   *Ibid.,* and James C. Abegglen and Etori Akio, "Japanese Technology Today," *Scientific American,* January 1981, p. J-12.

THE PROPENSITY FOR INNOVATION

Table 4. Japan: Patent Rights, 1980.

A. Patent Rights Owned by Major Companies (in cases)

Rank		FY 1980	FY 1975
1.	Hitachi	39,734	35,494
2.	Matsushita Electric Industries	33,411	20,819
3.	Toshiba	23,848	20,754
4.	Mitsubishi Electric	14,097	10,306
5.	Sumitomo Chemical	11,222	10,068
6.	Sanyo Electric	11,029	8,013
7.	Teijin	9,560	n.a.
8.	NEC	9,300	7,600
9.	Mitsubishi Heavy Industries	8,468	5,790
10.	Nissan Motor	7,309	3,116
11.	Toray	6,269	7,278
12.	Honda Motor	6,048	1,991
13.	Nippon Steel	6,000	5,200
14.	Hitachi S&E	5,754	1,932
15.	Asahi Chemical	5,300	4,000
16.	Fuji Photo Film	5,200	3,800
17.	Takeda Chemical Industries	4,810	4,565
18.	Fujitsu	4,484	4,565
19.	Furukawa Electric	4,436	2,718
20.	Ricoh	4,068	2,645

B. Patents Owned and Applied by Major Companies Listed on Stock Exchanges

		(in cases)
Patents owned	392,926	290,707
(Of which, established abroad)	(107,706)	(74,051)
Patent applications	157,879	100,626
(Of which, applied abroad)	(19,399)	(17,513)

Coverage: 566 firms for patents owned and 560 firms for applications.
Source: *Japan Economic Journal*, Tokyo: 30 June, 1981.

But Japan's technological trade is no longer a one-way affair. Although Japanese imports of technology continued to increase throughout the 1970s and the cumulative balance of royalty payments continued to exceed receipts, during the last half of the 1970s imports of technology rose by a modest 10.8 percent while exports of technology shot up 140 percent. The most sensitive index of Japan's actual technological trade position is the ratio of receipts to payments for new licenses. By this criterion, Japan emerged as a net exporter of technology in 1972 when receipts were 125.9 percent of payments under new contracts. Since then this ratio has continued to rise, and in 1977 reached an impressive 214.9 percent. [12] By 1980, according to a survey conducted by *Nihon Keizai Shimbun,* the 650 companies listed on the Tokyo Stock Exchange were earning US$425 million annually on technology exports while paying only US$350 million a year for imported technology.

Such a rapid rise in the technology trade balance is a reasonably pragmatic measure of the qualitative and quantitative improvement in Japanese innovation.

II. The Role of Government

The conclusions to be drawn from these data are significant. For considerably less money, Japan is getting far more in technology than other countries. During the last half of the 1970s, Japanese patent applications were three times those of West Germany, yet the expenditures of Japanese manufacturers as a percentage of sales were half those of West German industry. And, in addition, these results have been obtained at much lower cost to the public treasury.

To repeat a point made above, Japan's rapid technological progress and the low rate of government funding of R&D are not wholly coincidental or unrelated phenomena. Quite to the contrary. Technological division of labor, as it is practiced in Japan, is a deliberate policy consistent with broader governmental objectives to optimally manage scarce resources. Although this practice does not derive from any particular general theory of political economy, it is wholly congruent with the prevailing pattern of industrial policy-making in Japan. It is also consonant with the basic Confucian notion that each human institution, like each person, has its distinctive competence and that the general good is best served if each does what it does best. Let the innovators innovate — is the heart of the matter.

As it is the unique function and purpose of enterprise to create wealth through the management of technology and its requisite resources, efficiency and productivity are the criteria of performance. The more efficient the enterprise in the management of innovation, the greater the wealth creation and the lower the necessary tax take to meet the expanding need for those social services and infrastructural facilities which only government can provide. Thus, total taxes are only about 20 percent of GNP in Japan, compared with over 30 percent in the United

---

(12)   Science and Technology Agency, *Annual Report on the Introduction of Foreign Technology for Fiscal 1978,* Tokyo: November 1979.

States, and more than 40 percent in most West European countries.

Since it is not particularly sensible to take from corporations in the form of taxes in order to provide funding for R&D in the form of subsidies, other more efficient instruments are employed for channeling financial resources into higher value-added, higher-technology production. By so doing, the margin of error and waste in government is reduced and the stultifying tendency to over-regulate industry is abated.

General economic policies tend to be consistent with and serve to encourage innovation in industry.

Tax Policies

Tax policies, again, are a case in point. In order to encourage diffusion of semiconductor applications, bringing costs and prices down the learning curve as rapidly as possible, in the early years of color television the Japanese government allowed reductions in high commodity taxes on all color television sets incorporating microelectronic devices.

Since commodity tax reductions operated as a benefit to consumers, they had the initial effect of stimulating demand. But the indirect effects on production were far-reaching: competition among manufacturers to outdo each other in ''going solid-state'' enhanced the flow of innovations in both semiconductor devices and final equipment design; increased use of integrated circuits had the effect of reducing the number of electronic components from more than 1,200 in 1972, to fewer than half that number in 1977; and this in turn increased the possibilities of automatic component insertion, which by 1978 had risen to 80 percent of assembly and testing operations. The accumulated effect was just short of a major revolution in television receiver production technology, reducing the number of man-hours per set in assembly from 4.11 in 1973 to approximately 1.15 in 1978, while substantially improving the reliability of the end product.[13]

Depreciation, tax-free reserves and tax credits are devices used freely to promote investment in new product technology and production systems, and in many instances two or more of these incentives may be used to promote the same objective. Not only do these general provisions tend to strengthen the enterprise and enhance its propensity to innovate. Since they do not require that the government make judgments on the merits of particular investments, they achieve this objective with minimal

---

(13)   The Boston Consulting Group, *An International Comparative Study of Production Economics in Consumer Electronics,* London: November 1978. See also The Boston Consulting Group, *Alleged Export Subsidies to the Japanese Consumer Electronics Industry,* Tokyo: n.d.

political friction and bureaucratic obstruction. As industrial policy, in general, is founded on a clear and common understanding that new investments will tend to incorporate and stimulate technological advance, fiscal and monetary policies tend to be supportive of these objectives. Dangers of overall systems inefficiency are largely obviated by leaving decisions as to appropriateness of technology to the entrepreneur and the market place, keeping government agencies free to make and coordinate general policies.[14]

In this process, the distinctive competencies of both government and business enterprise are strengthened. Since both government and business have as their goal pushing forward the frontiers of technology to maximize wealth creation, there tends to be an identity of basic purposes which serves as a sound footing for cooperation and mutual support rather than conflict and adversarial postures.

In electronics, the Japanese government concentrated its attention on the development of industrial policies which would assure the long-term balanced development of the industry. Those policies set forth broad objectives to encourage production best suited to the economic realities of Japan, allowing the individual enterprises discretion in everything which depends on the circumstances of time and place, since only the enterprise concerned in each instance can develop the kind of feedback systems which are needed to manage technology and adapt strategies accordingly.[15]

Firms were able to use their knowledge and resources effectively, since they were able to count on the consistency and continuity of public policies and action. To provide firms which followed the general guidance of those policies with the necessary financial resources, Japanese authorities devised financial policies which have ensured a smooth and steady supply of funds at low cost through normal banking and capital market institutions. At the same time, education policies were adapted to the formation of human capital required for rapid growth through innovation in key industrial sectors.

Incentives were devised to induce even those firms which had been privileged suppliers to the military before World War II and remained principal suppliers of government public corporations in the postwar period, to diversify into activities and product lines where success depended on competitiveness in the open market. Leaders in advanced

(14)   Joseph A Pechman and Keimei Kaizuka, "Taxation," in Hugh Patrick and Henry Rosovsky (eds.), *Asia's New Giant — How the Japanese Economy Works*, Washington, D.C.: The Brookings Institution, 1976, pp. 317-382.
(15)   OECD, *The Industrial Policy of Japan*, Paris: 1972, pp. 11-31.

electronics, such as Fujitsu and NEC, which developed initially as suppliers of telecommunications equipment to the government, found the most interesting growth opportunities in more competitive lines of semiconductors, computers and home electronics. Strategies of diversification and integration in highly competitive product areas in turn enhanced the competitive advantage of their telecommunications equipment as well. During the 1970s, these two highly innovative companies emerged as major challengers of the most advanced and powerful firms in the world electronics industry, competing effectively for global markets with giants such as IBM, ITT and Texas Instruments.

III. The Propensity to Innovate

To stress that innovation in the Japanese electronics industry has not been due mainly to direct government support for R&D activities is not meant to imply that the role of government in the development of the electronics industry has not been important. The emphasis Japanese industrial policy has given the electronics industry since 1957 has itself clearly played a critical role in Japan's moving to technological parity in advanced electronics, and to its leadership in semiconductors, consumer electronics and other sectors of the industry.

But that role is at once more subtle and less interventionist than it is usually represented abroad. The technological imperative which has provided the underlying momentum of Japan's electronics revolution is not the creation of industrial policy-makers. Rather, industrial policy is the formal expression of that imperative, which springs deep within the economic condition of Japan, and from its surrounding international environment.

Cultural heritage has combined with economic reality in a complex matrix of motivations to imbue Japanese entrepreneurs and enterprises with an extremely high propensity to innovate; industrial policy deriving from those same motivations supplements this propensity with overall goal selection, guidance and appropriate assistance.

It would be commonplace to observe here that Japanese culture is the rich and distinctive expression of centuries of innovation, a heritage that long predates Japan's modern era if this fundamental reality were not so often overlooked in discussion of Japanese propensities for creativity. The Japanese capacity for imitation, so evident and important to the process of rapid modernization, could not have given rise to such abundant idiosyncrasy. Rather, historically, imitation and innovation have been the woof and warp of cultural and technological change in Japan, and their combined effect has been to hasten as well as enrich the process of change.

Ready and rapid acceptance of change, if the change met a felt need and if the practice or its product was recognized as superior to those prevailing, has been one of the remarkable traits of Japan since the first contacts with outside civilizations. The present enthusiasm with which workers accept continual improvement in electronic production technology, and the eagerness with which Japanese business and consumers grasp successive waves of new products and production systems, is a conditioned reflex bred of a culture which over time has demonstrated a continuing propensity for both imitation and innovation.

Technological assimilation, adaptation, improvement and innovation were recognized imperatives from the outset of Japan's decisive move to modernize and have been perceived as synonymous with modernization. Since Japan has no natural resources, the pace of modernization would perforce be determined in large measure by the extent to which Japanese through industrial enterprise could add value to imported raw materials, through application of the most appropriate and advanced technology, and then sell them on domestic and world markets. That technology was most appropriate which would maximize the return to Japanese production, with the lowest material, capital and labor inputs.

The extent of wealth creation in this process would, of course, determine the rapidity with which Japanese industry would catch up with those of the advanced industrial countries of the West. And for the first century of Japan's modern era, this was a primordial objective of nation and enterprise alike. In this catching-up process, world markets were as important as worldwide sourcing of materials, and the competitive marketplace rather than central planning, more especially in the post-World War II period, has been the ultimate arbiter of Japan's choice of technology as well as the tempo of technological change.

Now that Japan has, in the aggregate, attained technological parity with the West, the propensity of Japanese industry to innovate has by no means diminished. Having caught up, changes in Japan's international environment have given rise to a veritable explosion of innovative activity.

Present-day Competition

The reasons for this renewed burst of innovation are readily apparent. Now that Japanese industry has overtaken most other advanced countries, it is faced with both increasing competition and rising protection in its major export markets; and the only effective response is through innovation in high technology.

Moreover, it is in the nature of things that the chaser should be chased once he moves into the lead. New industrial countries, especially neighboring countries of Asia, have been making rapid and extensive inroads into markets in which Japanese industry had previously developed strong market positions. Since there is no way foreign markets can be protected by politically contrived barriers, the only effective response to these new sources of competition is increased competitiveness through higher technology. Since competition between Japanese industry and those of newly industrialized countries is mainly in world markets, a defensive strategy is not a real option for Japanese enterprise. Equally important, as the newly industrialized countries are important markets for Japanese firms, it is clearly the better part of wisdom not to risk retaliation against restrictive practices directed against the exports of those countries. Even if on economic grounds protection were an effective option, for political reasons Japan cannot afford to discriminate against the products of new industrial countries in Asia or elsewhere.

Innovation, therefore, is more imperative than ever before, and especially important is innovation in electronics as one of the basic technologies of the present and future generations of information industries.

Still other factors conspire to add compulsion to this innovative imperative. Especially, the changes of Japan's overcoming energy constraints depend very heavily on the extent to which, and the rapidity with which, Japanese firms can shift production from energy-intensive heavy industries to knowledge-based industries for which electronics is a basic technology. Energy conservation also depends very much on increased production efficiency, which in turn requires more sophisticated electronic systems, controls and devices. Uncertainties and increasing cost of materials supplies can also be offset, at least in part, by greater reliance on less resource-intensive electronic technology. And, no less important, the outcome of the continuing two-front war against inflation and pollution depends to no small degree on developments of advanced electronic technology.

If these conditions are not unique to Japan, the response very often is. Reasons for this lie at the core of the Japanese economic system.

In the first place, the competition among Japanese enterprises to overcome obstacles or grasp new opportunities through technological advantage is fierce. With relatively high fixed costs of both labor and capital, because of peculiarities of Japanese industrial organization, each Japanese manufacturer is committed to the inexorable pursuit of higher value added in production. In high growth industries such as electronics, the firm that fails to innovate, to steadily improve product and production technology, soon finds itself out of the running. Managers are forced to give highest priority to the management of techology to assure to the enterprise the essential flow and development of both men and money, of both human and financial resources. The result is a sustained high level of innovation and a concomitantly high level of investment in improved production systems.

It is precisely this ferocity of competition in the home market, not the collusion of some monolithic "Japan, Inc.," which hones the cutting edge of Japanese competition abroad.

The intensive competition between Japanese videocorder makers is a case in point. Sony and Japan Victor, along with others using their respective technologies, were linked in what seemed to be mortal combat in the home market during the last half of the 1970s and early 1980s. The result has been a continuing flow of improvements in videocorder technology, spurring the further development of videodisc systems.

The combined effect of competitive strategies at home and abroad has been to intensify the technological imperative and amplify it globally, exacerbating the propensity to innovate to a level that has defied foreign challengers. And since economies of scale and learning are critical factors of competitive power in the marketplace, only those Japanese competitors with innovation, production and marketing strategies which optimize global economic realities remain in the race.

These kinetic forces which give to Japanese innovation its powerful forward thrust are further enhanced by geographic and demographic conditions within which the central creative processes take place. Concentration of Japanese manufacturing and innovation in the Kanto-Kansai (Tokyo and Osaka) conurbation has had the effect of accelerating innovation not only to meet the changing demands of an increasingly affluent urban population, but because the rates of interaction between supply and demand, between innovators and the market, and between innovators themselves, tend to be more intensive. Feedback mechanisms in such an environment tend to be highly efficient, especially when the population is as homogeneous as that of the Japanese. Diffusion of technology tends to be accelerated, due to a broader demonstration effect and to higher density of forward, backward, lateral and diagonal linkage networks. Since innovation in advanced technology typically involves entire systems, rather than a single innovator, the existence of complementary skills and capabilities is an essential precondition for an increasing amount of innovative activity, especially in the electronics industry.

Agglomeration of Japanese industry tends to strengthen its propensity to innovate in yet another way. If one firm in the electronics industry introduces new technology, putting downward pressure on prices, as maintenance costs with existing equipment rise, firms throughout the industry are under pressure to replace existing technology with new. The action-response time-frame is virtually immediate. And since the Japanese electronics industry has been expanding rapidly, firms in the industry, assuming they are producing near capacity, are under constant pressure to improve their production technology in order to gain or retain market share.

Continued production with "obsolete" technology, a strategy that was adopted by profit-maximizing color TV manufacturers in the United States in the late 1960s and early 1970s, is thus hardly conceivable in Japan where value-added maximizing firms have more-or-less uniform cost curves and are under constant competitive pressures to scrap old equipment and build with new up-to-date technology. The premium on innovation in such a highly competitive environment is paramount.

Finally, but more important, this complex matrix of factors which endows Japanese industry with its exceptional propensity to innovate is buttressed by a clear and common recognition that the need for innovation is overriding. In the place of technophobia which plagues other advanced industrial societies, there is a broad consensus on the fundamental notion that only through improvement of the technological level can Japan's basic resources — human and capital — be made more productive. Continuing innovation to shift these resources to higher levels of wealth creation, thereby assuring steady improvement in living standards and the quality of life, thus becomes the main responsibility of management.

IV. The Propensity to Manage Innovation

Japanese companies characteristically seek to attain maximal growth through the management of technological change; only the mastery of increasingly higher levels of technology can assure the creation of added increments of wealth needed to obtain continuing improvement in the common good. Not accidentally, the Japanese management system has a high propensity for the effective management of innovation.

In striking contrast with the United States, where much of the important innovation activity is undertaken by new entrepreneurial firms financed by venture capital, in Japan technological leadership comes from within the largest enterprises and major industrial groups. Although smaller entrepreneurial firms such as Sony, Casio, Kyocera and Honda play an important role in pushing forward the frontiers of new technology, the main thrust in advanced technology has come from large-scale enterprises in the steel industry, shipbuilding, the automotive and electrical machinery industries.

For each technology there is an appropriate corporate size, degree of integration, internationalization and critical financial mass. The firm must be large enough to carry capital commitments required for effective management of technological change, in the long run. Similarly, the innovative firm must be sufficiently integrated and diversified to assure optimal management of that technology which constitutes its distinctive competence. And to be able to effectively manage any technology, a firm must be large enough to control the optimal market required to sustain a competitive level of innovation, which for much of the electronics industry is a global market.

Integrated Manufacturers

One of the major reasons why the Japanese electronics industry has succeeded in overtaking U.S. leaders in semiconductor technology is that there are no semiconductor manufacturers in Japan. Semiconductors are developed and produced by large integrated electrical-electronics manufacturers in Japan, not by small-to-medium scale specialized semiconductor merchant houses characteristic of the U.S. industry. The large integrated Japanese firms can devote 22 percent of sales to R&D in ingegrated circuits year-in and year-out, regardless of variations in the business cycle; similarly, they can invest heavily in plant and equipment needed to implement rapidly changing mega-technologies such as microelectronics. Specialized U.S. semiconductor manufacturers can do neither and have therefore lost their technological leadership.

If it is true that innovation can be a goal of enterprise only if it does not prejudice a minimum level of earnings, that financial threshold tends to be much lower in Japan than in the United States or Europe. As a result, between firms of equal size and similar structure, Japanese firms are capable of a substantially higher commitment of resources to R&D activities. This explains why Fujitsu, much smaller in size than IBM, was able to invest heavily in computer manufacturing and overtake the American giant, even though its total financial resources were much more modest.[16]

It is also true that product diversification and vertically integrated manufacturing systems have had a substantially greater synergistic effect

---

(16)   For a fuller discussion of the effects of differing financial systems on the development of the semiconductor industries in the United States and Japan see Gene Adrian Gregory and Akio Etori, "Japanese Technology Today: The Electronic Revolution Continues," *Scientific American,* October 1981, pp. J 16-18.

on technological management in Japan than in the United States or Europe.

Quite clearly, the fact that Nippon Electric is Japan's leading manufacturer of telecommunications equipment and its third largest computer maker is especially relevant to its leadership in microelectronics. Not only has cash flow generated in the sale of telecommunications equipment tended to sustain a high rate of investment in both computer and semiconductor R&D, plant and equipment and market development. By giving the management of technology highest corporate priority, NEC has assured that innovation and production in each of these major product areas is strengthened by innovation and production in each of the other areas. And since all three technologies must be optimized on a global scale, NEC has organized for worldwide marketing and manufacturing in all three — communications equipment, computers and semiconductors — to obtain the full synergistic effect of their linkage.

Cooperative Modes

While virtually every major electrical machinery, telecommunications equipment and computer manufacturer in Japan has obtained somewhat analogous results in the management of advanced electronics technology, heavy demands of large-scale R&D in successive generations of semiconductor devices and computer mainframes surpass the resources of the largest integrated companies. Merger of various companies, or the spin-off and amalgamation of particular divisions, to obtain further scale economies tends to be incompatible with the organizational arrangements for technology management within the large manufacturing concerns themselves and would seriously prejudice the network of horizontal and vertical relationships within the respective industrial groups (*keiretsu*).

In place of takeovers and mergers, special ventures between existing manufacturers are formed for the management of particular production technologies. Or, alternatively, *ad hoc* cooperative research undertakings such as the VLSI Research Association are used to obtain the appropriate critical mass for each particular R&D undertaking. Similar associations have been formed for developing new computer generations, electric automobiles, biomass energy technology and other mega-technologies which require resources beyond the means of individual firms for their development.

Within each enterprise, innovation is also managed cooperatively by research teams. *Ad hoc* teams of engineers and research scientists are formed for major projects such as Sony's Trinitron television picture tube, Japan Victor's VHS videocorder or Hitachi's laser beam videodisc system. Since innovation is the work of groups, although there are many inventions in Japan, there are few inventors. Due recognition is accorded innovative genius within the enterprise, and occasionally individual brilliance may be visible from without, but inventions are normally identified with companies rather than with individual inventors.

The management of cooperative modes of innovation is facilitated by the continuity of management and technicians in the enterprise, and by the identity of their interests and goals with those of the enterprise. This continuity, and the experience in cooperative endeavor which it makes possible, are the keys to efficiency in R&D activities of Japanese corporations, and together they constitute the solid foundations for efficient cooperative research undertakings.

These features of the Japanese management system enhance the propensity of Japanese firms to manage technology efficiently and effectively, and account in large part for the observable shift of innovative activity in the electronics industry, as well as in others, to the Western Pacific.

V.   The Human Factor

However criticial the propensity to promote, finance and manage innovation may be in the equation of technological change, ultimately the availability and quality of human capital is a vital determinant. The most appropriate organizational structures, managerial systems and monetary arrangements will not suffice for sustained innovation in the electronics industry, or any other sector operating on the frontiers of high technology, in the absence of an adequate supply of highly trained engineers and an educated work force. And here, once again, the Japanese electronics industry has a decided advantage.

Formal Education

While the Japanese educational and in-company training systems have continued to meet requirements for trained engineers relatively well, in the United States the shortage of engineers and technically educated professionals has reached what industry leaders have woefully termed "catastrophic" proportions.

Throughout the 1970s, 20 percent of all baccalaureates and about 40 percent of all master's degrees in Japan were granted to engineers, compared with about 5 percent at each level for the United States. As a result,

Table 5. U.S./Japan Electrical and Electronic Engineering Graduates (1969–1979)

	United States		Japan	
	B.S. Only	B.S., M.S., Ph.D.	B.S. Only	B.S., M.S., Ph.D.
1969	11,375	16,282	11,035	11,848
1970	11,921	16,944	13,085	13,889
1971	12,145	17,403	14,361	15,165
1972	12,430	17,632	15,020	16,052
1973	11,844	16,815	16,205	17,345
1974	11,347	15,749	16,140	17,419
1975	10,277	14,537	16,662	18,040
1976	9,954	14,380	16,943	18,258
1977	9,837	14,085	17,668	19,257
1978	10,702	14,701	18,308	20,126
1979	12,213	16,093	19,572	21,435

Sources: Engineering Manpower Bulletin (USA); Ministry of Education (Japan)

by the end of the decade, the number of engineering degrees granted in Japan surpassed those granted in the United States, even though the population of the United States and the total output of its electronics industry were both approximately twice that of Japan.[17]

But the contrast in the field of electronics and electrical engineering is even greater than these general statistics reflect. As early as 1970 Japan was granting more bachelor of science degrees in electrical and electronic engineering than the United States, and by 1979 the number of graduates in this field in Japan was as much as 60 percent greater than that of the United States. On a per capita basis, in 1977, Japan had almost three times as many electrical and electronic engineers as the United States, more than four times the total number in Great Britain, almost six times that of France, and approximately 70 percent more than West Germany.

And the gap continues to grow. While the total number of engineering graduates in the United States has been growing at a compound annual rate of 2.4 percent, the number of electrical and electronic engineering graduates declined at an alarming compound rate of 2.9 percent during the 1970s. In Japan, on the other hand, the number of engineering graduates grew from 1967 through 1979 at a substantially higher compound rate of 7 percent annually, and the number of electrical and electronic engineers increased at the even higher rate of 7.2 percent a year. (Table 5) While in France and the United Kingdom the average annual

(17) "Government Report Hits Education Gap," *Electronics,* 6 November 1980, pp. 93-96.

Table 6.   Graduates in Engineering and in Electrical / Electronics Engineering

(per million inhabitants)

	France	Japan	U.K.	U.S.	USSR	W.Germany
1965						
Engineering	20	82	32	—	—	16
Elec./Elec. Eng.	157	323	145	262	—	56
1970						
Engineering	34	133	46	85	248	11
Elec./Elec. Eng.	181	508	205	313	833	43
1975						
Engineering	28	162	45	67	269	48
Elec./Elec. Eng.	179	646	213	308	930	193
1977 (1)						
Engineering	33	185	46	66	271	109
Elec./Elec. Eng.	186	711	259	314	997	438

Note: (1) 1979 for Japan; 1978 for U.S.S.R.

Sources: *International Financial Statistics Yearbook,* 1980; *Economic Statistics Yearbook,* 1980 (Japan); *Statistic Abstract* (U.S.A.); *Statistiches Jahrbuch; Annuaire Statistique de la France;* APN (U.S.S.R.); British Embassy in Tokyo.

growth rate in electrical and electronic engineering graduates surpassed that of the United States during the 1970s, the rates of increase fell far short of the Japanese. Only West Germany surpassed the Japanese performance, with the total number of engineers growing at an exceptionally high 19.2 percent annually and the number of electrical and electronic engineers advancing at a compound rate of 17.5 percent a year. (Table 6)

In-house Development

But educational statistics sketch only a part of the picture, albeit the important relief. In Japan, companies invest much more in training and retraining than elsewhere, and the implications of this are far-reaching indeed. It is not uncommon in Japanese industry to shift engineers from one sector to another, providing extensive retraining in new professional fields. During the 1970s, in-company training was especially focused on shifting mechanical engineers to electronics, hardware technicians to software-oriented technology, and from blue collar to white collar employment. Massive retraining was a *sine qua non* of rapid industrial structural change; and rapid industrial structural change was imperative to managing the new environment of Japanese industry created by successive monetary adjustments, oil shocks and wage increases.

Here, again, appropriate organizational size and systems are critical for the effective management of technological change. In Japan, since lifetime employment is the rule and labor mobility between firms is low, especially in large, diversified and vertically integrated firms operating in high-technology fields, companies can afford to invest heavily in human resource development. In the United States, where human resources are a temporary contractual appendage to the critical financial assets of the enterprise rather than a permanent and essential constituency of the corporate entity, such an investment in human resource development would be tantamount to managerial irresponsibility in the allocation of company assets.

Moreover, there was a time, back in the 1960s, when the U.S. electronic industry leaders located near the citadels of engineering and physical sciences higher-education in New England and Silicon Valley boasted that mobility of technical personnel was one of the major advantages of the industry. Hit-and-run innovators, like honey bees, assured cross-pollination among companies thereby speeding the process of technological diffusion. Early high-flyers of the semiconductor industry, such as Fairchild Camera and Instrument Corp. of Mountain View, California, seemed to measure their technological prowess in the number of top engineers who left them to establish their own start-up ventures or were recruited by marauding competitors. Indeed, many, if not most, of the leading semiconductor makers in the latter 1970s had their roots in the fertile soil of Fairchild's early experience.

By the early 1980s, all that had changed. With the loss of much of its best talent, Fairchild's brilliance faded considerably. Unable to sustain its pace of innovation and plagued with capital shortages, company control passed into the hands of a French conglomerate. Technological leadership had long since passed on to Intel, Advanced Micro Devices (AMD) and Zilog, and thence to NEC, Hitachi and Fujitsu.

In a remarkable change of corporate practice, if not philosophy, semiconductor makers in Silicon Valley, Arizona and Texas responded to the shortage of engineers with measures to conserve vital human resources. Firms were noticeably more reluctant to lay off personnel with each downturn of the market. Mobility of skilled personnel has been recognized as one of the principal problems of the industry, increasing costs and lowering both productivity and quality in the face of mounting Japanese competition.

The entry of Japanese semiconductor manufacturers into world markets, coming at a time when the rapid growth of the U.S. industry was threatened by a major imbalance of supply and demand for engineers and production personnel, focused management attention on the critical importance of human capital development. It became increasingly apparent that the immediate economics of short-term profit-maximization, based in part upon flexible hiring-and-firing practices intended to minimize the variable cost of labor, was tantamount to corporate suicide in the long term. Management has been forced to focus on longer-term horizons and to reconsider corporate responsibility to employees, or cede the field to Japanese competitors more adept at the development of human resources for the management of advanced microelectronics.

Although the response to this new business environment has varied from company to company, an increasing number of U.S. electronic firms are joining Advanced Micro Devices Inc. president Jerry Sanders in the view that "If you believe in the future, it is self-serving and shortsighted to lay off employees."[18] There are also those who confess what is becoming painfully apparent: personnel mobility has become incompatible with profit maximization, even in the short run. Quite clearly, if the U.S. electronics industry is to compete effectively with Japan, and with new industrial countries such as Korea, Taiwan, Singapore and Brazil, it will be necessary not only to increase the attractiveness of employment but to make those changes in corporate philosophy and management systems which assure the stability of employment necessary for perpetual in-company training of existing personnel. These changes are, of course, wholly justified solely in terms of social justice and harmony. But they are also absolutely imperative for the optimal management of advanced microelectronics technology. Yet there are still those who loudly proclaim the incompatibility of high technology and human welfare.

In Europe, where the electronics industry has now grown as rapidly as in the United States and Japan and some university systems have responded relatively well to the needs of industry for engineers, the problems of retraining within the enterprise assume special import. Since the shortage of engineers is not as acute and the profession remains rather rigidly stratified, there is strong resistance to changes from one branch of engineering to another. And for quite different reasons, labor unions tend to discourage the mobility of skilled production personnel within the enterprise, ossifying organizational structures and obstructing technological changes which affect job classification and work assignment. So long as these rigidities prevail, there is little possibility that the European electronics industry will ever seriously rival the Japanese propensity to effectively manage innovation.

It is worth noting here that the effectiveness with which Japanese electronic firms manage technology is in no small degree due to the

---

(18)   Howard Wolff, "The New West: An Industry on the Move Changes Where and How it Does Business," *Electronics,* August 28, 1980, pp. 88-104.

prevalence of experienced engineers among the top managers. If the primary objective of the firm is to manage technological change, it follows quite logically that management must have a keen understanding of the technologies which are being managed. Firms which have any intention at all of living up to slogans such as "Hitachi Supreme in Technology" or Sony's "Research Makes the Difference," must have men among the top management with strong technical backgrounds. Indeed, at Hitachi most of the company's directors, beginning with the president, have engineering experience. They are, in turn, supported in their commitment to innovation and technological change by more than 3,000 research engineers with a budget in 1980 of US$554 million, or 3.7 percent of consolidated sales. For computers and semiconductors, however, R&D spending was running at a prodigious 13 to 22 percent of sales. Similarly, at Sony, engineers have been prominent among the top managers from the outset of the company's now legendary history.[19]

In the American and European electronics industries, where management of assets, finance and legal affairs tend to preempt the management of technology in the ranking of corporate priorities and the claim on corporate resources, financiers and lawyers tend to rise to the top while engineers remain on tap. Especially in the United States, where intercorporate mobility of managers is still the rule, top managers are frequently recruited from outside the firm without regard for experience in managing the company's particular technological competencies. It is also true that managers tend to change more often than in Japan, where continuity of relationships is important, not only to assure the stability of enterprise, but also because only through such continuity can the management of technology benefit from experience.

In the electronics industry, continuity of working relationships is especially critical to the management of technology since the experience factor is the key to competitive power in virtually every sector of the industry, and particularly in the production of microelectronic devices.

The explanation of the higher Japanese propensity for innovation is not to be found mainly in the work ethic, but rather in the experience of working together in the management of technology. Security of employment as well as continuity of management, two closely related features of the Japanese enterprise system, are *sine qua non* of that salutary state of affairs.

To complete the picture, it would be amiss not to stress that the

---

(19)   See the author's report on R&D in the Japanese electronics industry, "Getting Smaller to Get Bigger," *Far Eastern Economic Review,* October 23, 1981, pp. 68-71, as reprinted in Gene Gregory, *The Japanese Electronics Industry,* Tokyo: Sophia University, Institute of Comparative Culture, 1981, pp. 107-121.

strength of Japan's human resources devoted to the management of technology is not limited to the quality and numerical strength of the engineering corps. As remarked by Mr. Konosuke Matsushita to a foreign visitor he was showing through a television factory: "That man over there," he said pointing to a worker on the assembly line, "is my most highly prized manager." His point was well taken. Given the intensity of Japanese primary and secondary education and its emphasis on mathematics and science, the quality of workers on the shop floor is second to none in the world. Not only are they most probably among the world's best educated workers, but they are in all likelihood the most homogeneous and highly motivated, as well as being the least conflict-prone, of those in any country. It is this high quality of the labor force which makes possible Zero Defect and Total Quality Control production systems that have transformed quality into the most important ingredient of Japanese competitive power. Likewise, it is the quality of this labor force which makes possible the rapid introduction of robotics in Japanese industry without serious problems as a result of labor displacement. The consistency of goals of the highly educated Japanese workers with those of the firms where they work assures continual up-grading of their skills, sustaining a high propensity for rapid technological change.

Conclusion

It is a historic coincidence of no mean significance that the electronic age should be accompanied by a shift in innovative activity from the Atlantic to the Pacific, and that Japan should emerge as the country with the greatest propensity to innovate at the highest levels of advanced technologies. For the past two decades, Japanese industry in general and the Japanese electronics industry in particular, have been the pacesetters in patented inventions. By 1974, the aggregate level of Japanese technology had surpassed that of the United States and was therefore the highest among advanced industrial countries.

Clearly, this innovative activity has been sustained by a complex system of institutional arrangements, management practices and human resource development. To conclude that the shift in electronics innovative activity to Japan is attributable mainly to any one of these factors, or to attempt to divorce the propensity to manage effectively innovation in high technology from the realities of Japan's cultural heritage and economic condition, would be to indulge in reckless oversimplification. Those who would yield to the temptation simply to clone one aspect or another of Japanese experience with the hopes of obtaining a similar level of innovative activity and productivity are most likely to be disappointed with the genetic results.

The educational system which has given high priority to equipping Japanese for fruitful participation in innovative processes at all levels of the industrial spectrum, as an instance in point, is at once an expression of government policy to foster these industries and a long-standing tradition of rapid cultural and technological change. Moreover, a heritage of creativity and a clear recognition of the utility of technology, as well as its relevance for Japanese political independence, has given to engineering particular social significance. As a result, social energies and institutions have been focused on the management of technology, not through central governmental planning but by a prevailing consensus on the purposes of technological change which derives its main stimulus from human needs manifest through an open competitive marketplace.

In keeping with this consensus, Japanese enterprise is therefore constituted not mainly for the management of assets or the maximization of profits, but rather to maximize wealth creation through the creative management of technology. To assure the optimal performance of this system, Japanese financial institutions and practices are notably designed to give optimal support to innovative activities, thereby raising the threshold of sustainable technological virtuosity to a remarkably high level. Rapid technological change is in turn spurred by high investment, gently but surely guided by industrial policy to assure the timely transfer of resources from outmoded and declining industries to growth sectors with high value-added. But the underlying momentum for this process of technological change is not provided mainly by government funding, it is rather induced by the powerful push-pull effect of the world's most rigorous competitive business environment, and supported by a management system having a high propensity to develop human resources and organize them for cooperative management of technological advance.

CHAPTER 4

Managing Technological Change

IN THE EARLY 1950s the elctronics industry was very much an Atlantic affair. Few national leaders paid much attention to it, except in wartime when radio communications and radar became vital to national interests, and what might pass for electronic equipment was hardly identifiable in the foreign trade accounts of the advanced industrial nations. In most countries, effort was devoted to the design of communications systems, whose terminal equipment and transmission networks constituted the measurable whole of the industry. Much of the effort went to assure autarchical exclusivity to national manufacturers, and what could not be attained through technological design was achieved by intricate cartel arrangements. Quietly, between themselves, national communications organizations and their main suppliers agreed on what equipment to manufacture, at what prices; and equipment manufacturers on both sides of the Atlantic contrived, through patent pools and marketing arrangements, to satisfy those requirements. Consumers took what was available from their respective national industries, and manufacturers exported only to their own country's preferential markets.

Now all that has changed. Electronics has moved to the forefront of industrial activity, replacing steel as the keystone of the industrial system and the symbol of national economic prowess. Development of the electronics industry has thus been granted priority on the agenda of industrial policy-making in advanced and developing countries alike. New technology has obliterated old monopolistic structures. The epicenter of electronics manufacture and innovation has shifted convulsively from the Atlantic to the Pacific Basin, with the emergence of Silicon Valley and Japan as the bipodal foci of a protean technological revolution. Among the momentous aspects of this revolution has been the emergence of Japanese manufacturers as leaders in new product development and in production systems which combine advanced microelectronics and highly

Excerpted from "Japanese Technology Today: The Electronic Revolution Continues," *Scientific American*, October 1981, Supplement on Japanese Technology Today.

developed human resources with cooperative institutional arrangements to set consummate standards of quality, reliability and efficiency for the industry. A country reputed for the ability of its industry to imitate modern technology has now emerged as major source of innovation in advanced electronics.

Three signal developments in July 1981 underscore this revolutionary change in the configuration of the global industrial and technological order. During high level consultations on security matters, the U.S. government asked Japan to provide advanced very large scale integration (VLSI) technology to enhance air and anti-submarine defense capabilities. A week later, it was revealed in Tokyo that General Motors had requested the Toshiba Corporation to help in developing an electronic control system for automobile engines; compared with GM's best device using nearly 20 large-scale integration elements, the Toshiba system performs as well or better with only four or five more highly integrated circuits. At the same time, it was learned that General Electric was negotiating with Hitachi, Ltd. for extensive technological assistance in the manufacture of VLSI circiuits in pursuance of its strategy to make a fullscale reentry into the semiconductor business, from which it had withdrawn ten years previously. According to informed sources, GE was seeking not only Hitachi's advanced technology, but the formation of a joint VLSI manufacturing company in the United States.

Evidence that Japan was overtaking the American lead in integrated circuit (IC) technology had been mounting ever since Japanese semiconductor manufacturers came from nowhere in 1976 to take a 40 percent share of the world market for 16K random access memories (RAM), the then-ultimate in the art of packing circuits onto the 31-square-millimeter standard chip. In 1978, Fujitsu Ltd. was the first to announce the commercial production of 64K RAM circuits, which pack 4 times as many circuits on the standard chip. The Japanese industry thereupon grasped the lead in global competition in this new generation of devices, with Hitachi, NEC, Toshiba, Matsushita and Mitsubishi following in force, all bidding with Fujitsu for a share of the action.

Some industry observers have predicted that Japanese makers will take anywhere from 60 to 70 percent of the world market, given their ability to deliver reliable parts in quantity; and even U.S. industry executives have stated that it will be difficult to hold the Japanese industry's world market share below 40 percent.

Any lingering doubts that Japan had caught up with the American semiconductor industry in key areas of advanced semiconductor technology were dispelled early in 1980 when representatives of Matsushita, NEC-Toshiba and Nippon Telephone and Telegraph's Musashino

Laboratory provided a solid-state-electronics conference in San Francisco with detailed descriptions of memory chips that could store still another four times as many bits of information — 256K — as the 64K RAM chip. At a time when U.S. semiconductor firms were wrestling with 64K RAM production problems, Japanese manufacturers were gearing up for the next generation of 256K dynamic RAMs. Although there is no hard evidence that Japanese semiconductor makers will begin production before 1984-85, indications are that market demand, rather than technical restraints in manufacturing, will dictate the timing for introduction of these new tightly packed VLSI microcircuits.

It is in this transition to the stage of VLSI production that the current lead of Japanese manufacturers will assume critical importance. Their advance in 64K RAM production, and the strong market position which this assures Japanese makers, signifies more than just another in a series of industrial triumphs dating back to the 1950s when Sony introduced the first commercially produced transistor radios. This combination of production and market prowess is expected to play a pivotal role in determining the strength of the Japanese industry throughout the VLSI era, which will extend through the 1980s and well into the 1990s.

The probable consequences of this structural change in the global electronics industry are of epochal import. If the Japanese electronics industry has been so successful in product and production innovation during the 1960s and 1970s — a period when it was largely dependent on

Fig. 1   VTR Price Trends in the 1980s (in 10 thousand yen)

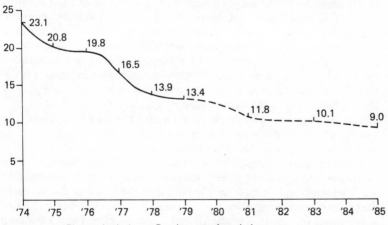

Source: Japan Electronics Industry Development Association

81

foreign sources of technology — there are reasons to expect that leadership in advanced microelectronics technology will enable Japanese firms to move even further out in front, pioneering the substitution of electronics for mechanical systems in a broadening range of products, introducing entirely new product generations and gaining global market share in the process.

Reasons for this expectation are not based upon logic alone, however compelling that may be. In this case, syllogism is dramatically buttressed by past experience as well as by the observable dynamics of technological and industrial change.

Extraordinary Measures

Especially significant is the Japanese record of industrial policy-making in this critical sector. As early as 1957, policy-makers identified the electronics industry as a priority sector for development, passing into law what was appropriately called Extraordinary Measures for the Promotion of the Electronics Industry. At the time, few Japanese firms could be classified as electronic equipment manufacturers, and not one among them had distinguished itself as an exporter of electronic products. Total production of the industry, such as it was, had reached barely ¥60 billion (US$167 million) at the end of fiscal 1955, and growth rates had not yet reached the point of takeoff.

The record since then is now legendary. The Japanese electronics industry emerged as the world's leader in the production of color television sets and a broad range of home entertainment products, electronic organs, electronic watches, electronic calculators, numerical controls for machine tools, robots, automotive control systems, medical electronics and virtually every component used in these various types of equipment. By the mid-1970s, Japan's output of computers, communications equipment, office machines and semiconductors was second only to that of the United States. Opto-electronics was emerging as a new field for Japanese special development, and an increasing number of products such as video taperecorders (VTRs) were the creatures of a newly discovered — or rediscovered — genius for innovation.

Significantly, during the last half of the 1970s, when other industries were languishing in the doldrums of cyclical recession or suffering from structural depression, the output of the Japanese electronics industry continued to develop at a fast pace. The total value of its production, second only to that of the United States, reached ¥8,683 billion in 1980 — or almost 150 times that of 1955 — making electronics one of

Japan's major industrial sectors, on a par with the automotive industry.

Present parity appears, however, to be preliminary to tomorrow's preeminence. Since with each successive generation of more highly integrated circuits the diffusion of electronic technology is extended to new products and applications, permeating one sector of the industrial system after another, the electronics industry will soon surpass all traditional industries in size and diversity.

During the post-oil-shock recession years of 1975-1978, for example, output of electronic products increased 50 percent, compared with a 33.5 percent rise in automobile production. At the same time, the automotive industry became an important new market for electronic products. With the increasing importance of fuel economies, electronically controlled fuel injection systems were developed using new generations of microcomputers. And to save both energy and labor in manufacturing processes, the automotive industry invested massively in robots and other automatic-production equipment with microcomputer controls.

Fig. 2   VTR Output in the 1980s (in 100 million yen)

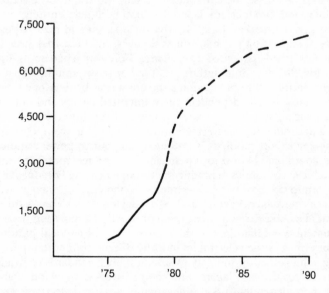

Source: Japan Electronics Industry Development Association

83

If not to the same extent, the steel industry — which electronics is rapidly replacing as the basic sector of Japan's industrial edifice — also became an important market for electronic equipment during the latter 1970s.

"Today, computers are serving to systemize all sorts of work in steel-making, ranging from control to energy-saving to labor-saving," Sumitomo Metal Industries' Executive Vice-President Toshio Ikeshima notes. "To date we have invested ¥30 billion in computers. In addition, we annually spend about ¥6 billion for leasing of computers plus. 8.4 billion in other related expenses. The sum amounts to 1.2 percent of annual sales. On the other hand, the annual amount of benefit from all this is calculated to be ¥50 billion, of which 40 percent is from labor-saving (reduction of workforce), 23 percent from improvement in product yield, and 37 percent from improved efficiency, energy-saving, reduction of inventories and other factors."

Long Range Vision

As obvious as these benefits might seem, diffusion of electronics technology and products in Japan has not been left entirely to the mechanics of the market. In 1974, after the oil crisis had made clear the necessity of reassessing Japan's industrial policy, the Industrial Structures Council published its revised *Long Range Vision* of Japanese industry, designating the electronics industry as the key sector around which virtually the entire future production structure was to be developed. Since Japan is almost wholly dependent upon imported energy and other raw materials supplies, rising crude oil prices and the uncertainties which followed in commodity markets seriously endangered the competitiveness of many key export products. Continued competitive power required a massive, coordinated effort to allocate and use resources more efficiently, and to achieve economies in production through increased productivity as well as improved quality. As electronics technology is "energy-saving, resource-saving, labor-saving and space-saving," in the words of Michiyuki Uenohara, senior vice president of NEC Corporation, its rapid development was catapulted to the highest levels of national priority.

Not only was the eelectronics industry of central importance for increasing value-added in industry to sustain higher standards of living in an ever more affluent Japan, but competitiveness in world markets became more dependent upon innovation in advanced electronic export products and the most rapid possible diffusion of efficiency-inducing electronic technologies throughout the industrial system. In the apt

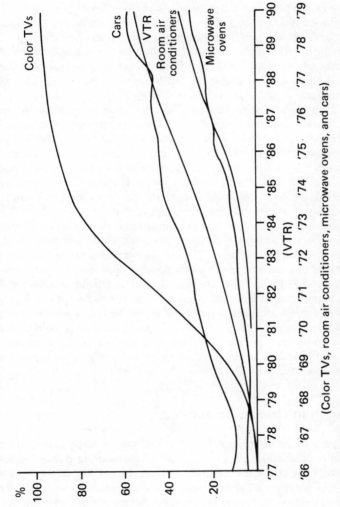

Fig. 3   Comparative Diffusion Rates of VTR and Other Consumer Durables

(Color TVs, room air conditioners, microwave ovens, and cars)

Source: Japan Electronics Industry Development Association.

words of Singapore's Premier Lee Kuan-Yew: "There is nothing like the threat of hanging in the morning to concentrate one's attention the night before." It was just such a perception of danger that gave a massive added thrust to the dynamic development of the Japanese electronics industry in the post-oil-crisis period.

But crisis was not the sole impetus to the now-renowned cooperative VLSI research and development project which was launched in 1975, following the directions indicated in the new industrial "vision." Nor was necessity itself sufficient cloth from which to fashion success in the intended bid for technological leadership in advanced electronics technology.

In the two decades since its birth, the Japanese electronics industry had, of course, accumulated a vast stock of technology. Innovation had come early and often from the industry, largely as the result of cooperative research and development — albeit mostly in applied technology — within enterprises and industrial groups. It was mainly because of this intense pursuit of product and production innovation that Japan had become the world's leader in consumer electronics and was moving out front in other product lines as well.

If this is becoming increasingly apparent even to the casual observer of the Japanese scene, the dynamic processes which have made it so remain somewhat more elusive. The critical interactive factors which determine that process, and which go far towards explaining the Japanese advance in VLSI technology, are not, however, difficult to identify. The basic ingredients of progress in microelectronics technology in Japan — simply stated — are: men, money and enterprise functioning in a fiercely competitive business environment. These, in turn, have fashioned particular corporate strategies and structures which serve to accelerate the pace of technological change and assure the efficient management of scarce resources.

Strength in Human Resources

History is eloquent witness that the electronics industry is one that will spread on a global scale in the quest of high-calibre human resources most adept at efficient enterprise. In brief, highly educated people who work well together are the *sine qua non* of electronic technology. Herein lies the principal explanation of Japanese ascendancy to leadership in consumer electronics production, the rapid overtaking of the American industry in LSI production, and the advantage so far attained in VLSI technology. It is likely to be the prime determinant of the role of Japanese firms in the industry for decades to come.

Given the paucity of Japan's natural resources, the affinity of Japanese industrialists for a knowledge- and skill-intensive industry such as electronics is derived from no special insight. It is plain common sense. Just as comparative advantage was originally obtained by wise use of abundant scources of educated low-cost manpower in the earlier stages of Japan's industrialization, the prospect for continued technological and industrial progress in its advanced stages is to be found only in technologies and production which require abundant sources of educated high-cost manpower.

Electronics technology is appropriate to Japan not only because of the relative scarcities of natural resources and the abundance of manpower, however. The quality of human resources makes the difference. Traditional arts and crafts, as well as the light industries typical of Japan's pre-World War II economic development, relied heavily upon dexterity of hand, concern for detail, a penchant for intricacy and a quest for perfection. Skills which found their supreme manifestation in the carving of ideographs on individual grains of rice, were ideally suited to the tedious tasks of designing and producing high-density microelectronic devices.

For a country totally reliant on its human resources for survival and welfare, education commands primacy of place among societal values. It follows, too, that for industry in such a country, education — the development of human resources — is of foremost concern. Not by accident does Japanese industry recruit workers directly and systematically from high schools, the fast-growing and most technically advanced firms

Table 1. VTR Market Growth in the 1980s

(in thousands of units)

		1979	1981	1983	1985	1990
Domestic demand	Fresh-buying demand	480	1,360	1,780	2,310	1,480
	Carryover & add. buying demand	—	40	120	230	1,090
	Total	480	1,400	1,900	2,540	2,570
Diffusion ratio (%)		459	10.4	19.3	30.0	51.9
Exports		1,720	2,650	3,800	4,300	3,500
Cross demand		2,200	4,050	5,700	6,840	6,070

Source: Japan Electronics Industry Development Association.

hiring the best graduates from the best schools. Nor is it accidental that managers and engineers are just as systematically recruited at graduation from university. And since the reputation and strength of educational institutions depends to a great extent on their ability to place their graduates in positions with leading companies, there is competition among schools to prepare students for lifetime employment in the best industrial, financial and commercial enterprises.

Since the electronics industry offers the best opportunities for human knowledge and skill development, as well as for rapid advance in companies having the fastest growth, students show a strong preference for work in electronics, and electronics manufacturers can select the best students. It only remains for educational institutions to make the selection process possible, and the most fruitful, for both students and employers.

It is not surprising, therefore, that Japanese universities graduate about 50 percent more electronics engineers than universities in the United States, which works out to approximately three times as many per capita. In the information industry *par excellence* — electronics — people are the critical ingredient, the more so the more knowledge-intensive electronics technology becomes. There is, therefore, a keen, and quite unique, recognition in Japan that microelectronics and its progeny — computers, automated equipment, robots — do not in fact replace people. Electronics makes people more essential. it renders the development of their most important attributes — brainpower and cooperative endeavor — imperative. In software alone, Japanese industry sees the need for highly trained manpower as insatiable.

"If we have to continue to increase the number of software engineers at the present rate," Toshiba's manager of technology development Sakae Shimizu observes, "all the people on the planet will have to be turned into software engineers in a few years. This, of course, is impossible, and so we have to develop software technology to avoid this." And that will take more, more-highly-trained engineers.

Obviously, given the pace of technological change in microelectronics with a new generation of IC devices emerging every two to five years, the requisite perpetual development of human resources cannot be left to the universities, however efficient they might be in providing raw, inexperienced recruits. It is in meeting this vitally important need for continuing education or training that Japanese industry excels over all others.

Lifelong employment makes perpetual and intensive investment in human resources possible. Japanese companies can provide training to their employees without fear that they will leave the firm in response to

Table 2. Outlook for Consumer-Electronics Production in the 1980s

Units: in thousands
Value: in billions of yen

		1976	1977	1978	1979	1980	1981	1983	1985	1990	Average annual growth rate (%) 1985/1978	1990/1985	1990/1978
Color TVs	Units	1,115	987	888	941	930	940	950	950	950	1.0	0	0.6
	Value	7,681	7,008	6,173	6,180	6,175	6,430	6,897	7,315	8,480	2.5	3.0	2.7
B/W TVs	Units	540	534	505	392	361	320	307	280	180	− 8.1	−8.5	− 8.2
	Value	1,035	1,064	973	689	650	595	610	596	452	− 6.8	−5.4	− 6.2
Taperecorders	Units	2,961	2,919	2,793	2,863	2,809	2,752	2,609	2,476	2,256	− 1.7	−1.8	− 1.8
	Value	5,044	5,079	4,716	4,305	4,354	4,431	4,540	4,680	5,166	− 0.1	2.0	0.8
Combination stereos	Value	3,919	4,076	4,137	3,700	3,912	4,299	4,937	5,575	6,741	4.4	3.9	4.2
Radios	Units	895	909	793	514	420	371	349	335	314	−11.6	−1.3	− 7.4
	Value	590	526	467	277	233	212	212	216	234	−10.4	1.6	− 5.6
Car audio	Units	1,871	1,879	1,957	1,953	2,044	2,067	2,096	2,099	2,022	1.0	−0.7	− 0.3
	Value	1,868	1,994	2,056	2,144	2,353	2,477	2,713	2,931	3,438	5.2	3.2	4.4
VTRs	Units	29	76	147	215	306	408	583	715	701	25.4	−0.4	13.9
	Value	607	1,310	2,113	2,881	3,978	4,814	5,888	6,464	6,996	17.3	1.6	10.5
Video discs	Units					17	25	90	140	190		6.3	
	Value					145	198	625	820	950		3.0	
Other	Value	1,700	1,531	1,407	1,624	1,768	1,912	2,142	2,319	2,632	4.3	2.6	5.5
Total value of consumer electronics		22,408	22,538	21,970	21,800	23,568	25,368	28,564	30,916	35,089	5.0	2.6	4.0
(Domestic supply & export)		(14,079)	(14,625)	(13,627)	(14,770)	(15,008)	(15,710)	(17,591)	(18,182)	(18,677)	(3.9)	(0.5)	(2.2)

Source: Japan Electronics Industry Development Association.

offers of higher remuneration by competitors. Workers, engineers and managers, for their part, can devote their full energies to self-development, knowing that they will have an opportunity to use their acquired knowledge to good advantage — their own, their company's and their nation's — for the whole of their working careers. And, of course, just as technological innovation in the electronics industry is the key to competitive advantage, technological skills are a vital element in career advancement, even in companies where seniority sets the order in which employees mount the corporate ladder.

Lifelong employment adds a further differential factor to the equation of human resource development in Japanese industry, a factor which has special import for prowess in microelectronics production. In the economics of all semiconductor device production, experience is a critical

Fig. 4   Japanese Consumer-Electronics Production in the 1980s (in 100 million yen)

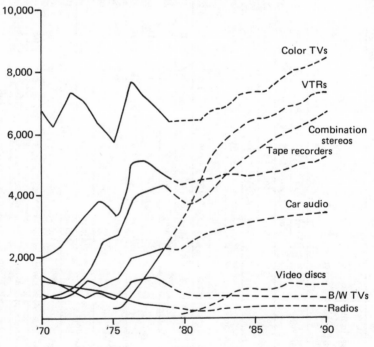

Source: Japan Electronics Industry Development Association

ingredient of competitive power. Each time a company doubles its production of a given device, cumulatively over time, it does the job better, achieving cost reductions of from 20 to 30 percent. It is important to note that the experience involved here is collective: the experience of a group working together at a given task. It follows, therefore, that those companies in which the members work together the longest, have the most cumulative experience and the most cooperative systems of working together, are likely to move down the experience/cost curve faster and further than their competitors.

But, however appropriate the institutional arrangements, they do not work by themselves. Like a symphony orchestra, a company's performance depends ultimately on the quality of its director. The finest musicians badly directed cannot render great performances. Likewise, the most highly trained human resources cannot produce quality electronic products without expert direction.

Fig. 5   Facsimile Equipment Production in the 1980s (in billions of yen)

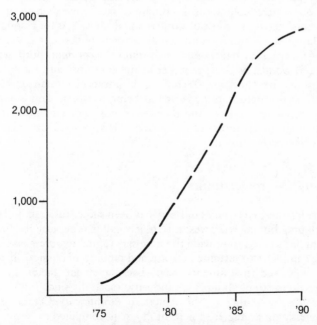

Source: Japan Electronics Industry Development Association

91

This would seem self-evident, and would not merit attention here were it not for the fact that in many advanced countries, the managers of manufacturing enterprises are more concerned with managing assets than in managing people and technology. Management tends to be impersonal; people and technology are expendables, traded in the marketplace of corporate assets by wizard-conglomerators. But manufacturing is people working together to optimize the results obtainable from a given stock of technology and to develop new technologies which will yield better results from those common endeavors. This being so, Japanese manufacturing firms tend to be managed by engineers who have had a lifetime of experience in working together with colleagues in the same enterprise, and one is usually chosen to manage the corporate team for his skills in leading cooperative effort in the management and development of the company's technological stock, skills he has manifested over a 25-30 year period. As a result, Japanese electronics companies have been less prone than some others to lose their technological way. They are likely to be more responsive to changes in technology, and to changes in needs for technology. As Hideo Sugiura, executive vice-president at Honda Motor, puts it: ''Our philosophy of technology puts needs above anything else — technology is subordinate to needs. Needs here means not only users' needs but those of workers and dealers.'' As for Honda, so for most Japanese electronics companies: corporate philosophy is a vitally important aspect of management which states the common purposes of the men and women working together in the enterprise and the role of technology in that endeavor. Technological advance, human development and social progress are perceived as being consistent and essential aspects of corporate purpose. And the effect tends to be synergistic, providing the main thrust for rapid technological change in future-oriented industries such as electronics.

Money and Microelectronics

If the Japanese electronics industry has been successful in mobilizing and developing human resources, it is in no small part because the financial system has worked to provide the necessary capital resources on terms which are conducive to sustained growth and rapid technological change. High savings and investments have been a major factor in the phenomenal growth of all Japanese industry, especially since 1960; from 1960 to 1977, according to the American Productivity Center, gross private investment averaged 35 percent in Japan, compared to 25 percent in West Germany and about 17 percent in the United States. Moreover,

Table 3. Demand for Electronic Computers in the 1980s

(in billions of yen)

Items \ Year	1978	1981	1983	1985	1990	Average growth rate (%)		
						1985/1978	1990/1985	1990/1978
Gross demand (gross supply)	10,212	16,680	20,940	26,000	46,300	14.3	12.2	13.4
Domestic demand	9,516	15,060	18,510	22,670	39,890	13.2	12.0	12.7
Overseas demand (exports)	697	1,620	2,430	3,330	6,410	25.0	14.0	20.3
Domestic production	9,102	15,260	19,260	23,960	42,710	14.8	12.3	13.7
Imports	1,111	1,420	1,680	2,040	3,590	9.1	12.0	10.3

Source: Japan Electronics Industry Development Association.

financial resources have been available to Japanese manufacturers on terms and conditions which are particularly conducive to continuing investment over longer time frames. And for no industry has adequate and appropriate financing been more important than for the semiconductor industry, where, as a U.S. industry leader puts it, ''You almost have to invest a dollar to sell a dollar's worth of integrated ciricuits.''

Fig. 6 Effective Demand for Electronic Computers (1965-1990)

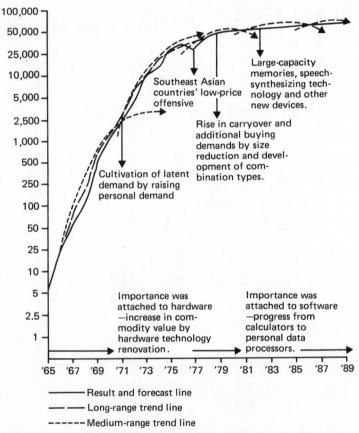

Source: Japan Electronics Industry Development Association.

94

In the years 1966-1970, when Japanese semiconductor manufacturers undertook heavy investments to begin IC production, their plant and equipment investments totaled ¥43.2 billion, or 59 percent of total IC sales in that period. At the same time, R&D outlays for the development of IC technology ran as high as ¥21 billion, or an additional 29 percent of revenues from sales.

While production of integrated circuits in Japan grew tenfold during the 1970s and the race for innovation intensified, the cost of silicon wafer fabrication lines rose five or six times and R&D costs increased proportionately. A recent Ministry of International Trade and Industry (MITI) survey shows that between 1973 and 1978 Japanese IC makers reinvested an average of 17.8 percent of their sales in plant and machinery. Then, in fiscal 1979, total capital spending of the nine major IC makers increased by 77.5 percent compared with the previous year, rising to ¥109.4 billion, or more than twice the total amount invested by the industry in the startup 1966-70 period. In fiscal 1980, investment of the nation's top 11 semiconductor manufacturers rose another 52.2 percent to ¥169.5 billion, most of which was for new integrated circuit production facilities. In the meantime, R&D expenditures by IC manufacturers rose 100 percent from ¥17 billion in 1973 to ¥38 billion in 1978, and continued advancing at a high rate into the 1980s as the industry geared for entry into VLSI production.

By way of comparison — one which explains the closing of the gap between the American and Japanese industries — U.S. semiconductor manufacturers invested on the average only about 12 percent of their gross revenues from sales of integrated circuits between 1973 and 1978; between 1979 and 1980 U.S. manufacturers increased their outlays by just over 20 percent. The reasons for the higher rate investments in Japan are to be found in lower costs of capital and in the availability of bank loans which are relatively insensitive to irrational money market forces and the epicycle of savings in the business cycle. The typical Japanese semiconductor manufacturer's cost of capital in 1979 was 9.3 percent, or about half the 17.5 percent cost confronting U.S. counterparts. Moreover, the Japanese capital market's demands in the form of profit margins or return on investments were more in keeping with long-term financial requirements and performance of industry. Profits of NEC, the leading Japanese IC maker, were 1.2 percent on a consolidated basis in 1978, compared with 5.5 percent for Texas Instruments and 5.6 percent for Motorola. During the same period, Fujitsu's consolidated profits of 2.6 percent were a small fraction of IBM's 14.8 percent. These lower returns on sales deliver an entirely satisfactory rate of return, however, on the highly leveraged stockholders' equity in the Japanese firms.

But greater availability and lower cost of capital are not the only reasons why Japanese manufacturers are able to invest more heavily in semiconductor technology and production facilities. All major Japanese semiconductor makers are also highly diversified, vertically integrated electronic equipment manufacturers, including in their various product mixes a wide variety of home appliances, data processing, telecommunications, automotive electronics and medical equipment. Semiconductor sales account for less than 20 percent for NEC Corporation and far less than 10 percent for other major makers; this compares to approximately 30 percent for Texas Instruments and Motorola and over 70 per-

Table 4. The Electronic Calculator Revolution, 1965-1990

Year	Gross production quantity (in thousands)	Ratio to preceding year (%)	Exports (in thousands)	Ratio to preceding year(%)	Imports (in thousands)	Ratio to preceding year (%)
1965	4	—	0.8	—	—	—
1966	25	625	6	750	—	—
1967	63	252	20	333	—	—
1968	163	259	74	370	—	—
1969	454	279	237	320	—	—
1970	1,423	313	731	308	—	–
1971	2,040	143	1,267	173	—	—
1972	3,866	190	2,612	206	9	—
1973	9,960	258	6,366	244	54	600
1974	15,453	155	10,215	160	145	269
1975	30,040	194	25,727	252	88	61
1976	40,426	135	35,192	137	43	49
1977	31,835	79	28,168	80	458	1,065
1978	42,319	133	32,549	116	677	148
1979	49,500	117	36,000	111	1,600	236
1981	52,000	—	38,000	—	2,000	—
1983	60,000	—	47,000	—	2,200	—
1985	63,000	—	49,000	—	2,500	–
1990	66,500	—	52,000	—	4,000	—

Source: Production — MITI "Dynamic Statistics of Production."
Imports & exports — Finance Ministry "Customs Clearance Statistics."

cent for specialized semiconductor manufacturers in Silicon Valley. Because of their vertical integration, Japanese makers can draw on cash flow generated by downstream products to sustain heavy capital requirements and R&D costs of the IC operations. At the same time, highly diversified downstream production provides a ready in-house demand for IC's, which reduces risks of capital outlays, assures scale economies of production and enables firms to obtain learning-curve economies more rapidly.

The net result — in an industry where capital costs are high and product life cycles of successive generations of integrated circuits are quite short — is some very important advantages for diversified Japanese manufacturers over cash-hungry, manpower-short specialized U.S. manufacturers. Economies of scale and learning, plus lower capital costs, enabled Japanese manufacturers to sell 16K RAMs in the U.S. market at 20-30 percent below the prices asked by American makers, according to

Fig. 7   Electronic Calculator Output and Prices (1965-1990)

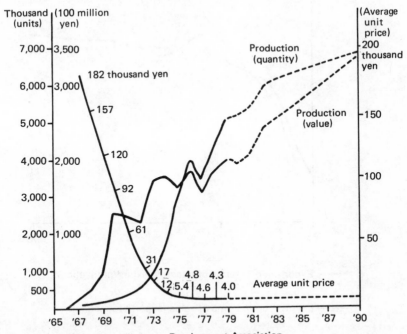

Source: Japan Electronics Industry Development Association

97

testimony before the U.S. Senate Finance Committee. During the post-oil-crisis recession, while U.S. semiconductor makers cut back sharply on capital spending, Japanese manufacturers could continue to invest in new plant, equipment and technology.

The more rapid product diffusion as a result of integrated production, combined with these advantages, gives Japanese makers greater flexibility in responding to changing market needs, competitive presures and technological change. Thus at the end of fiscal 1981, seven Japanese companies will be manufacturing a total of 2.5 million 64K dynamic RAM chips monthly, or about three times the output in June 1981. This mon-

Fig. 8   Electronic Calculator Export Rates (1965-1990)

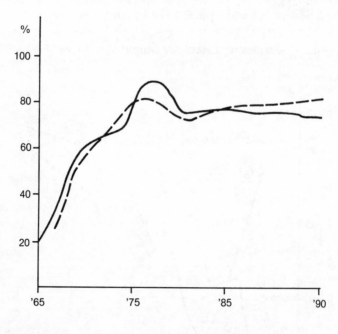

———— Export ratio in terms of quantity (export/production)

— — Export ratio in terms of value (export/production)

Source: Japan Electronics Industry Development Association.

thly production volume compares with the total world production for all of 1980 of a mere 440,000 64K dynamic RAM chips.

Such dramatic increases in production have far-reaching implications for downsteam applications and upstream supply of materials and machinery. While the diffusion rate of 64K dynamic RAMs remains still largely a matter of conjecture, the heavy investment by Japanese makers in new plant and equipment to produce this new generation of devices has already been a boon to makers of IC manufacturing machinery and specialized materials suppliers, catapulting an entire new generation of export industries into world markets which until now have been virtually the exclusive domain of American manufacturers.

The Enterprise and Technological Change

Formidable as are the advantages which accrue separately from a finely tuned industrial policy, superior human resources and high capital investment, it is the capacity of the Japanese enterprise to combine these forces in cooperative endeavor which gives it its ultimate synergistic power. Yet, although it is widely appreciated that there are many aspects of Japanese corporate culture which give to the management system it has produced some especially effective features worthy of emulation, the fundamental difference between Japanese and Western enterprise which will ultimately determine the rate of progress in high-technology industries such as electronics is generally given scant attention.

If, for example, Japanese electronics manufacturers have succeeded in attaining higher productivity than their American or European counterparts, it has been found that this competitive advantage cannot be co-opted simply through higher investment in robots and other automated equipment. There are firms abroad which are as automated as those in Japan, but they usually do not get the same results. Similarly, although Japanese semiconductor makers have obtained strong competitive advantage from higher quality products, the American and European firms have failed to obtain the same high degree of reliability in the end product.

The distinguishing feature of the Japanese corporate enterprise which has made possible the combination of men and money for higher productivity, better quality and more rapid technological change is its cooperative mode of management, as opposed to the zero-sum management systems characteristic of most Western enterprises. This cooperative management system, in turn, is derived from the absence of those dichotomous adversary relationships which pit capital and labor, management and workers, against each other in eternal conflict within Western enterprise.

Ultimately, the Japanese cooperative management system and its underlying philosophy is an expression of fundamental corporate purpose. Unlike Western enterprise, the primary purpose of which is seen as profit maximization, the Japanese enterprise has as its principal objective the creation of wealth, adding value through production. Since from added value must come all rewards to capital and labor, the "social partners" in the Japanese enterprise share a common interest in maximizing its performance.

Workers, managers and shareholders are all duly informed through regular detailed statistical reporting services of the performance of their company, as well as of the performance of competitors at home and abroad, in terms of value added in production. All recognize that if they take more than their share from the wealth created, either someone else must pay or the strength of the enterprise will be dissipated as a result of inadequate investment. Also, there is common recognition that higher investment in new technology and new production facilities means higher productivity, higher value added, and ultimately higher rewards to all members of the enterprise in an equitable sharing of the results of future performance.

To attain this objective, resort to external organizational power is unnecessary. Since there is general agreement and shared interest in the purpose of enterprise, company unions dedicated to those same purposes are at once adequate and desirable as appropriate instruments of communications and negotiation between manager and workers. Workers, as well as management and shareholders, have a stake in quality control and in the mechanization of production, and will voluntarily devote much of their own time and effort to those ends without immediate or direct compensation. Not surprisingly, therefore, today's labor union leaders in most companies are likely to be found among tomorrow's managers.

The implications of such an enterprise system for the electronics industry is immediately apparent. Not only does the resultant cooperative management mode assure higher quality and productivity, which provide the fine cutting edge of competitive power, but it provides the appropriate environment for maximum expression of creative forces within the enterprise. Innovation, therefore, flows naturally, producing a steady stream of new electronic products. Moreover, since rapid technological diffusion within the enterprise is essential for maximal wealth creation, application of microelectronics where savings of labor, energy, materials, space or time are possible, meets little resistance. Quite to the contrary, reason dictates that it be promoted vigorously as a common good.

At the same time, since higher value-added increases the tax take, the state has a vested interest in enterprise efficiency and rapid technological innovation. It is also true, of course, that the gross national product is nothing more than the sum of the value-added by all enterprises, which means that there is an identity of public and private interest in enterprise performance.

There is nothing unnatural or contrived in the manifest cooperation between government and business in the joint promotion of the electronics industry. The resultant higher value-added is necessary for the attainment of the respective purposes of both. And, as it is recognized that higher value-added by industry is the prerequisite of higher wages, there is a broad consensus among all social partners of the inherent wisdom of devoting all needed resources to the rapid development of the electronics industry.

This broad consensus gives to the continuing electronics revolution in Japan its special quality, and its dynamic thrust which makes possible much greater efficiency in innovation as well as in production.

Market Size, Strategies and Structures

If men, money and enterprise are critical ingredients in the development of the Japanese electronics industry, fierce competition at home and abroad has been the force fashioning corporate strategies and structures to obtain full advantage of these basic assets. Although mainstream economic theory suggests that capital intensiveness, large fund requirements, and rapid technological change which characterize the industry should lead to oligopolistic market structures, this has not been the case in the Japanese electronics industry. Rather, the nature of industrial groupings in Japan and the rapid diffusion of electronic technology in advanced, newly industrialized and developing countries throughout the world have made the global electronics marketplace the scene of increasing competition.

Essentially a large-scale technology, requiring the mobilization of massive resources for development, electronics requires optimal markets for fullest utility. The market for most electronic products is, perforce, inherently global, and the application of basic electronic technologies will tend to seek those products for which there is the highest income elasticity of demand in the global marketplace. To compete in this market, large-scale organization is imperative, not only to assure maximal economies of scale and learning in production or to sustain the heavy outlays necessary for continued innovation, but because global structures are necessary to sell the output of large production units and to manage information flows which enable the fine-tuning of innovation and production to catch the full force of market demand on each successive product upswing.

Leading Japanese electronics manufacturers are, therefore, characteristically large-scale enterprises. Although there is a vast number of smaller electronics firms in Japan, most of these are subcontracting production which is inherently small-scale and has been less conducive to mechanization in the past. Others fill special niches, some in global markets which require custom or batch labor-intensive production technologies. In the main, however, major electronics firms in Japan are more diversified and highly integrated than their competitors abroad.

This is especially remarkable in semiconductor manufacture. Japanese firms competing in this sector are significantly larger in total sales and assets than their U.S. counterparts, roughly two to four times larger than Texas Instruments and Motorola, and much larger than National Semiconductor, Fairchild and Intel in both sales and assets.

But, as important as corporate size and the extent of diversification and vertical integration are for funding capital needs, these structural features have other equally important advantages. Not only do they greatly facilitate financing of capital-intensive microelectronic and computer operations, they also speed the process of technological diffusion and enhance the reactive mechanisms which keep technological change in tune with market demand. These in turn have the synergistic effect of priming the processes of downstream product innovation, thus providing an immediate in-house market for new microelectronic devices, new materials, new automated machinery and new software.

Thus "mechatronics" has emerged in Japan as a new field of technological management. Committees within enterprises, within industry associations, and within and among government agencies, systematically examine the possibilities for application of new generations of microelectronics throughout the broad but shrinking range of mechanical devices. Once again, the results tend to be synergistic. On the one hand, added impetus is given to the market for new microelectronic devices, making possible earlier market entry and more rapid return on investment, which in turn make continuing technological change the more feasible and timely. On the other hand, the broader and more rapid introduction of microelectronics in mechanical equipment saves vast amounts of energy, materials, manpower, time and money — all resources in scarce supply. The effect is to reduce cost, improve functions and enhance the reliability of end products — with all this means for competitive power in world markets for sophisticated high value-added products.

102

Not only does this spur investment in advanced electronics at home, but, with the increasing sophistication of technology and markets and given the mounting forces of protection, major Japanese electronics companies have shifted their strategies to increasing investments in overseas production facilities. Following a decade of investment in offshore facilities in low-labor-cost developing countries, after 1975, Japanese consumer electronics equipment makers began manufacturing color television receivers in the United States and Western Europe. More recently, since the end of the 1970s, some of these same firms, along with communications equipment and computer manufacturers such as Fujitsu and NEC, have begun to move integrated circuit manufacture to both advanced and developing countries to better meet the demand of local markets and avoid trade conflicts which were experienced over color television receivers.

Table 5. Japanese Overseas Color Television Production

	Hitachi	Matsushita	Mitsubishi	Sanyo	Sharp	Sony	Toshiba
Hong Kong				x			
Indonesia				x			
Malaysia				x	x		x
Philippines	x	x		x			
Singapore	x				x		x
South Korea		x					
Taiwan	x	x		x	x		
Thailand	x			x			
Australia		x		x	x		
New Zealand				x			
Canada		x		x			
United States	x	x	x	x	x	x	x
Argentina				x			
Brazil		x		x	x		
Mexico	x						
Spain		x		x			
Great Britain	x	x	x	x		x	x
Nigeria				x			

Source: Annual Reports

103

Early in 1981, Japan's four largest semiconductor manufacturers — NEC, Hitachi, Toshiba and Fujitsu — began strengthening their production operations in the United State to cope with the coming VLSI era. In July, Mitsubishi Electric revealed that it, too, was considering erecting a VLSI production facility in the United States, as a joint venture with Westinghouse Electric. In each of these facilities, production will focus immediately on the 64K dynamic RAM, consolidating and adding to the substantial power the industry has already obtained in the market for this new generation of devices and setting the stage for the timely introduction of the first generation of VLSI devices.

Similarly, in Western Europe, these major semiconductor makers are following the strategies and experience developed in color television manufacture with manufacturing facilities strategically located for optimal results in regional markets. NEC, Fujitsu and Hitachi have all finalized plans for European production and will soon be producing 64K dynamic RAMs to strengthen their position in Great Britain and throughout the Continent.

In sum, the Japanese electronics industry has strong structural and strategic advantages calculated to assure a position of leadership in world markets for a wide range of microelectronic devices and the new generations of electronic products which will emerge during the remainder of the 1980s and in the 1990s. Major Japanese firms have the necessary size to manage effectively and competitively largescale electronic technologies. Their degree of diversification and vertical integration is matched by few competitors throughout the world. And they have the added power which comes from appropriate global strategies and structures in virtually all product areas.

To a very large degree, the success of Japanese electronics manufacturers depends on this global outlook, supplemented by an implicit if not always explicit commitment to adapt both strategies and structures to the realities of world markets in order to optimize available material and human resources. As markets change, technology changes to meet new needs, new resources are developed and experience gained. Strategies and structures are being altered accordingly to optimize new realities.

Optimal advantages from large-scale electronics technology are clearly not to be obtained through national technological independence or superiority, but by the development of global objectives pursued through what Dr. Michiyuki Uenohara, senior vice-president of NEC and long-time general manager of Central Research Laboratories at NEC, likes to call "symbiotic cooperation."

Within an increasingly interdependent technological system, firms and nations are challenged to find and develop new modes of cooperation which will permit full exploitation of production economies. Ultimately, of course, the main stimulus for this internationalization of production is competition. But competition and symbiotic cooperation (within enterprises, national markets or the world marketplace), rather than being opposites, are very much two sides of the same coin. The very nature of electronics technology is such that competition must inevitably beget new patterns of cooperation. The autarchic option is tantamount to technological isolation and economic suicide.

Especially in consumer electronics, Japanese manufacturers have in the past played a key role in setting the mode of both competition and cooperation, and they are likely to continue to do so for the foreseeable future. In response to international competitive forces and national priorities in host countries, Japanese consumer product makers have extended technical assistance to, and invested in, the electronics industries of developing countries throughout Asia, the Middle East, Africa and Latin America. Moreover, through competition, followed by cooperation, Japanese makers have contributed to the restructuring of consumer electronics industries in Western Europe and the United States, with remarkable results in increased productivity, quality and export potential.

But as they have advanced technologically and their weight in the global electronics industrial system increases, Japanese electronics manufacturers have been called upon to devise new organizational approaches which, while assuring the benefits of cooperation, are designed to achieve a higher level of cooperation in world markets, including, of course, the Japanese market itself.

This will become steadily more important since, for the reasons advanced above, the Japanese electronics industry will continue to progress more rapidly than most others in advanced countries. Social and economic institutions, cultural values, and a pervasive recognition of economic dependence all conspire with geopolitical imperatives to assure a remarkably high structural adaptability to the opportunities of technological change and contingent changes in comparative advantage in the marketplace.

Since industrial structural change is likely to be more rapid in Japan than in other advanced countries, where ossification of structures is induced by social trends toward oligarchies and by the reduced capability of governments to manage economic and social policies, it follows that the focal point of innovation and invention will continue to shift gradually to Japan — especially in key technologies such as electronics.

There are clear indications that this shift is well under way. While R&D investment in electronics in advanced industrial countries has generally been declining in recent years, it has been rising steadily in Japan. More important still, during this period an increasing share of R&D expenditures by other industries in Japan has been devoted to applications of electronics and electronic capital equipment in their respective sectors.

CHAPTER 5

Mega-Research and Microelectronics

THE MICROELECTRONICS revolution that is already transforming virtually every facet of Japanese industry at breathtaking speed seems bound to advance to a new and even more accelerated stage as major Japanese electronic device and equipment manufacturers give new emphasis to high technology. Following the successful development of VLSI (very large scale integration) circuitry technology through a MITI-supported cooperative research effort, every Japanese electronics firm that expects to still be doing business by the end of the 1980s has stepped up expenditures on R&D; most of the large integrated electronic equipment manufacturers have reorganized their rapidly growing R&D establishments; and an increasing number of these same manufacturers are appointing technical experts to top management posts.

Investments in R&D by leading Japanese electronics companies, which on average have been doubling every five years for the past decade, continue to increase at an annual rate of approximately 15 percent (see Table).

At Hitachi, for example, R&D outlays rose from just ¥36,161 million in fiscal 1972 to 127,973 million in fiscal 1981, increasing from 2.97 to 3.81 percent of sales during the ten-year period. The two next largest integrated electrical/electronic equipment makers — Toshiba Corporation and Mitsubishi Electric Company — have been following the same pattern of investment in high technology development. In fiscal 1980, Toshiba boosted its R&D outlays to ¥74,300 million, an increase in one year from 3.4 to 4.0 percent of sales, while Mitsubishi raised R&D expenditures 20.9 percent to ¥52,000 million, or 4.3 percent of sales turnover.

Major appliance makers are also preparing to harness the full force of VLSI technology. Although Matsushita Electric's strength traditionally has been seen to lie in its sales power, the Osaka consumer electronics

Originally published as "Meeting the Foreign Challenge: Mega-Research Investment for Japanese Microelectronics," *Research Management*, May-June 1983.

giant has added substantial technological prowess during the 1970s, expanding its R&D activities at an annual rate of about 15 percent. From 1976 to 1980, outlays by Matsushita's 23 research laboratories rose from ¥59,400 million to ¥101,500 million, producing a whole new generation of integrated circuit applications, which will affect virtually every product of the company's 40-odd independent manufacturing divisions. Among diversified electrical/electronics equipment manufacturers, Matsushita now ranks second only to Hitachi in terms of absolute R&D expenditures.

At NEC, the leader in telecommunications and semiconductors and one of the Big Three computer makers, R&D spending also increased 15 percent in fiscal 1981 to ¥50,800 million, which was more than 2.5 times the outlays for R&D expenditures in 1977. And, although Fujitsu does not regularly publish data on R&D outlays, Japan's leading computer

R & D Expenditure by Japanese Private Industry, 1970-1979

	(Billions of yen)		
	1970	1975	1979
All industries	823	1,685	2,665
Manufacturing	761	1,537	2,447
Food	21	46	68
Textiles	14	23	33
Pulp & Paper	6	14	14
Publishing & Printing	3	6	5
Chemicals	175	322	490
Petroleum & Coal	10	17	26
Rubber products	10	29	46
Ceramics	18	42	73
Iron & Steel	36	89	120
Nonferrous metals	18	26	40
Fabricated metal	14	29	55
Machinery	72	116	186
Electrical machinery	97	167	312
Communications & Electronics Apparatus	130	289	383
Motor Vehicles	79	196	373
Other Transport Equipment	16	94	72
Precision Machinery	19	36	77
Other Machinery	22	53	75
Transport, Communications & Public Utilities	39	81	144

Source: Science & Technology Agency.

manufacturer reportedly spends 9.5 percent of sales on the development of new technology.

Global expenditures of Japan's integrated electrical/electronic equipment manufacturers tend to understate the level of investments in high-technology electronics, however. The particular strength of these vertically integrated and highly diversified firms is that their large total revenues make possible the concentration of substantial R&D resources in cash-hungry advanced electronics technology. This structural feature of the industry enables makers to devote as much as 22 percent of integrated circuit sales to R&D in microelectronics, compared with four to six percent for total R&D spending in most of these firms. As a result, total R&D spending on integrated circuit research by the 11 top Japanese makers averaged approximately 18 percent annually during the 1970s and continues to run at high levels as makers prepare for production of 256K bit dynamic RAMs (random access memories) in 1983 nd 1 megabit RAMs in the second half of the 1980s. Even those manufacturers highly specialized in a limited range of integrated circuits are now devoting substantial sums to VLSI research. Sanyo Electric, which commands over 70 percent of the market for audio amplifier chips and supplies about 40 percent of its own integrated circuit needs, is a case in point: during the four years 1981-1985, this diversified appliance manufacturer will spend more than ¥8,000 million (US$40 million) on VLSI research alone.

Although there remain vast yet undeveloped microengineering frontiers to be explored before the full potential of VLSI silicon technology is obtained, a critical task now is the development of applications technology. As the timing of the introduction of new VLSI devices already developed by several Japanese makers depends largely on market demand, it has become important to develop downstream applications for the new devices. Competitive power for VLSI in global markets will ultimately depend on how rapidly makers can obtain the combined advantages of high yield (low rejects) as well as scale and learning, and these in turn will be functions largely of the rate of diffusion of the new devices through product application.

In this process of diffusion, Japanese integrated circuit manufacturers have a special and often decisive advantage in the diversification of downstream equipment manufacture — including a wide variety of computers, communications equipment, robots, medical equipment, office equipment, home appliances, and audio-visual equipment — which make possible rapid and extensive integrated circuit applications. Through their networks of central laboratories, specialized development laboratories, and factory applications engineering, Japanese electronics manufacturers are able to speed the diffusion of new semiconductor technology, obtaining at once important economies of production of new semiconductor devices and important competitive advantages in downstream product innovation.

The Propulsive Force of Competition

The main source of this innovative momentum is provided by intense competition. Despite the large scale of integrated Japanese electrical/electronic equipment manufacturers, which would normally lead to an oligopolistic market, competition among these firms within the Japanese market itself is fierce. And since the Japanese market for integrated circuits is larger than that of all Europe combined, its competitive pace tends to gear Japanese industry to a rapid pace of innovation. Added to this primary competitive thrust are two secondary, and substantial, forces: the race with U.S. semiconductor, computer and office equipment manufacturers for world markets on the one hand, and, on the other, the increasing competitive threat of South Korea, Taiwan, Hong Kong and Singapore in global consumer goods markets. Since VLSI circuits make possible drastic reductions in size and cost of many industrial and consumer electronic products, competitive power in the marketplace for such products will largely depend upon the timely diffusion of VLSI devices. This entails not only the redesign of the products themselves, but very often major modifications in production processes.

At the outset, when VLSI devices are relatively expensive, the main opportunities lie in industrial applications. Size and costs of computers will be reduced to a fraction of those at present, while performance and functions will increase greatly as a result of introducing logic circuits with several tens of thousands of gates, and dynamic random access memories with capacities of 265K or 1 megabit. At the same time, large-capacity read-out memories (ROM) using megabit-class VLSI are speeding the development of voice input-output and image input-output systems, as well as *kanji* printers. Research laboratories of Japanese electronic equipment manufacturers are therefore reviewing virtually every industrial electronic product for possible advantages to be obtained through the application of new VLSI technology.

Likewise, consumer electronic products and their manufacturing processes are being subjected to the same kind of technological reassessment. Home entertainment equipment, household appliances, cameras, automobiles, toys and games will all undergo major transformations, and a whole generation of new devices such as home facsimile, personal computers and word processors will appear as soon as VLSI circuits are available at low prices. Or, more accurately, as soon as these products have been redesigned or developed for the marketplace and demand becomes effective, the volume production of VLSI devices which is necessary to bring their costs down will become possible.

The implications for equipment and integrated circuit manufacturers are far-reaching. Depending upon the rate of innovation and investment in new production systems, market shares built up over long years can vanish in the short span of a year or two. For some makers, of course, this means that the timely introduction of new products and capital expenditure in production facilities strategically located around the globe can bring rapid improvements in market share in an equally short time span.

The recent surprise by which Japanese semiconductor makers took 40 percent of the world market for 16K dynamic RAM devices and an even larger share of the market for 64K dynamic RAM devices is but a foretaste of things to come. Similarly, the sudden reversal of market positions of personal computer manufacturers in Japan, where Japanese makers increased their combined market share from 20 to 80 percent in the short period of a year or so, will no doubt be repeated in a vast range of product areas and in many markets.

Superior Mass Production Capabilities

Japanese manufacturers were able to move rapidly to obtain large shares of world markets, despite early leads by U.S. manufacturers, because of superior mass production capabilities and higher quality achieved through more abundant and lower cost capital, more abundant supply of engineering talent, and a more appropriate management system. And these advantages are likely to serve Japanese firms equally well in future direct confrontations in competitive bids for market share. But, as recent experience demonstrates, such competitive advantages, in the absence of a strong lead in basic technology, tend to breed trade friction and the inevitable protectionist measures which follow. Moreover, competitive power based upon mass production capabilities peculiar to the Japanese business environment is not readily transferred to offshore production sites which are becoming increasingly necessary to abate those protectionist pressures.

A strong position in original technology — as earlier U.S. leadership in computers and semiconductors clearly showed and Japanese leadership in video tape recorders confirms — gives manufacturers more options, greater flexibility, and greater bargaining power with governments and potential joint venture partners in moves toward global rationalization of production and markets.

Quite obviously, at the macroeconomic level such higher technology results in higher value-added in production and therefore qualitative growth in GNP which is necessary for higher living standards. In addition, the more advanced electronic technologies clearly conserve energy, materials, space, and manpower. At the same time, they tend to enjoy counter-cyclical demand, easing recessionary pressures on the economy during downswings in the business cycle. And, no less important, they assure stability in foreign exchange earnings as well as bargaining power with "resource-rich" countries which will improve conditions of raw materials and energy supply.

Innovation, in both VLSI chip design and application, has therefore taken on a much greater importance than it had already attained in earlier generations of integrated circuit development. To speed the process of innovation, not only are Japanese electronics manufacturers spending more on R&D, as well as on capital equipment required for the resultant new production processes, but R&D activities are also being restructured within companies to assure greater flexibility of response to changing market demand and the increasingly fierce competition in basic electronic technology.

Some Industrial Examples

Since its founding some 70 years ago, Hitachi, Ltd. has placed heavy emphasis on R&D to decrease its reliance on imported technology. More as a statement of this objective than of actual fact, the adoption in the 1960s of the slogan "Hitachi Supreme in Technology" identified the primary focus of corporate strategies which have since been manifest in the company's position as one of the nation's most intensively research-oriented enterprises and in the consistent choice of technical experts at its helm. Research and development operations, which employ more than 10,000 research personnel in 13 laboratories, are diverse and wide-ranging, reflecting the comprehensive range of Hitachi's electrical and electronic products. But in the past five years, as total outlays for R&D mounted steadily, the company shifted its emphasis in allocation of funds and personnel sharply toward electronics.

While Hitachi's increase in R&D expenditures, to ¥127,973 million in fiscal 1981 from 36,161 million in 1972, has been nothing less than spectacular, the company's emphasis has been more that of a specialized electronics manufacturer than that of an integrated electrical/electronics company. If during the past decade total R&D spending at Hitachi has grown from 2.97 to 3.81 percent of sales, it has been largely due to heavy outlays for R&D in computers and semiconductors, which has been between 13 and 15 percent of annual sales of these products. Reflecting this change in corporate direction, capital expenditures during the past five years have shifted remarkably away from heavy electrical apparatus and consumer products to industrial electronics.

According to recent studies by Nomura Research Institute, Hitachi's capital spending in electronics has increased by about 20 percent annually over the past five years, with particular emphasis on microelectronics and its applications. And the appointment of Katsushige Mita, an engineer who has played a major role in the development of Hitachi's industrial electronics activities, suggests that this trend will continue. Before taking over the presidency, Mita made it clear that he wanted industrial electronics to account for at least 30 percent of Hitachi's total sales within five years time, up substantially from 19 percent in fiscal 1980.

The main thrust of Hitachi's R&D operations is indicated by the company's Central Research Laboratory, where emphasis is on basic and electronics research. Not only is Hitachi devoting a major share of its research resources to electronics, but emphasis is increasingly being placed on basic research. Although fully 30 percent of current R&D expenditures concern basic research, this percentage is slated to rise, reflecting the importance which the company now attaches to developing its own original technology. Parallel to the main thrust in electronics technology development, high priority basic research programs are currently underway in: consumer electronics; new materials and electrical machinery; solar power and nuclear fusion, as well as the improvement of coal-fired generating facilities; and new systems of technology that integrate the considerable range of Hitachi's technological expertise.

The combination of these technologies has resulted in such sophisticated systems as the first integrated production process computer control system to use optical fibers, water purification and waste water treatment control systems, an integrated distribution system, a highway monitoring control system, and building administration systems.

To develop new production systems which at once use these new basic technologies and are adapted to the manufacture of new products employing them, Hitachi has established a Production Engineering Research Laboratory, which among other achievements has kept Hitachi in the forefront of robot manufacture and application. In addition, there are independent research groups in each factory coordinated by special research centers which administer and promote major R&D plans that encompass a number of factories and research facilities.

Toshiba's Technological Offensive

At Toshiba Corporation, as part of a new overall management strategy to boost the company's innovative capabilities, a General Technology Committee was established in 1978 with then-executive vice president Shoichi Saba as chairman to formulate overall technology policies; the same committee was further charged with supervising the execution of technology strategies coordinating all Toshiba new product and production technology development.

To provide specific line execution coordination, a general chief engineer was appointed to each of the company's three broad product sectors: consumer products, industrial electronic products, and heavy apparatus. These top-level engineering executives, in turn, preside over chief engineers within the various specific product divisions of the sectors, strengthening coordination between the operational departments and assuring a system of controls which promotes rapid diffusion of new technologies in all the products and activities of the divisions. At the same time, this system provides a reverse feedback system which makes certain that new technological developments at the divisional level are diffused through the corporation's operations to obtain optimal advantages of innovation wherever they are developed within the corporation.

Accordingly, company systems and structures were revised to mount a three-pronged technological offensive with several basic goals.

• At the company's Research and Development Center, additional resources were allocated and organizational changes were made to concentrate on the development of materials and basic technologies common to the company's broad product spectrum and various production systems. Greater emphasis was also given to the development of advanced basic technologies which will serve as a foundation for future generations of products and production systems.

• The second prong of the offensive entailed the establishment of individual sector/division laboratories to develop new products and applied technologies that utilize the basic technologies developed by the Research and Development Center. In October 1978, the Consumer Products Engineering Laboratory and the Nuclear Engineering Laboratory were established to develop new household appliances and nuclear power equipment and systems. Then, in April 1979, a Semiconductor Device Development Laboratory began operations to develop advanced integrated circuitry and other semiconductor products for the industrial electronic product sector. These were followed in April 1980 with the inauguration of two additional new applied research facilities: the Medical Equipment Engineering Laboratory and the Electronic Device Engineering Laboratory, the latter concentrating on the development and application of electron tubes.

• As the third prong in the new R&D strategy, a Manufacturing Engineering Laboratory has been assigned the task of developing new technologies which will increase the efficiency of the production process, raise productivity and product quality, and obtain further savings in materials and energy.

Emphasizing the overall corporate commitment to this technological thrust, Shoichi Saba, the chairman of the General Technology Committee, was promoted to president of the corporation in 1979, and a managing director in charge of engineering was appointed to head the Committee. Under his direction, a Technology Planning and Coordinating Division now controls the development and diffusion of new technologies within the company and also keeps abreast of technological developments abroad, arranging for technological exchanges wherever advantageous.

At NEC, the sharp increase of R&D expenditures by 2.5 times in the past five years has spurred a similarly extensive reorganization of R&D activities. To obtain greater overall flexibility in research, in July 1980 NEC's management restructured the company's Central Research Laboratories in Kawasaki, spinning off six separate laboratories: basic technology research, optoelectronics research, C&C (computers and communications) systems research, software product engineering, resources and environment protection research, and a scientific computer center. These six laboratories, under the general supervision of the Research and Development Group, have clearly defined goals for the development of new basic information technologies which will provide the main innovative force of the future.

Research in optoelectronics and advanced software programs is given special emphasis in this new organizational arrangement, signalling important developments in these two critical fields where Japanese technology still lags behind U.S. leaders. At the same time, the new structure of NEC's technological activities is intended to integrate developments in these respective fields into a powerful, comprehensive whole with particular stress on the intertwining of technologies and their application in the company's "C&C" approach to information equipment business.

115

To promote rapid diffusion of new technologies, greater precision in production and wider automation of all operations, NEC has formed a new Production Engineering Development Group. Based on the company's recent achievements in the development of automatic assembly systems for communications equipment, manufacturing and testing systems for magnetic disk equipment and high-speed automatic bonding of large-scale integrated circuits, this group will expand its activities through a specialized Production Automation Development Laboratory, which supplements the work of the Production Engineering Laboratory and the Production Facilities Development Division. Operations of all three of these research units are coordinated by a Production Engineering Planning Office responsible for formulating overall corporate production technology strategies.

Given the increasingly high demands which higher outlays on R&D activities make on the company's human resources, the company buttresses its new technological strategy with three facilities for in-house technical training in advanced electronics for engineers and technicians: the NEC Technology Education Center, the NEC Institute of Technology, and the NEC Technical Training School. With its life-long employment system, NEC backs its belief that "human resources are the company's most important assets" with heavy investment in extensive educational opportunities in advanced technologies, software, production control and related management fields.

A Company Run By Engineers

NEC has long been a company run by engineers, some of whom are outstanding international leaders in their field. Both chairman Koji Kobayashi and current president Tadahiro Sekimoto have played key roles in the development of global satellite communications, making NEC a world leader in ground station design and installation. But in recent years the elevation to the corporate board of directors of the company's top R&D executives, Michiyuki Uenohara and Tomihiro Matsumura as senior vice president and associate senior vice president respectively, is evidence of the mounting importance of a commitment to advanced technology which now absorbs more than 5 percent of the company's turnover.

This pattern of intensified investment in R&D and shift of resources to industrial electronics by major heavy electrical and communications equipment makers has its parallel in developments at the leading integrated consumer applicance makers such as Matsushita, Sanyo and

Sharp, where R&D organizations have been restructured especially to assure rapid development of VLSI technologies and their application to consumer products. At the same time, each of these major home appliance makers is accelerating the process of diversification into office automation equipment, a process which began at Sharp in the late 1960s with the development of the first electronic calculator. Each of them is also investing heavily in solar energy technology, having added research and development of amorphous solar cells to their solar energy projects.

If the mounting wave of R&D activity by integrated electrical/electronic equipment manufacturers and their continuing shift of resources to the development of new electronics technology are paramount forces in the incipient VLSI revolution, they are but symptomatic of a prevalent burst of innovative activity throughout the industry. Specialized research laboratories at Fujitsu Fanuc, Sony, Japan Victor, Pioneer, Omron, Canon, TDK, Kyocera, and a legion of smaller manufacturers of new materials, components, instruments and machinery, add further innovative force to the development of new high-technology electronic products and production processes.

The upshot of this explosion of R&D activity, as Matsushita Electric made quite apparent at its special technology exhibition in Chicago in 1981, will be a widespread transformation of home, office and factory, making the 1980s the age of the semiconductor. Japanese industry is clearly determined to demonstrate the full measure of its creative power, and all indications are that this creativity will find its finest and foremost manifestation in the ingenuity with which manufacturers utilize silicon chips to transform more and more products.

CHAPTER 6

The Great Engineering Gap

PLAINLY, THE ASCENDANCY of major Japanese industries such as steel, car manufacture, shipbuilding and electronics is closely related to the performance of Japan's human resources. Maybe Japanese do work better together; but the explanation is not simply that they are better organized, or equipped with more or superior tools. Their working ability is also in large part the result of a life-long education system that has raised the quality of knowledge for a large and critical number of the population to levels unattained anywhere else in the world.

On average, the Japanese are more technologically literate than people in other industrial countries. On this solid foundation, Japan has built a finely tuned, intensive and life-long education system, which is especially designed to develop and train engineers and technicians for productive careers in the high growth sectors in successive stages of the nation's economic development.

In typically systematic fashion, the Japanese begin this vital undertaking at the beginning. At the high school level, the average Japanese student is already more advanced in mathematics and science than his peers elsewhere. At higher levels of education, throughout the 1970s, 20 percent of all baccalaureates and about 40 percent of all master's degrees in Japan were granted to engineers, compared with about 5 percent at each level in the United States. More important still, once the engineer is employed in a major Japanese company, especially one operating in high technology, he is assured of continuing state-of-the-art education to keep him abreast of the latest developments.

In stark contrast to the U.S., where material rewards and professional distinction are most accessible to lawyers and MBAs, the engineering profession in Japan has high social status and attracts the best and most achievement-oriented young people. Senior managers of Japanese industry tend to emerge through the engineering function, and a high percentage of chief executive officers come from the engineering ranks.

Reprinted from "Why Japan's Engineers Lead," in *Management Today*, May 1984.

Even more fundamentally, understanding technology is a cultural factor in Japan that transcends these more mundane considerations. If technology is to be developed as the essential instrument of human material progress, the knowledge and skills of the engineer must necessarily be at the head of this process. As a result, the engineer, not the financier or the professional manager, has come to represent the spirit of enterprise in Japanese society.

Not surprisingly, in purely quantitative terms the availability of well-trained engineering personnel in Japan has been growing more rapidly than in the United States, West Germany and France for over a decade. During the 1970s, employment of scientists and engineers in R&D alone rose by 60 percent in Japan, against less than 20 percent in the United States, 25 percent in France and 50 percent in West Germany. In the context of the total labor force, these relative growth rates become even more significant: over the same period, the total U.S. labor force increased by 24 percent, the Japanese by 6 percent and the French by roughly 5 percent. In West Germany, the labor force actually declined by 6 percent.

But the total number of Japanese engineers employed in production has been growing even faster. By 1980, Japanese industry employed 35 engineers per 10,000 population, against only 25 in the United States, reflecting a higher aggregate technological level in Japan. From the engineering ranks had risen not only half of all directors of Japanese industrial enterprises, and an even higher percentage of the upper management cadre, but also the élite civil servants — half of them hold engineering-related degrees. Significantly, too, while only one in 10,000 Japanese is a lawyer and three are accountants, in the United States there are 20 lawyers and 40 accountants for every 10,000 of the population. It is not especially surprising that U.S. industry is confronted with serious shortages of engineers, even in rapid-growth sectors such as electronics.

By 1980, the shortage of electronic engineers, according to Intel president Andrew Grove, had reached 'catastrophic' proportions. Japanese universities, by this time, were graduating 60 percent more electrical and electronic engineers than U.S. institutions, just the reverse of the situation a decade earlier. On a per capita basis, since 1977 universities in Japan have graduated annually almost three times as many engineers in this critical field as in the United States, more than four times the total in Great Britain, almost six times that of France and approximately 70 percent more than West Germany.

And the gap continues to grow. While the total number of engineering graduates in the United States has been increasing at a compound annual rate of 2.4 percent, the number of electrical and electronic engineering graduates declined at an alarming 2.9 percent during the 1970s. In Japan, meanwhile, the number of engineering graduates grew from 1967 to 1979 at the substantially higher compound rate of 7 percent a year, while the number of electrical and electronic engineers increased at the just faster rate of 7.2 percent a year. Although enrolments in U.S. universities showed some improvement in this sector in the early 1980s, the cutbacks in engineering school faculties and facilities that accompanied earlier declines in numbers have not been reversed.

But the most important aspect of the shortage is qualitative, not quantitative. U.S. industry, and electronics in particular, has too few engineers well-grounded in the basics of engineering and committed to long-term careers in one company, two shortages that are notably absent in Japan. The first lack is especially serious. Far from being a matter of simply improving physical facilities and increasing the size of the faculty, or of carrying out much-needed reforms of university engineering curricula (all these difficult enough in themselves), the problem runs far deeper. There has, in fact, been a general deterioration of the entire education system in the United States, and to a lesser extent those in other advanced countries, while that of Japan has continued to improve.

The education of engineers necessarily begins with the very first educational experiences. Operating on this commonplace wisdom, the Japanese have given higher priority to basics than other industrial countries. Nearly two-thirds of all Japanese begin their formal education at the age of four, in kindergarten, while only a third of American children of this age are in school. In primary and secondary schools, Japanese spend a third again as many days in school as their American contemporaries. Japanese students at all grade levels tend to spend more hours at their studies after school as well. In Japan, according to a recent American study, first-graders (six-year-olds) surveyed spent an average 233 minutes each week on homework; in the United States it was a mere 79 minutes. By the fifth grade (age 10) the average weekly homework was up to 368 minutes in Japan, but only 256 in the United States.

This is not at all because the Japanese are duller or slower learners. On the contrary, numerous recent studies indicate that the mean IQ of the Japanese has risen spectacularly from the turn of the century, and that, at least among the younger generation, it is now the world's highest — some 11 points above the U.S. and West European averages.

A British psychologist, Richard Lynn, reported in 1982 that fully 77 percent of the Japanese younger generation have higher IQs than the average American or European. Lynn concluded that a far greater proportion of the Japanese population has a high IQ, with about 10 percent of all Japanese scoring more than 130 — the levels generally found among professional groups such as research scientists, engineers and doctors — compared with 2 percent of Americans and Europeans.

Among the dramatic changes in Japanese society which have contributed to this remarkable increase in IQ, improvements in the educational system and approaches to learning have the most obvious significance. Recent studies by the University of Michigan's Center for Growth and Development sustain this conclusion. Tests of cognitive ability and specific mathematical achievement, conducted by psychologist Harold Stevenson and his colleagues in the United States and Japan, indicate that it was the educational system, rather than nutritional standards or hereditary genetic factors, which enabled the Japanese to perform better in the testing. The correlation was especially evident in the case of mathematics. Higher Japanese performance in mathematics tests was in more or less direct proportion to the relative amount of time spent studying the subject. Mathematics consumes 25 percent of the first-grade classroom time in Japan; only 14 percent in the United States.

If there is any link between IQ and industrial performance, there are far-reaching future implications in having a more advanced educational system in a knowledge-intensive society which is becoming increasingly dependent on technological innovation. The superior quality of the Japanese educational system seems certain to assure that Japanese industry will play an increasingly innovative role in future decades.

Comparative international tests conducted by UNESCO, showing that average Japanese scores in mathematics and science are significantly higher than those of any other country, give further support to this conclusion. Analyses of these findings show that mathematics instruction at junior high and high school levels follows the pattern of primary schools, advancing faster in Japan than in other industrialized countries. The difference is astonishing. As a result, Japanese students normally have more maths by the ninth grade (14) than most U.S. high school graduates. Calculus, probability and statistics are taught at the high school level in Japan, but half of all U.S. high school students take no mathematics at all after the age of 15.

A similar pattern prevails in other fields of science. As a consequence, estimates Stanford University researcher Thomas P. Rohlen, the average Japanese school leaver has as much basic knowledge as the average American college graduate. What's more, although high school is not compulsory in Japan, at least 95 percent of all young people enter high school and about 90 percent graduate; the comparable figure for American youths who took their diplomas at the beginning of the 1980s is under 75 percent.

Contrary to American experience, the Japanese have not sacrificed quality in the process of making their education system universal. A telling point is the recruitment of teachers. In Japan, where the competition for teaching positions is rigorous, only university students with good academic records can apply, while in the United States or United Kingdom, teachers are normally drawn from the less academically able, as measured by scholastic aptitude test scores. Japanese teachers are correspondingly well paid in comparison with other professions; in contrast, low salaries for science and mathematics teachers have driven large numbers of American teachers to better-paid positions in industry and have discouraged college students from entering the teaching profession at all.

All this educational emphasis goes some way to explain the success of Japanese workers and the rapidity with which Japanese firms have overtaken Western leaders in industry after industry. Higher technological literacy tends to speed the process of innovation and change, while reducing the incidence of technophobia. But there is yet another factor which magnifies the effects of the knowledge acquired by individual students. The intensive socialization they undergo throughout their education, and especially in their high school years, prepares students for cooperation in the use of their learning. This socialization process also accustoms the average Japanese to intelligent and meticulous work habits, which are especially useful on the factory floor and in engineering studies.

Since engineering is perceived as a socially useful function, assuring the wealth of the nation and attracting high social status, engineering studies in universities are at once popular and prestigious. Accordingly, the number of students taking engineering courses in Japanese colleges and universities at the beginning of the 1980s was approximately 368,000, or 20 percent of the total enrollment. Remarkably, this was more than seven times the number of students studying pure sciences, the social relevance and practical utility of which have not been so readily apparent.

By comparison, Britain at that time had 36,000 engineering students enrolled in universities — not even one-tenth the Japanese total, although Britain outranks Japan in the number studying physical sciences. While Japanese universities were turning out some 75,000 engineers a year at the turn of the decade, the number failed to reach 9,000 in Britain, where science graduates averaged 25 percent. More surprisingly, the number of engineering graduates in the United States, about 72,000 a year, was also smaller than that of Japan; however, physical scientists, although in much lower demand, were graduating at a rate eight times faster than in Japan. The U.S. university system has actually been graduating significantly fewer engineers than Japan, despite having five times as many students in all.

The lower number of Japanese physical science students has some rather compelling economic explanations. At the beginning of the 1980s, around 5,000 scientists with advanced doctoral degrees were unable to find jobs. For many, the most they can hope is to continue their work as unpaid researchers at universities. To meet their priority targets, neither Japanese industry nor government requires the theoretical training of these scientists. Engineers, not scientists, provide the technological and managerial backbone of Japanese industry.

But the educational statistics sketch only part of the Japanese picture. Especially in engineering, formal education is the beginning, not the end, of the learning process. Rapid technological innovation coupled with radical structural change in industry has added greatly to the importance of continuing education. And in Japan, companies rather than universities provide the bulk of this post-graduate training.

In Japan, companies invest much more in training and retraining of engineers than elsewhere, with far-reaching implications. It is not uncommon, for instance, for Japanese firms to shift engineers from one field to another, providing extensive retraining in new professional areas. During the 1970s, in-company training focused especially on shifting mechanical engineers to electronics, hardware technicians to software and blue-collar workers to white-collar employment. Massive retraining was an imperative of the industrial structural change that was necessary in the wake of successive monetary adjustments, oil shocks and wage increases. The effect was not lost on the institutional and organizational facilities required for continuing training; since then, they have honed the competitive cutting edge of high technologies for the 1980s.

Among the leaders in the development of formal corporate education, Hitachi maintains no less than five institutes wholly or in part devoted to the development of engineering capabilities. The most important among them, the Hitachi Institute of Technology, attached directly to the president's office, trained 5,000 engineers during its first 10 years. Its intensive one to three-week courses are designed to broaden the trainees' field of vision, as well as to upgrade and update their knowledge. The other four institutes include two technical colleges offering 15-month courses for technical school graduates, which the company claims are the equivalent of a four-year college or university engineering course. A productivity training center, wholly devoted to education in the arts of production efficiency, is not reserved exclusively to engineers, but rather diffuses the results of production engineering throughout the organization. And the Hitachi Comprehensive Management Research Center in Abiko, Ibaraki-ken, places special emphasis on the training of executives, from newly-appointed section chiefs through to departmental and divisional managers of all enterprises affiliated to the Hitachi *kigyo keiretsu* (industrial group of companies).

All freshmen technicians are gathered at the Hitachi plant in Ibaraki-ken for a full two-week intensive training session which is then followed by three-day 'lodge-in' sessions — Friday to Sunday — every other weekend for the first two years of their employment. In addition, a consistent and systematic on-the-job training program gives engineers and technicians the practical know-how which can only be developed through hands-on experience after assignment to a particular job. This means that those in supervisory positions in the factory must give special importance to developing their teaching and job-assignment skills in order to advance the development of their subordinates. All this, not including the costs of on-the-job training, required an outlay of ¥10 billion (US$42 million) a year at the beginning of the decade — which is sufficiently large to classify Hitachi as a major educational institution. Qualitatively, too, that is not far from the truth; an engineer joining the company can look forward to a life-long program of education to assure the continual upgrading of his technological knowledge.

This same principle is the cornerstone of NEC's five-year-old Institute for Technology Education. By 1977, it had become painfully evident that the knowledge gap in the company's engineering was growing rapidly. A survey in that year established that, while 50 percent of company recruits in the period 1960-65 had felt that it took 11 to 12 years for a gap to open up between their knowledge level and the level of expertise required for their work, a similar percentage of the 1975 engineering entrants were aware of a gap after only three years. In fact, most new entrants considered that, after two years, their university knowledge was no longer sufficient for them to carry out their jobs effectively.

The problem was not that university education had worsened, but that technology is evolving so rapidly that existing knowledge, including what is learned at universities, becomes obsolescent much quicker. In addition, however, the technical level and demands of Japanese industry have grown much faster than the ability of the educational system to meet them. Faced with the obvious consequences of such a knowledge gap for its wide range of high-technology production, NEC in March 1978 formed a top-level corporate task force headed by a senior vice president and a working group consisting of engineering department managers. They decided to establish the new Institute for Technology

Education under the direction of Dr. Yukimatsu Takeda, a top executive with long and broad experience in the areas of both telecommunications and microelectronics.

In its first four years, the new Institute awarded graduation certificates to 1,500 engineers. Entrance to the Institute is open to all NEC engineers with more than two years' service. But candidates are not selected by management, even though the curriculum is specifically designed to meet the overall company requirements for implementing its 'C & C' (communications and computer) strategy. Rather, the initiative must come from the individual engineer, who is automatically accepted, on the one condition that he has obtained the consent of his boss.

This prior approval is not required because the advanced training is an escape from work. On the contrary, attendance at the Institute requires extra effort on the part of the engineer-trainee, who must continue to meet his regular responsibilities, plus those required by the course of instruction. Professors at the Institute must also work harder. Most of them are NEC executives for whom teaching responsibilities have to be carried out over and above their regular duties. Reliance mainly on capable members of the company who know the relevant technologies and their application has two purposes: systematization of technology transfer within the company to obtain optimal advantage from the company's own knowledge resources, and further training of the instructors themselves, who must refresh and organize their knowledge for effective teaching.

The Institute's Principal Technology Study Program is an advanced engineering course which provides 160 trainees with a year-long instruction in specialized fields of technology in seminars limited to 10 trainees per subject. Seminars meet for a half-day every two weeks for six months, after which each attendant must write a technical paper meticulously prepared over the next six months under the direction of the instructor. A medium-level engineering program lasting a year begins where university education leaves off, to upgrade the engineer's general technological level. Five separate courses of study are offered, with lectures one day a week for a year. Other special seminars on requested subjects are offered to trainees at various levels for periods varying from 10 to 20 days per course. Institute programs supplement, rather than replace, the engineering, management and administrative training provided at the divisional or departmental levels, capping NEC's life-long education program. "In fact," says Dr. Takeda, "work in the age of information technology is permanent, life-long education. The enterprise has become essentially as much an educational institution as it is a place of work."

In Japan, corporate technical training has become a surrogate for the university graduate school of advanced engineering. The difference is that post-graduate training is practical and carefully tailored to specific corporate objectives, giving the engineer the opportunity to combine practical learning with study of advanced theory in a way that is meaningful in terms of a clearly perceived career development. Unlike university graduate education work, this is open on a continuing basis to every engineer working in leading high-technology companies.

In sum, at all levels of human capital formation, the Japanese education system seems to be responding more effectively than its competitors' to the economic needs of society and the individual alike. Ultimately, success in the competition between rival knowledge-intensive companies, as well as national economies, will depend on the quality of human resources. For Japanese, education, motivation and innovation are essential ingredients of success. The available evidence suggests that the nation's ability to sustain creative technological advance by highly motivated, trained engineers and workers is assured. Right down to the present, the highest priority has historically been given to what is perceived as the single most important task: maximizing the nation's sole resource, human potential, through education designed to meet the needs of the times. Japan's remarkable attainments in engineering education make its leadership in high technology during the coming decades look a very safe bet.

CHAPTER 7

Finance for High Technology

EVER SINCE THE Japanese semiconductor industry appeared out of nowhere in the late 1970s to grab a reported 70 percent of the world market for 64K RAM chips, it has become increasingly apparent that Japanese work better in part because they are better financed. More and cheaper money equips Japanese workers with more and better tools, which accounts for much of their higher per capita productivity and rapid technological advance. At the same time, a stategic shift of resources, with appropriately-timed finance allocated to higher value-added production, has accounted for as much as 20 percent of the total productivity increase over a decade.

Among high-technology industries, few are more sensitive to the availability and cost of capital than semiconductors. Ten years ago, a semiconductor factory could be built for US$1 million; in 1984 that pays for just one electron-beam etching machine used in VLSI production, and a complete up-to-date facility costs upwards of US$20 million. Although output has soared during that period, capital needs per dollar of sales are said to have risen by more than 50 percent over the period and continue to climb at an increasing rate.

Remarkably, while the U.S. frontrunners were forced to restrict their outlays for the development of new technologies and production capacity during the recession following the first oil crisis, Japanese semiconductor makers continued high-level investment in both R&D and new production facilities. As a result, as the boom in demand for 16K RAM chips gathered momentum in 1978 and 1979, Japanese manufacturers had the capacity to supply the market. U.S. makers, unable to meet the rising demand, were faced with the painful alternative of losing market share or buying in Japan themselves to meet their customers' requirements. This was also necessary, in some instances, because the Japanese could deliver much higher quality electronic devices at the same prices that U.S. manufacturers charged for standard products.

Reprinted from "Japan's Winning Capital Asset," *Management Today*, June 1984.

The scenario was repeated during the recession following the second oil crisis, which came just at the time the industry was gearing up for heated competition in the next generation of semiconductor devices — the 64K RAM chips — signaling the arrival of the VLSI (very large scale integration) era. Before the U.S. industry, squeezed by record high interest rates and a general shortage of capital, could get its act together, Japanese suppliers had established a strong lead in world markets. R&D was more advanced in some critical fields, new production facilities — including the world's largest semiconductor plants — were in place, and Japanese semiconductor makers had nationalized output in Europe and the United States before Silicon Valley firms were able to bring the new devices to market in commercial quantities.

Now a third and even more gripping act of this drama has moved on stage. Although four Japanese manufacturers have prototypes of the advanced 256K RAM devices, with two more ready to join them in the wings, only two U.S. players are among the *dramatis personae*. And, in a play within the play, in good Shakespearean tradition, Japanese companies have begun to sell advanced supercomputers in a market that even IBM has not seen fit to enter, while seeking revolutionary technological advances through their Fifth Generation Computer Project.

The story is by no means confined to the high-flying electronics industry, however. After two decades of massive investment in automobile and steel production facilities and their related technologies, by 1978 Japanese manufacturers had emerged with commanding capital-to-labor factor ratios. In that year, Japanese automakers employed approximately US$110,000 worth of plant, equipment and working capital per employee, compared with an average US$40,000 for the big four U.S. firms. Japanese steelmakers had invested even more heavily in what is usually regarded as a mature technological sector. Their US$200,000 worth of assets far outstripped the equivalent US$60,000 in current tangible assets and working capital available to the average U.S. steelworker. And these differentials are even greater when the Japanese industries are compared to their counterparts in the U.K., France and Italy.

Quite obviously, there is no way firms in these countries can compete effectively in home or world markets when they are unable to sustain appropriate levels of investment. Continued under-investment over long periods of time has emasculated companies in key industrial sectors. However well motivated, workers equipped with less than half the physical facilities cannot match the production capabilities of their better-endowed competitors. However well organized, firms which cannot sustain high levels of investment, especially at those stages when garnering market power is most critical, will be doomed to the role of

130

also-rans in the race against those which enjoy the necessary funding at advantageously low capital costs.

Curiously, the ability of the Japanese financial system to assure the necessary flow of funds to industry, under highly favorable terms and conditions, for quantum leaps forward in technology and productivity, has never received the attention it merits. Part of the problem, at least for some American observers, has been the conviction that the U.S. capital market, and hence the financial system, is the world's most advanced and efficient, and that the Japanese system is by these standards a primitive model. But the principal function of American capital markets has never been to finance industry, much less provide funds needed for technological advance; in the first instance, it is much more to provide a secondary market for stockholders to trade their certified claims against firms in which they have already invested. Nor has the finance of industry been the special forte of the U.S. banking system, which is predominantly oriented towards the medium-term financing of domestic trade.

In contrast, the modern Japanese financial system, evolved after the Meiji Restoration, was originally developed for, and remains highly specialized in, the financing of industry. At the very outset, the imperatives of national survival determined this priority. And the massive financial requirements of rebuilding Japanese industry after the 1923 earthquake and World War II reinforced this imperative.

The Cost of Capital

As a result, not only has the banking system itself been especially designed to meet the particular needs of industry, but both interest rate policies and the tax system have given high priority to the task of supplying capital to industry at the most favorable rates. And since the supply of capital must ultimately begin with savings, all relative systems — financial, tax and wage — have been consistently designed to foster the high rate of saving for which the Japanese are now remarkable. Supply side economic management began by assuring the flow of funds to industry at the cheapest possible costs.

Despite the obvious significance of this consistency, the efficiency with which the Japanese financial system has delivered funds to industry has attracted comparatively little attention. Relatively low labor costs, business-government cooperation, industrial policy and other easier explanations seemed to tell all. But had it been fully understood that the cost of capital to Japanese firms had been running well below that of American industry, at least since the beginning of the Income Doubling

Plan of the 1960s, the competitive strength of Japanese industry could have been more readily understood and anticipated.

According to one recent study by Dr George N. Hatsopoulos, the cost of capital services to Japanese companies has varied from half to a small fraction of that paid by American firms throughout the period since 1961. If these calculations are correct, in 1973 and 1974, U.S. companies paid about 40 times more for their capital services than their Japanese competitors, and in 1981 that multiple was still more than three.

In fact, using standard U.S. procedures for evaluating the cost of capital, Hatsopoulos found that the after-tax real cost of funds for profitable Japanese firms has actually been negative for the past 20 years or more. This highly auspicious arrangement has been possible, he argues, since most firms rely heavily on bank borrowing for their capital requirements; and while interest payments are deductible from taxable income, the inflationary appreciation of fixed assets is free of tax.

The overall real cost of business investment is no simple calculation, of course. Cost of capital depends both on its source (debt or equity) and how it is used (whether for financing fixed assets, inventory or accounts receivable). Tax codes differ in their treatment of capital raised respectively through debt or equity. And the use of that capital, too, is subject to different tax allowances, depreciation rates and other treatments which affect the overall cost differential from country to country.

Nonetheless, the real cost of capital for any firm is composed of two main ingredients: the basic cost of raising money, and the costs and benefits arising from the different ways in which the money is used. Since Japanese companies typically rely more heavily on debt financing, their capital costs are determined largely by the prevailing low interest rates. U.S. companies, on the other hand, which rely predominantly on equity for their external needs, are primarily dependent for the cost of their capital on the price-earnings ratio of their stock. The American firm which depends heavily on equity financing will pay less for its capital the higher its shares are priced in the market. Maximizing profits on equity is, therefore, the way in which the American manager minimizes the cost of capital.

While studies vary in their findings of the magnitude of difference, the consensus is that Japanese firms' capital costs are substantially lower than those of their U.S. counterparts. According to calculations released by the Electronic Industries Association, U.S. semiconductor manufacturers must allocate 13 percent of a product's selling price to the service of capital, compared to 8 percent by their Japanese competitors.

Dr Robert Noyce, vice-chairman of the most celebrated American chipmaker, Intel, estimated at the outset of the decade that the U.S.

semiconductor industry would have to raise some US$41 billion in new capital by 1990. If no new equity were raised, given a debt/equity ratio of 25/75, firms in the industry would have to increase after-tax earnings to 13.5 percent of sales to finance their capital requirements — or over three times the industry average during the decade from 1968 to 1977, when it was relatively free of Japanese competition. "In the final analysis," Noyce concluded, "international competition in semiconductors may be a battle of financing, and advantages in access to capital may be decisive in determining the world market share (of a manufacturer) by the end of this decade."

The problem is not, as some would have it, one of Japanese government support for high technology, which in semiconductors is substantially less than the U.S. government's total contribution to American industry. Rather, the crux of the matter is capital costs, which cut across the entire spectrum of industry. Clearly, this problem cannot be met by government subsidies. According to the Hatsopoulos study, for projects requiring 10 years of R&D expenditures, the required subsidy would need to be more than 500 percent of the companies' contribution.

As such rates are unthinkable, the logical conclusion is that the real cost of funds and capital prevailing in the United States may well preclude the development of many technologies which can be afforded easily in Japan. Indeed, this is exactly what happened in recent years at the applications stage for semiconductors, when Japanese firms led the development of solid state radios, television, calculators and watches.

Savings Outstrip West

That this is the more likely pattern for future technological and industrial development becomes clear on a closer examination of the reasons for the greater and more appropriate availability of capital in Japan, as well as its lower cost. In the first place, Japanese gross national savings have continued to outstrip those of all other industrial countries by substantial amounts, assuring a high level of capital formation. Throughout the 1960s and 1970s, Japanese savings varied from 32 percent to 40 percent of GNP. For comparison, savings in France and West Germany averaged about 24 percent during the past two decades, in Britain 20 percent, and in the United States an even lower 17 percent. Personal savings from 1976 to 1980 reveal an even wider disparity: Japan 20 percent, France 16 percent, West Germany 14 percent and the United States just 6 percent.

Although part of the difference may be explained by cultural and institutional factors, the large Japanese lead is attributable mainly to a variety of incentives specifically designed to induce savings. Interest on savings, contrary to practice in the United States, is tax-free up to a certain level, while consumer credit, which acts as a disincentive to savings, is deliberately kept low. Low consumer credit, in turn, reduces the upward pressure on interest rates. The salient point is obvious: the higher the rate of savings, the greater the supply of capital and the lower its cost. And Japanese savings are likely to remain the world's highest for the foreseeable future, sustaining a continued downward bias in interest rates.

Second, a broad agreement exists between the Japanese government and business circles that the tax system should be used to promote savings and investment. Complementing tax exemptions on income from personal savings, measures adopted to stimulate private investment include special tax-free reserves, accelerated depreciation on certain kinds of assets in which the government wishes to promote investment, as well as tax credits for R&D.

In addition, the tax code is very industry-specific, designed to influence resource allocation by encouraging investment in higher value-added advanced technology industries. And the overall burden of taxation is relieved by the lower rate of inflation, serving to spur further the rate of saving and investment and to direct capital into industries targeted for rapid growth.

Third, unlike its counterpart in the United States, the Japanese government does not seek to solve all major economic problems by tinkering with the price of money. Rather, the primary determinant of interest rates has been the imperative of assuring an adequate and low-cost supply of capital to industry. Low interest rates, added to the tax deductibility of interest payments, have served as incentives for companies to rely largely on banks for their capital requirements.

As a result, for most major Japanese firms, debt financing is available when required, regardless of the debt-equity ratio. Funds are readily provided by banks at low cost when the most long-term and risky investments must be made. Although debt/equity ratios between 75/25 and 85/15 are common in Japanese firms, the level of debt may well expand to five to 10 times equity during periods of expansion.

A fourth factor is that some of this debt is financed on 'soft' terms through the governmental Fiscal Investment and Loan Program, which allocates some US$100 billion annually. While the bulk of these funds is directed to small and medium-sized enterprises, agriculture, housing and regional development, included in this program is the budget of the Japan Development Bank which, at the beginning of the 1980s, was lending around US$600 million annually to high-technology ventures. Significantly, in the past two years, the JDB has included foreign firms investing in high technology production in Japan among the beneficiaries of this preferential lending.

In the past, therefore, equity financing as a means of raising new capital has played a minor role in Japan. More often, new stock issues at below the market value have served as a source of income for existing stockholders. But as share values of debt-financed Japanese firms have tended to rise, the cost of equity financing has declined steadily. Ironically, the opposite trend has prevailed in the United States, where firms are more heavily reliant on equity financing. High interest rates have driven share values down, increasing the cost of raising new equity capital. Thus, while real capital costs rose from 15 percent to 20 percent in the two decades from 1961 to 1981, much of this increase was due to the stock market's fall, pushing up real equity costs. In 1981, the nominal pre-tax cost of equity was well above 30 percent, compared to 'triple A' bond rates of about 12 percent.

The consequences for Japanese industry of its favorable supply of capital are auspicious. Most important, Japanese innovation will continue to be fueled by a flow of capital at low real cost. Since the economic value of innovations ultimately depends on the ability of industry to invest in new plant and equipment to put them into action, it is very sensitive to both the greater availability and lower cost of capital. As Japanese firms will continue to be more capable of shifting resources steadily to higher value-added technologies, an increasing range of technologies will inevitably be developed more profitably in Japan than elsewhere.

Since the cost of labor continues to rise sharply in relation to capital costs, the capital-to-labor factor ratio will increase steadily at relatively high rates. With this ratio continuing to grow more rapidly in Japan than in Europe, a new international division of labor, already taking form, is likely to emerge. The more capital-intensive industries will develop in Japan. Factory automation will continue to outstrip that of other industrial countries, since, with lower capital costs, Japanese firms will be able to sustain the pace of diversification which is necessary to assure the redeployment of labor in new jobs with a higher quality of working conditions.

Japan's Capital Asset

Lower capital costs will also assure Japanese firms the necessary funds for internationalization of production at a quickening pace. Equipped with higher technology, the Japanese tend further to rationalize production, globally moving output to locations where access to markets and lower cost non-financial production factors are readily assured. Higher stock values and appropriately lower capital resources will combine to improve options for international expansion through acquisitions, wholly-owned subsidiaries or joint ventures.

The net result, then, of a continuing flow of low-cost capital will inevitably be to improve the competitiveness of Japanese firms in world markets. Value-added productivity increases will outstrip those of competitors abroad, not only because of increased innovation, more investment in advanced manufacturing systems, and rational global investment strategies, but also because of the ability to sustain the high level of industrial structural change which the latter entail.

The unknown factor in this equation is, of course, how long growing pressures at home and abroad for a liberalization of interest rates and fuller internationalization of Japan's capital and money markets can be resisted. If the Japanese market were opened to foreign borrowers and all restrictions were removed, the demand for funds in Japan would increase, with resultant upward pressure on interest rates. At the same time, Japanese investors would normally seek higher returns in foreign markets, lessening the supply of capital at home with similar effects, and bringing interest towards equilibrium.

This would create radical problems for the financial system and the economy as a whole. Most important, it would limit the ability of MITI and the Ministry of Finance to direct funds into targeted industries to assure the direction and pace of industrial structural change. And this would inevitably have negative effects on the competitiveness of Japanese exports. At the same time, higher interest rates would increase the burden of the public debt, which in itself suggests that progress toward liberalization is not likely to command much enthusiasm in official circles.

There is, of course, one other possibility, which, at least at present, seems equally remote. If the United States and other OECD countries were to succeed in restoring the lower interest rates, inflation rates and equity risk premiums that prevailed during the 1960s, the real cost of capital services would drop, substantially reducing the Japanese competitive advantage. This would require, at the very least, revamping tax systems to encourage savings and investment, along with measures to discourage improvidence among consumers and reduce deficits.

In the United States such a program would entail radical changes from the present high interest rate policy and the overvalued dollar which it sustains, to an abiding concern for the competitive power of American industry at home and abroad. Presently, and for the foreseeable future, both major U.S. political parties appear to have more pressing concerns, while the European countries show almost no sign of achieving the monetary or technological coordination required. The capital-backed ascendancy of Japanese industry will, therefore, in all likelihood go unchallenged for many years.

PART III.

Consumer Electronics

CHAPTER 8

The Protectionist Threat

CAN UNITED STATES television manufacturers regain their former pre-eminence in a massive restructuring of the industry following the phenomenal example of U.S. calculator manufacturers during the first half of the 1970s? And, if so, what are the implications for the Asian electronics industry? These are the muted, and largely unasked, questions that ultimately must be answered once the smokescreen of legal gobbledy-gook and diplomatic doubletalk is cleared from the battlefield in which the electronics industries of three continents are locked in contest.

On the surface, there are ominous signs of a resurgence of protectionism, sponsored by a new and powerful coalition of industry and labor. This is highlighted by the decision of the U.S. Customs Court of New York to overrule the Treasury Department refusal to impose countervailing duties on Japanese television sets which benefit from consumption tax exemptions and the almost simultaneous recommendation of the U.S. International Trade Commission (ITC) that higher customs duties under the so-called "escape clause" be given an "injured" U.S. television industry.

Both the Customs Court decision and the recommendations of the ITC were based on the inherently protectionist 1974 Trade Act, which considerably strengthened the provisions under which imports into the United States can be restricted — either because the imports are regarded as unfairly competitive or simply because of their adverse impact on American industry.

But the action brought by Zenith Radio Corporation in the Customs Court and the petition of the ITC by the consortium of 11 labor unions and five firms from the industry — the Committee to Preserve American Color Television (COMPACT) — must be seen in the context of a much broader legal counter-offensive against television sets from Japan, Taiwan, South Korea and Mexico.

Reprinted from "Science: America's Ultimate Weapon," *Far Eastern Economic Review*, 1 July 1977.

In January 1976, GTE Sylvania Incorporated, also a member of COMPACT and a petitioner in the latest ITC escape clause action, had already filed a complaint with the ITC alleging that 13 Japanese television companies had engaged in unfair acts in the importation of portable color television receivers. The complaint alleged violation of Section 337 of the Tariff Act of 1930, a highly-protectionist piece of legislation born of the hysteria following the 1929 crash.

GTE Sylvania was joined in this petition by the Philco Corporation, whose trademark GTE Sylvania purchased in 1974 when Philco discontinued television receiver production. Subsequent to the Sylvania-Philco complaint, the ITC initiated its own investigation of Japanese television marketing practices covering 14 areas of alleged unfair acts.

Both of these proceedings were suspended during the commission's consideration of the COMPACT escape clause petition, which was made in September 1976 after a three-fold increase in color television imports gave the Japanese 30 percent of the U.S. market. But they remained on the ITC books, and were taken up again as soon as the commission's recommendations for relief for the industry had been acted upon by the President.

The Carter Administration's answer to the ITC was to have its peripatetic Special Trade Representative, Robert Strauss, conclude an agreement in May with the Japanese to restrict imports of television sets, and under a proposed consent order the ITC may drop its investigations of unfair pricing practices in return for agreement by Japanese manufacturers to submit to detailed surveillance of their exports to the United States over a five-year period.

In addition to the proceedings before the ITC and the Customs Court, there is continuing anti-dumping surveillance by the Treasury Department and a major anti-trust suit pending in a U.S. District Court brought by Zenith in 1974 against 21 Japanese manufacturers and their U.S. marketing subsidiaries. Seeking almost US$1,000 million in damages, as well as an injunction against continuing alleged violations of the law, Zenith charged that Japanese producers have engaged in a varie-

U.S. Television Imports

	1975	1976
	(US$ million)	
Japan	263	615
Taiwan	157	200
South Korea	20	34
All other	8	37
TOTAL	448	886

ty of anti-trust violations, including an unlawful cartel and conspiracy in restraint of U.S. foreign and interstate commerce, and conspiracy to monopolize such trade in violation of the Sherman Act.

Apart from enormous legal fees and the cost of multiple investigations and court actions, what specific impact this will have on imports of finished or unfinished television sets and sub-assemblies from Japan, Taiwan and Korea is by no means clear.

Higher Prices

If the U.S. Customs Court ruling is upheld and a 15 percent countervailing duty is imposed on Japanese color television imports, this — combined with the "voluntary restraints" the Japanese industry has accepted — would cut imports of color receivers by as much as 50 percent from current levels. The U.S. industry has sought to reduce Japanese color television imports to 1.2 million sets a year from the record 2.96 million of 1976.

Prices have already begun to reflect these changing supply conditions. Anticipating possible extension of countervailing duties to other electronic products as well, U.S. importers of radios, hi-fi equipment and television sets from Japan have been forced to raise prices to cover eventual retroactive duty payments.

Had the ITC tariff recommendations been accepted by the Carter Administration, the added retail cost would have been an estimated US$131 million in the first year as a result of an assumed 50 percent pass-on of the increased import duties and an increase in retail prices of 75 percent of the higher tariff rate.

Under the "imaginative" orderly marketing agreement which Strauss negotiated in its stead, some 4 million U.S. consumers will have to pay more for their color sets during the three years from now until the end of the decade of the '70s. Added to the price rises already provoked by the Customs Court decision, this gives a conservative measure of the cost to the U.S. economy of protectionist measures already taken.

If the President had accepted the ITC recommendation of a tariff rise on all television receivers, it would almost certainly have meant substantial changes in the nature, as well as the volume, of television imports from Japan, South Korea and Taiwan.

At this juncture, however, nothing is certain. The Treasury Department has already appealed against the Customs Court ruling and said it will carry the challenge all the way to the Supreme Court if necessary. The chances of reversing the lower court decision, Special Trade Representative Robert Strauss emphatically stated to newsmen in Washington, cannot be considered better than 50-50. The court's decision took into consideration not only existing U.S. statutes, including provisions of the 1974 Trade Act, but also the so-called "grandfather clause" of the General Agreement on Tariffs and Trade (GATT), under which signatories accepted the GATT provisions insofar as they were not inconsistent with existing domestic legislation.

The Treasury Department will have difficulty in sustaining its position that exemptions from or remission of indirect taxes on exported products do not constitute bounties or grants as defined under the GATT agreement. In the meantime, Japanese companies will have to post bonds or letters of credit covering the full amount of increased duties on imported sets. While in itself not a costly requirement — about US$7.50 per US$1,000 free-on-board value — it introduces an element of uncertainty and risk which will be a deterrent to imports from Japan. In the ensuing confusion, not surprisingly, many importers are canceling contracts and others are raising their prices to cover themselves in the event that duties are ultimately levied.

The agreement which Special Trade Representative Robert Strauss hammered out with the Japanese government in May 1977 for voluntary restrictions of exports for a period of three years has enabled the Carter Administration to reject the higher import duties recommended by the ITC. But even within the Administration there are those who believe that the Japanese were forced to pay too high a price to fend off protectionist pressures.

To provide what Strauss has termed "short-term relief to some short-term problems," the Japanese government agreed to limit exports to 1.56 million complete color television receivers and 190,000 unassembled units — a total of 1.75 million sets. In return, to head off congressional pressures to have Japanese acquisitions of U.S. electronic companies in the United States reviewed and disallowed, the agreement struck with Tokyo expressly encourages further Japanese investments in U.S. color television production facilities.

Relieving Pressures

In addition, Japanese manufacturers with production facilities in the United States will have no limitation on imports of components for television sets or on kits which lack nine basic parts, including picture tubes. Anything more complete than this will be treated, for the purpose of the restraint agreement, as a finished set.

144

Since Sony makes its own Trinitron color tubes in San Diego, and both Matsushita and Sanyo buy tubes from U.S. manufacturers, this aspect of the agreement fits the existing economic exigencies of rational television manufacture. Neither is the exclusion of other components, selected to assure a minimum 40 percent U.S. input into the final product, likely to pose any serious problems for these manufacturers. For other Japanese television manufacturers, this aspect of the agreement will probably hasten the decision to invest in U.S. manufacturing plants.

While this will somewhat relieve the competitive pressures on U.S. television producers, with an opportunity to regain the lost share of the market, it poses a threat to major private-label mass-merchandisers such as Sears Roebuck. Sears, which sells approximately 9 percent of all color televison sets in the United States and depends heavily on Sanyo and Toshiba as prime contractors, appears to be in danger of temporarily or permanently losing what it thought was a secure long-term source of supply.

Certainly, if the U.S. industry gets its way, the links between U.S. mass merchandisers and Asian suppliers will be considerably weakened.

The COMPACT labor-industry group has launched its campaign to override the President's rejection of ITC-recommended tariff imposition, claiming that the agreement completely disregards the seriousness of damage the U.S. color television and related industries have suffered.

Since the agreed level of imports would still be 60 percent above the 1971-75 average that the U.S. industry has suggested as a suitable limit, the COMPACT group will pressure Congress to sustain the ITC's recommendations. However, few informed observers in Washington rate very highly its chances of success.

With a Japan-U.S. "orderly marketing" agreement fully in force, settlement of the legal problems confronting Japanese imports will have only just begun. Some industry observers consider the pending Sylvania-Philco complaint alleging unfair practices in the importation of portable color television sets from Japan to be potentially one of the most important to be considered by the commission. It is conceivable that through this case the ITC could develop a wholly new method whereby protectionist interests could attack imports and obtain remedies equivalent to embargoes on specific products.

The complaint is that Japanese manufacturers are selling their products to the United States at below-cost prices, and further are receiving subsidies from the public treasury. On the face of it, the complaint would appear to be subject to the Anti-Dumping Act and the Countervailing Duty Law, but the U.S. television industry has lost cases previously brought under both statutes.

The industry now argues that practices complained of in the present case fall within the purview of the ITC under Section 337 of the Tariff Act of 1930. If the ITC accepts this logic, the U.S. industry will have a broad new device with which to attack import competition. Quite apart from the ultimate decision on such cases, a problem arises from the procedures under Section 337. Unlike other ITC proceedings, they fall under the Administrative Procedures Act, and therefore have all the complexities, expense and delay of court litigation — along with the few touches unique to ITC proceedings.

GTE Sylvania and four of the Japanese manufacturers concerned — Toshiba, Hitachi, Sharp, Sanyo Electric — have agreed to propose consent orders to the ITC to settle GTE Sylvania's charges. Under the consent orders, the Japanese companies would agree they would not violate U.S. law in future and would provide the ITC with details about their exports to the United States.

Excessive Intervention

While the proposed orders did not establish that the Japanese had in the past engaged in what Sylvania alleged were predatory pricing and other unfair practices, they would bar such practices in the future and preclude an agreement to fix prices or restrict sales. Each Japanese company would have to report annually, by screen size, its unit volume of production, revenue, and costs for all color sets sold in the United States or exported to it for resale. The orders further provide for inspection and audits to assure compliance. The ITC could hold hearings to determine any violations of the consent orders and could impose penalties that would include exclusion of a violator's products from imports into the United States.

Whether the Japanese have simply agreed not to do something they were already not doing is a question which this procedure would, in fact, beg. The danger is that the public may be misled by such action to believe that the consent order is tantamount to admission of past guilt. The consent action may be the best way to avoid the high cost of long litigation, but it is important that it be described clearly for what it is and not allowed to fuel runaway protectionism or serve as another means of continued harassment of importers by excessive and almost random intervention by the government agencies.

As in all such matters of law, the pyrotechnics of public confrontation and the debate of fine points of jurisprudence are likely to mask the fundamental questions. The recent investigations and decision of the ITC demonstrate the extent to which rhetoric can obfuscate reality and politicize industrial and trade decision-making.

Attention was focused on the two points of law involved. All that had to be shown was that imports of television sets had increased and that this increase was a substantial cause of serious injury, or a threat of such injury, to the U.S. industry. Since the Commission took as its time-frame the period 1968-76, its decision was a foregone conclusion. Imports during that period, mainly from Japan and Taiwan, had increased by 188 percent and rose even more sharply in 1976; during this period the number of U.S. television manufacturers had fallen from 18 to 11.

Three other manufacturers had been acquired by foreign multinationals: Magnavox by Philips; Motorola's television production facility by Matsushita; and a majority interest in Warwick by Sanyo. Moreover, during the recession year of 1974 and 1975, a significant number of firms — as many as eight of the 11 manufacturers in 1974 — were operating at a loss.

Investigations also confirmed that there has been a significant decline in the average number of persons employed in receiver assembly in recent years and that there have been significant temporary layoffs and shortened work weeks. The commission acknowledged that in the short span of five years 1971-76, the U.S. television industry underwent a major structural reorganization.

The entire U.S. industry shifted from relatively labor-intensive tube technology to the manufacture of solid-state sets which makes possible the introduction of intensive automation in production lines. This, in turn, raised cost efficiencies, increased yields, upgraded quality and reduced maintenance. Labor became a relatively less important factor in the total cost of production.

But in its judgment, the ITC found that imports were a more important cause of the decline in employment from 36,654 production workers in 1971 to 23,388 in 1975, although the reduced workforce produced approximately the same number of receivers in 1975.

The commission was also apparently not obliged to include in its calculations either the amount of production U.S. firms shifted to offshore manufacturing facilities during the period 1968-76, or the effects of such transfers on corporate profitability. The commission dismissed the fact that Zenith and RCA — which between them share 43 percent of the U.S. television market and supply a considerable portion of television imports from offshore facilities — continue to show healthy profits, and in 1976 reported earnings at all-time highs. Also ignored was the general improvement in television operations in General Electric and the record earnings for 1976 of Magnavox with new management skills, technology and capital infusions from North American Philips.

Significantly, the ITC chose to ignore the fact that the U.S. industry leaders had kept pace with competition through heavy investment in technology, automation and quality control. Instead, it focused on small, non-integrated assemblers whose days were inevitably numbered, regardless of imports, by basic changes in technology requiring increasingly large volume production.

Several other factors suggest that the U.S. television industry is not the invalid which the ITC investigations indicate. It is remarkable that no major U.S. television manufacturer was associated in the claim of injury under the escape clause provisions of the 1974 Trade Act.

Far from being infirm, or faced with the threat of extinction, as recent rhetoric claims and the ITC recommendations appear to confirm, there is substantial evidence that U.S. television manufacture is on the verge of a comeback which promises to be even more significant than that of the calculator industry. Not only does television blanket a broader range of potential applications, unlike the zero-based comeback of U.S. calculator manufacturing, but the home industry still has close to 70 percent of the color market and accounts for a substantial share of black and white sales.

Vertical Integration

The dramatic U.S. comeback in calculators clearly demonstrates how the U.S. lead in technology can change the industry and markets in a very short time. In 1970, U.S. calculator imports from Japan accounted for 40 percent of a US$224 million market — a share acquired in just four years. A year later, Japanese calculator imports represented almost 60 percent of units sold in the United States, and approximately 45 percent of the total market value.

But this was their peak year, in terms of market share. By 1974, Japanese calculator imports had dropped to 21 percent of the value of U.S. calculator consumption, or an estimated US$750 million; the number of electronic calculators shipped by integrated U.S. manufacturers was estimated at over 8 million, with a value of approximately US$500 million — a dramatic recovery in the same time it had taken Japanese imports to take a commanding lead.

In 1971, several developments converged to produce a radical change in the competitive advantage of the U.S. industry vis-à-vis Japanese and other Asian producers. The vigorous price and performance improvement in American MOS/IC technology, along with innovative Japanese marketing, had already established clearly that the electronic calculator was a product with a high elasticity of demand, its market broadening substantially as prices declined.

Then in 1971, several U.S. integrated circuit (IC) makers introduced single-chip four-function calculators. Labor costs as a percentage of total production cost dropped further, and Japanese firms lost the advantage previously obtained in the assembly of transistorized and multi-chip IC calculators. This change was intensified and accelerated in mid-1972 by the entrance into the United States and world markets of the first U.S. semi-conductor manufacturers with their own line of pocket calculators.

This action restored to the U.S. calculator industry the vertical integration lost with the disappearance of the electro-mechanical calculator, and added a further dimension to the competition between U.S. and Japanese firms for U.S. market shares.

Helped by import surcharges and the revaluation of the yen in relation to the dollar since 1971, U.S. firms quickly reversed the Japanese industry's advantage, using large-scale production and aggressive pricing policies based on the "learning curve" concept which they had been employing in the semi-conductor market.

Essentially, this concept is based on the premise that manufacturing costs decline by a certain percentage each time the cumulative number of manufactured units doubles. What happens, in fact, is that the more units manufacturers produce, the more efficient is their production. Thus, companies can determine their pricing schedules with finely-tuned synchronization of future market growth and production efficiencies. Largely as a result, U.S. semiconductor manufacturers have captured a large share of the U.S. and world markets since 1972.

Object Lesson

The global structural effects of technological change in calculator manufacture has been revolutionary. With the emergence of the solid-state calculator, the entire electro-mechanical industry was replaced by a new generation of manufacturers in a few years; wedded to old technology, none of the electro-mechanical calculator manufacturers was able to make the technological change to enhance its position in the market. Most failed entirely to adapt, and disappeared.

With the introduction of MOS/IC (metal on semi-conductor/integrated circuit — in effect a miniature computer) technology and the new strategy and structures it requires, great pressure is continually exerted on individual producers to increase the efficiencies of manufacturing and distribution operations to remain profitable in a mass market. These pressures have produced successive shake-outs of structurally weak or relatively inefficient assemblers and manufacturers. Others have retreated to the more profitable and less-competitive upper end of the price spectrum.

The evidence suggests that the object lessons of this experience have not been lost on other sectors of the consumer electronics industry; the impact which fundamental technological change accompanied by aggressive marketing had on the U.S. calculator industry is very likely to be repeated with the development and the diffusion of IC technology. Television receiver manufacture is now undergoing just such a far-reaching structural reorganization, after several years of rapid technological change.

In the 1960s, about all that was necessary to assemble television sets was a conveyor belt and a plentiful supply of trainable female workers. Since the conveyor belts and such test equipment as was necessary were easily transportable, production tended to move to the lowest-cost labor source. Other factors such as accessibility by transport, local supplies of components and services, and relative efficiency of government administrations — plus more investment incentives — were among the important secondary considerations determining the location of plants.

Labor cost differentials were so high that it was economical for some U.S. makers to close their existing plants and move production entirely to Taiwan or Mexico. Partly as a result of this movement, U.S. imports of complete black and white receivers exceeded domestic production throughout 1971-75. Moreover, when black and white receivers produced in the United States from incomplete receivers assembled in offshore facilities are subtracted from total U.S. producers' shipments, both the quantity and value of such shipments are reduced to an even lower proportion of the total.

The shift to solid-state technology in television manufacture has already changed this pattern significantly. Solid-state technology has simplified the circuitry of more complex color television receivers, as well as monochrome, with substantial savings in labor. But most important, the new technology introduced the possibility of more intensive automation. Automatic sequencers, automatic insertion of components in printed circuit boards, wave-soldering equipment and computer-controlled automatic test equipment have eliminated much more of the need for manual labor.

A single machine can insert components into a printed circuit board at the rate of 72,000 pieces per hour, compared with a hand rate of 300 pieces per hour. It would require 240 workers to achieve the 72,000 pieces per hour insertion rate, yet automatic insertion machines can be operated by as few as 11 workers. "Once you automate, the question of whether you're paying 25 cents or US$3 an hour becomes an awful lot less important," Zenith chairman John Nevin noted in explaining his company's changes in strategy and structure.

Wager on Revival

But the advantages of automation do not stop with direct cost efficiencies. Automated processes also reduce the number of rejects, upgrade quality and cut maintenance costs. With labor increasingly less important in the total cost of production, and with the other advantages that accrue from automation, the U.S. industry is now capable of becoming completely cost-competitive with Japanese and Taiwanese producers.

To be sure, automation costs money — which is another way of saying that television assembly has become a capital-intensive process. From this it follows that the critical mass of a production unit of an automated plant is much higher than labor-intensive assembly using old technology. While small-scale labor-intensive assembly can continue to compete, manufacturers now have the option of large-scale automated production. And in the case of television receivers — which because of their bulk are relatively costly to ship — substantial economies can be obtained by moving the plant to the market. Furthermore, location of assembly plants in the home market tends to reduce risks as well as the drain on managerial resources for U.S. companies.

Some of the implications for the Asian industry of this change deriving from solid-state technology was demonstrated by Zenith's transfer of production of 16-inch black and white sets to the United States from its plant in Taiwan because it found that it could achieve savings by manufacturing close to the market.

For Zenith, this move was part of a major structural reorganization which began with the switch to an entirely new line of solid-state sets in 1972. Zenith management rethought every activity the company was involved in, discontinued some, and relocated others. Having decided that Zenith's main business is home entertainment, the company got rid of a lot of sophisticated electronics and military equipment production which were spawned during the Cold War and the space race.

Resources since then have been concentrated on expansion and automation of consumer electronics production, related research and development, and a more aggressive advertising and marketing effort. In 1974 alone, capital spending on new plant tripled to US$47 million.

Zenith's wager on a revival of the U.S. television industry is echoed at RCA. In his keynote address to an industry conference, RCA's William Boss, chairman of the Electronic Industries Association's consumer electronics group, expressed the industry's confidence that "the video industry is now poised for unparalleled growth that will be fueled by a continuing supply of new products and new features." He went on to predict a 1977 retail market for video-related products reaching US$4,000 million. This hardly sounds like an industry on the verge of extinction.

Significantly, the optimistic view is clearly held by Japanese manufacturers, as well as by Philips of Holland, which has recently invested in color TV manufacture in the United States. Of course, their recent acquisitions have substantially enhanced the competitive strength of the U.S. industry by the infusion of new capital and technology from abroad. Matsushita is modernizing the production facilities it acquired from Motorola, introducing advanced production techniques developed in Japan and other parts of the world.

Sanyo's acquisition of the Forest City, Arkansas, plant of Warwick — a prime contractor of the Sears Roebuck store chain — was finalized in December 1976. Sanyo will invest substantially in new plant and equipment and hopes to double the number of workers at the plant by mid-1977. More Japanese manufacturers are expected to follow suit. The Mitsubishi group, Hitachi and Toshiba are all involved at various stages of plans for manufacture in the United States.

Not surprisingly, the first company on either side of the Pacific to understand the structural changes in the industry that would logically flow from solid-state technology was Sony — the pioneer in solid-state television receiver manufacture, having first produced and marketed solid-state black and white television sets in 1960 and color in 1968. By 1971, Sony had already built its wholly-owned plant in San Diego to produce Trinitron color television receivers and tubes, and is now adding the finishing touches to a video recorder plant in Alabama.

Sony president Akio Morita, who was quick to grasp the importance of changing economics in the industry, said in 1973 that Sony would make no further investments in production facilities in Japan. Both economic and political considerations dictated transfer of manufacturing to its major markets.

The ITC recommendations to increase import duties to 25 percent on both color and monochrome receivers, whether complete or incomplete, was expressly intended to spur the readjustment of the industry. Since the increased duty would offset the cost advantage of offshore production, it is likely that it would tend to encourage the return of assembly from Taiwan and Mexico to the United States.

Rapid Changes

And to give further stimulus to this process, the commission intentionally excluded sub-assemblies and components from the increased levies, an exception that was included in the "voluntary" restraint agreement reached with Japan. Higher duties on complete and uncompleted sets would serve as added inducement to manufacturers to invest in more automated assembly while continuing to rely on offshore sources for more labor-intensive component and sub-assembly supply.

If the resultant turn-around in the U.S. television industry may not be as dramatic as that achieved in calculators, the significance of this development is likely to be greater because of the central role of television in entertainment, information and data communications. The industry has tremendous potential. New television functions and applications include interactive communications, household computer terminals, and as a facsimile terminal and component of a home video system. And the evolution of solid-state technology renders television production ideally suited to U.S. capabilities.

Rapid changes in the function of television which are already in the pipeline will require increasingly sophisticated circuitry and at the same time be characterized by shorter life cycles. These multi-functional units — perforce larger console or table models — will utilize more ICs but will be assembled in larger-scale highly-automated production units close to their market.

As solid-state technology evolves, the impact of technological change on television manufacture seems likely to follow the pattern of the calculator industry. Television manufacturers, whether they are vertically-integrated manufacturers or assemblers of purchased components, new entrants or established industry leaders, in Singapore, Seoul, Chicago, or Chigwell in England, will find the pace of change quickened, the capital intensity of the industry increasingly higher, and the demands on resources intensified.

Final assembly will tend to move closer to the market and the source of technology, while more labor-intensive production of components and sub-assemblies which have a low ratio of transport cost to total production cost will continue to be located offshore from the world's major markets, where labor costs as well as purchasing power are high. The broader implications of this development for Asian industry are clearly far-reaching.

The restructuring of television production in Asia does not depend entirely on the impact of semi-conductor technology change, of course. Among other things, the growth of the Asian market for television receivers will become an increasingly important determinant of production location, scale and profitability in the region. Significantly, although the immediate purpose of Grundig's recent investment in television production in Taiwan was to supply European markets with monochrome television receivers, the company's longer-range view and primary motive for the investment was to establish a base from which to develop a share of fast-growing Asian markets.

But the number of firms in the Asian television industry is likely to decline rather than expand in coming years; substantial financial resources will be critical to the success of new entrants, vertical integration will become an increasingly important ingredient, and the link between mass merchandisers and manufacturers will tend to become more permanent and direct.

Since established brand-name television manufacturers in the United States and Europe have been reluctant, unwilling or unable to supply mass merchandisers from home production, a natural alliance has developed between Asian consumer electronics manufacturers and large U.S. and European mail order and discount houses, department and chain stores. This alliance is now threatened by changing economics of television production and the parallel development of protectionist measures.

Critical Size

Not all Asian manufacturers can follow the example of Sanyo to acquire production facilities near their major customers. Nor is it certain what the nature, timing or phasing of changes in this global production-distribution network will be. But structural changes will proceed at an increasingly rapid pace, just as they did in the calculator industry and in the recent boom-bust cycle of citizen band radio and television games.

If the Asian industry retains its momentum in miniaturization, the division of labor on a screen-size basis that has persisted between Japanese industry and the U.S. and European industries may continue. But the exploits of U.S. semiconductor manufacturers in the production of handheld calculators suggests that this pattern may also change in the television industry.

While their response will necessarily be different, Japanese and offshore Asian manufacturers will have to adjust to these rapid changes. The experience of the decades of the 1960s and '70s suggests that these adjustments will be made dynamically and with a resultant closer integration between industries and markets in Asia and the United States. The U.S. television industry has the technological, financial and managerial resources to play a leading role in this process. But this is unfortunately not the case in Europe, which is likely to be the focus of the next round of painful readjustments in industries' policies and strategies in response to changes in technologies and in world markets. By and large, European consumer electronics industries do not have the global reach of either the U.S. or the Japanese industries; nor, with the exception of Philips, do they have the same degree of vertical integration.

The European industry, fragmented largely on national lines, will have difficulty achieving the critical minimum size needed for large-scale automated production. If structures are not broadened through mergers, acquisitions and investments across borders, integrating the industry more fully into the global system, the chances are that European markets will become more protectionist than they are already — with more restrictions on imports from Asia.

CHAPTER 9

The Legal Conundrum

WHILE CONTROVERSY over the 1978 imposition of massive dumping duties on Japanese television producers continued to bubble at boiling point, the U.S. International Trade Commission (ITC) turned the investigatory heat up higher with its ruling at the end of September that microwave ovens imported from Japan were being sold at less than fair value in the United States, causing injuries to domestic industry.

Based on the ruling, the ITC asked the Treasury Department to institute a formal investigation into the prices of the ovens. Since the Japanese manufacturers of microwave ovens (electronic cooking devices which heat food by the application of ultra-high frequency energy) are also major television manufacturers which have been beleaguered by a continuing barrage of investigatory actions, the response from their side was immediate and curt.

Sharp Corporation president Akira Saeki dismissed as groundless and incomprehensible the charges, originally made by the U.S. Association of Home Appliance Manufacturers (AHAM), that Japanese microwave oven manufacturers had been dumping in the United States. Nonetheless, the Japanese industry made it known that the ITC ruling was being taken quite seriously. While preparations were undertaken to support their denial of dumping charges, the leading suppliers to the American market announced plans to begin manufacturing ovens in the United States to hedge against uncertainties such as recent Treasury Department retrospective dumping duties.

The U.S. industry's latest anti-dumping action is symbolic of a wider problem of neo-protectionism not only in the United States but also in Europe to some extent. As the experience of the U.S. television industry shows quite clearly, domestic producers are able to slow imports and inflict heavy damages through continual harassment by administrative investigations and court action. These procedures have become a standard

First published as "The Profits of Harrassment." *Far Eastern Economic Review*, 26 October 1979.

weapon employed by U.S. corporate strategists to protect their market shares against competition from imports.

This harassment strategy entails the transformation of the legal department of large U.S. corporations into a profit center. With a broad delegation of powers, corporate legal departments systematically comb the statute books for harassment opportunities. Adjudication and petitioning of administrative agencies is then combined with political action by corporate lobbyists and supported by strident public relations action. In this way the targeted judiciary and executive agencies are submitted to maximum pressure from Congress and the media.

Moreover, it has become standard operating procedure for firms employing these strategies to team up with labor unions in harassment actions. The objectives of the action are clear cut. Imports are to be impeded to protect market shares, price rises are effected both by increasing the costs of operation in the U.S. market and through the imposition of countervailing or dumping duties. Price cutting is to be checked wherever possible. Equally important, changes in organizational structure of the industry and market are to be resisted by employing the full measure of potential legal obstacles and political opposition to foreign investments.

Tactics employed in pursuit of these objectives are quite common to U.S. legal practice. "Trial by ambush," using some obscure or archaic statute to sandbag the competition, is supplemented by "trial by avalanche," which ties the competition up with endless deposition-taking, requests for documentation and interrogations that are repetitious, often irrelevant, intrusive and sometimes downright nonsensical.

In recent years, since the advent of the Japanese electronics industry as a major force in the U.S. market, there has been a continuing barrage of investigatory and other legal proceedings directed against the entire range of consumer electronic products and electronic components from Japan. The history of litigation in the courts, petitions for administrative regulatory action and pressures brought to bear on the U.S. Congress make abundantly clear that the multiplicity of actions brought to harass imports of electronic products from Japan reflects collective and individual corporate action. This is not only anti-competitive but appears to constitute an abuse of available legal processes to protect often high-cost, inefficient production at the expense of the American consumer and Japanese manufacturers.

The various avenues and legal instruments used for this purpose run the gamut of national security issues, the Anti-Dumping Laws of 1921 and 1916, the Countervailing Duty Act of 1979, section 337 of the Tariff Act of 1930, the escape clause provisions of the Trade Act of 1974 and that old standby, the Sherman Anti-Trust Act.

THE LEGAL CONUNDRUM

Transistorized pocket portable radios, transistorized portable television receivers, and portable taperecorders, among other electronic consumer appliances, were first mass produced and marketed on a commercial scale in Japan. So, too, was the electronic calculator. And very early in its development Japan emerged as the leading producer of solid-state color television receivers, microwave ovens and a wide range of active and passive components used in consumer electronic equipment.

Hardly had Japan achieved its first major successes with electronic products in the U.S. market before the harassment began. On September 17, 1959, the U.S. Electronic Industries Association filed an application with the U.S. government alleging that imports of portable transistor radios from Japan posed a threat to national security. The investigation which this petition provoked was finally dismissed on May 29, 1962, some two and a half years later, but only after the Japanese government had consented to impose an overall quota on export of radios with three or more transistors.

At first glance, this would seem to have been a successful attempt to restrict imports to hold the market for an obsolete product — the vacuum tube table radio. And in the short run it appeared to have just this effect. However, it also appears to have had damaging long-term results for the U.S. radio industry and explains at least in part, the present lacuna in the U.S. industry's capability for radio production, either at home or abroad.

After the apparent success of this first protectionist effort, a steady stream of actions of a similar nature followed, with the Anti-Dumping Act of 1921 and the Anti-Dumping Act of 1916 as the major legal instruments. Of the 23 investigations initiated under these statutes, 11 were dismissed after the investigations failed to disclose any dumping sales or injury where such practices were found. In three cases, investigations were discontinued on a finding of commercially insignificant quantities of export sales reflecting dumping margins.

In the monochrome and color television case, the only commercially significant instance of a dumping finding on the imports of electronic products from Japan into the United States, action — or the lack of it — by the Treasury Department over a 10-year period became the subject of controversy.

The anti-dumping complaint which prompted the investigation into TV sets from Japan was initially filed in 1968. In the autumn of 1969, the U.S. Customs Service advised that little or no dumping margins had been found on television imports from Japan and recommended that the secretary of the treasury terminate the investigation; provided that the Japanese manufacturers involved furnished the necessary assurances that there would be no less-than-fair value sales of television sets in the future.

Despite this advice, the Treasury Department reversed its policy five months later, and indicated that price assurances would not be a satisfactory basis for terminating the investigation. Furthermore, the U.S. Customs then proceeded with another fundamental change in policy: dumping determinations would be made henceforth only on a country-wide basis. This meant that if any Japanese television manufacturer was found to have dumped its products, all Japanese television manufacturers would be included in an anti-dumping action. Thus, national origin was made the sole test for inclusion of a producer in a dumping assessment if any single producer of a given country actually dumped. Accordingly, one firm found to be in full compliance with the statute and two other firms with minimal sales reflecting insignificant margins were included in the dumping finding issued on March 8, 1971.

Between 1971 and early 1977, the U.S. Customs Service sought to establish the difference between the value of imported Japanese receivers and the value of comparable receivers sold in Japan. After several attempts to verify massive documentation submitted by the Japanese industry, and further complications caused by widely varying rebate practices, in December 1977 the Treasury Department obtained approval from general counsel to use the commodity tax formula for setting foreign market television values. In mid-March 1978 the Treasury announced that instead of using the previously employed method of calculating margins in all dumping cases, it would resort to a completely new formula which would use as the standard of fairness those values used in Japan for calculating the Japanese commodity tax. Customs then proceeded to apply this new formula retroactively to unliquidated television entries going back to 1971, assessing importers some US$400 million in dumping duties covering the period up to April 1977.

Since then, after strong protests from the Japanese industry and government, the Treasury Department has agreed to re-study the TV ruling and assessments have been limited to the period up to June 1973. But even this amount, which comes to some US$46 million, is unacceptable to Japanese manufacturers, who have indicated their intention of not complying with attempts to impose the duty. There the matter rests.

In the meantime, however, an American group, Zenith Radio Corporation, provoked worldwide attention by attempting to force U.S. Government action imposing countervailing duties on imports of television sets. Joined in this action by separate complaints from Magnavox and the Electronic Industries Association, Zenith alleged that television sets and other consumer electronic products from Japan were unfairly benefiting from export subsidies resulting in trade distortions which were injurious to the American industry.

When the secretary of the treasury dismissed all three complaints in 1976, Zenith then appealed against the decision to the U.S. Customs Court of New York. The sole issue involved in the appeal was whether or not the Japanese tax system, which imposes excise duties only on home market sales of television sets, constitutes an unfair export subsidy. This notion, if accepted, compels the bizarre conclusion that exports in order to be fairly traded must bear an aggregate tax burden in an amount equivalent to all of the internal taxes imposed on the same product for home consumption.

In view of international practice (accepted in the United States) exempting exports from internal excise and commodity taxes, the adoption of this formula would have brought chaos in trade relations and most likely have precipitated a major trade war.

Although Zenith won a unanimous decision in the U.S. Customs Court, that decision was overturned by the U.S. Court of Customs and Patents Appeals and the latter's ruling was upheld by the U.S. Supreme Court in its 1978 decision. But the limits of harassment of Japanese television imports had by no means been reached.

In January 1976, GTE Sylvania filed another complaint, again alleging unfair Japanese government subsidies on television sets exported to the United States, this time under section 337 of the Tariff Act of 1930. The complaint, filed with the International Trade Commission, again raised the charge of dumping, despite the fact that there was already an appeal before the U.S. Customs Court on the subsidy issue and the same imports were concurrently undergoing appraisement under the Anti-Dumping Act.

Then, apparently not content that Japanese television imports were already subject to sufficient scrutiny, the International Trade Commission initiated yet another investigation alleging some 14 unfair trade practices coming within the scope of the anti-dumping, countervailing duty and anti-trust laws. Since the ITC's action was clearly duplicative, overlapping one or more legal proceedings then in progress in all three areas, suspicions inevitably arose that the U.S. government was a party to a conspiracy of multiple harassment.

But a most important action was yet to come. On September 22, 1976, U.S. television producers, including GTE Sylvania and jointly with 11 labor unions, filed, for the second time, an escape clause action on color television seeking the imposition of import quotas on the ground that the U.S. industry had been seriously injured by the effects of

Japanese competition. The first escape clause action, filed on monochrome and color TV sets in June 1971, had been dismissed in November 1971 on a finding of no injury.

Coincidentally, no doubt, the ITC made its recommendations to the president calling for the imposition of higher duties to assure protection of what it found to be an injured U.S. television industry at about the same time the U.S. Customs Court of New York voted to overrule the secretary of the treasury on the countervailing duty issues. But instead of following the ITC recommendation, President Carter responded by despatching special trade representative Robert Strauss to Tokyo to negotiate an Orderly Marketing Agreement (OMA) with the Japanese government, thus restricting the imports of color television sets into the United States. Under the OMA, signed in May 1976, the Japanese and U.S. governments agreed that color television exports to the United States from Japan would be limited to 1.75 million annually.

Once again, on the face of things, the multiple harassment appeared to have achieved its purpose. Imports of Japanese color television sets would be cut back at a critical point in time for the U.S. industry, giving Zenith time to revise its strategy for more highly-geared competition with Japanese makers and allowing other firms to regroup their forces. It remains to be seen whether the long-term results will have more positive results than the earlier curtailment of radio shipments.

Significantly, the first major move taken by Zenith after the signing of the color television OMA was to transfer much of its production offshore. Ironically, at the same time, Japanese manufacturers stepped up their investment in color TV production in the United States, in part for economic reasons but more especially to reduce the possibilities of a repetition of the kind of harassment action mounted against imports of complete sets in 1976.

Japan's switch from exporting to investment in production in the United States will undoubtedly reduce the incidence of anti-dumping actions as well as other claims of unfair trade practices, but it does not avoid another important avenue for legal harassment. During the 1970s, Zenith, in paticular, was especially aggressive in civil actions against Japanese firms for alleged anti-trust violations and in the use of political pressures on the Justice Department to impose penalties on Japanese manufacturers under anti-trust laws.

To block Matsushita's acquisition of Motorola's television division in 1974, Zenith took the initiative in a protest to the anti-trust division of the Department of Justice. Significantly, Magnavox joined Zenith in the protest, contending that if the Matsushita move was not stopped, it would result in a drastic restructuring of the U.S. television industry — a development in which Zenith, as price leader in the U.S. television oligopoly had much to lose. Zenith and Magnavox maintained that because Matsushita is the largest manufacturer of television components in the world, the combination of its existing manufacturing capability in Asia with Motorola's marketing assets would "be seriously anti-competitive."

When Zenith's own offer to buy certain Motorola plants was rejected as non-competitive and no other buyers came forward, the Justice Department authorized Matsushita's acquisition. A few weeks later, North American Philips, an affiliate of N.V. Philips Gloeilampen-fabrieken of Holland, made a tender offer to take over Magnavox. Apparently Magnavox had already begun discussions with Philips prior to the protest against Matsushita's acquisition of Motorola's television assets.

But even more remarkable than the move to stop Matsushita's entry into the community of U.S. television manufacturers had been Zenith's anti-trust suit filed in the Federal Court in Philadelphia charging that a number of Japanese television companies and their subsidiaries had violated American anti-trust and anti-dumping laws. Zenith's charges were based mainly upon the so-called check price agreements, which were cited as evidence of the existence of a Japanese conspiracy to sell television sets at unreasonably low prices. In fact, far from being a conspiracy to take over the U.S. market through predatory pricing, the check price arrangements were designed by Tokyo's Ministry of International Trade and Industry (MITI) to have exactly the opposite effect.

The check price agreements, which were openly and publicly made, were forced upon Japanese television producers and exporters by MITI as a consequence of pressure from the United States itself to control the growth of Japanese television exports to the United States. The imposed "agreements" prohibited Japanese producers and exporters from selling TV sets for export to the United States at prices lower than those sanctioned by MITI. They were minimum price agreements designed specifically to place a floor under export prices to protect the U.S. industry from the potential of too rapid growth of Japanese TV exports.

The U.S. Department of Justice has repeatedly declared its position that if the Japanese government in a legitimate exercise of its power to control exports leaving Japan were to direct such conduct as a check price agreement, U.S. courts would be highly likely to uphold the arrangement against an anti-trust challenge on the ground of a "foreign compulsion" or "act of state" principle. Moreover, as Assistant Attorney General Donald Baker pointed out in a letter to Senator Edward Kennedy on February 16, 1977, since it appears that the check price agreements were imposed to avoid or respond to anti-dumping charges in the United States, it would be quite anomalous to charge the Japanese with dumping if they price-cut in the United States, and with an anti-trust violation if they do not.

Zenith, however, retained special counsel to encourage the Senate Anti-Trust and Monopoly Sub-Committee to urge the Department of Justice into the action against the Japanese industry. As a result, the Justice Department was forced to conduct a preliminary investigation which involved the inspection of 35,000 pages of documents — fully 5,000 in the original Japanese — before confirming its position to the U.S. Senate Committee on the Judiciary on April 12, 1978. The investigatory action, which required six months and considerable public expense, as a result of Zenith's persistent harassment, added obviously to the mounting costs of legal services of Japanese firms operating in the United States, achieving little else.

Just how much the Japanese electronic industry has had to expend for legal services in the United States in defense against continuing, multiple harassment during the two decades since the end of the 1950s is not known. But apart from the legal counsel engaged by the Electronic Industries Association, every major Japanese firm operating in the United States must maintain its own legal counsel and support them with a massive flow of documentation and information on the industry, the company's activities in the United States and its operations in Japan.

The legal and administrative costs which result from multiple harassment adds to the rising bill passed along to the U.S. taxpayer. High legal costs which raise the cost of doing business in the United States must be covered in higher prices of Japanese producers. And, to the extent that these protectionist measures succeed in restricting imports or increasing duties, consumers are saddled with the ultimate penalty of higher-priced TVs and other electronic equipment.

This has undoubtedly brought some short-term benefits to U.S. manufacturers such as higher cash flow and profits. It has not, however, demonstrably improved their competitive position at home or in world markets. On the contrary, there is much evidence which suggests that the firms which engage in this kind of harassment become increasingly less competitive. Funds which might better be allocated to badly needed research and development, for example, are diverted into legal action to assure short-term profit objectives. Similarly, capital investments tend to be inhibited, rather than encouraged. In a very real sense, multiple harassment is the court of last resort for structurally weak enterprises in declining industries. And, unfortunately, indications are that they will not be saved in the final judgment.

A Case in Point

As with earlier anti-dumping approaches to Washington, the microwave oven case is based on questionable grounds. Following summary investigations conducted by the U.S. Customs Service (after receipt of a petition filed by the Association of Home Appliance Manufacturers), the Department of the Treasury stated that there is substantial doubt that imports of microwave ovens from Japan, allegedly sold at less than fair value, are causing injury to U.S. industry. This is something which must be shown before dumping duties may be assessed.

But there remains the knotty problem of establishing what is fair value, and the margin by which each shipment by each maker to each customer might have been sold under the fair value of each item. Based upon information supplied to the Treasury by the petitioner through a descriptive analysis of similar models sold in both the United States and the Japanese market, it appears that the margins of dumping could range from 27 percent to 107 percent.

Judging from previous experience, an accurate and rapid determination of the facts of the case is virtually impossible, however. As the U.S. Commissioner of Customs wrote in a letter to the Treasury in October 1977: "Aside from the question of the integrity of the information received, we are also of the opinion that the existing administrative procedures, necessitating the collection and analysis of vast amounts of commercial information before an anti-dumping appraisement can be performed, represents a perversion of the intent of the act, in that delays for unreasonable periods of time negate the remedial protection intended by Congress for the affected United States industry."

But the case for injury of the U.S. microwave industry is not at all clear. The petition cites a substantial loss of market share, price suppression or depression, underemployment, declining profitability, underutilization of capacity, declining capital investment and a static industry structure as evidence of injury sustained from imports from Japan.

That the U.S. industry, as a whole, has been having some difficulties is undeniable. But as the Treasury observed in its August 29, 1979, notice in the Federal Register, the information presented does not clearly establish that this state of affairs is due to imports from Japan, or that it is by reason of alleged sales at less than fair value.

In mid-1978 it became apparent that the seven-year boom in microwave oven sales in the United States was losing momentum. By that time at least two U.S. oven makers had already quit the business, and rumors were rife that other companies were preparing to follow. [Only Raytheon Corporation's Amana and Litton Industries' microwave cooking products divisions, which between them share more than half the market, were not complaining. Both manufacturers reported substantial increases in sales for the year.]

The performance of the U.S. industry has never been exactly brilliant — in fact, it would be kind to call it undistinguished. In 1978, there were 11 companies producing microwave ovens in the United States, employing only some 6,000 workers. Total output, which amounted to 1.4 million units, was valued at US$450 million. Imports, on the other hand, reached a total of 800,000 units for the year 1978, with an estimated value of US$160 million. Japanese makers, which supplied 98 percent of imports, held slightly more than a 30 percent share of the total market.

The claim of a loss of maket share by the U.S. industry depends largely upon the choice of base year and period of comparison. However, "although higher now than in 1975," the Treasury Department noted in August 1979, "the Japanese share of the market is now lower than either in 1972, when the product was relatively new on the market, or 1978. U.S. manufacturers' share of the market is below historic highs but above both 1972 and 1978 levels." Japanese makers' share of the U.S. market dropped from 49.6 percent in 1972 to 30.5 percent in 1978, Japanese sources indicate.

According to Sharp Corporation, the leading Japanese exporter of microwave ovens to the United States, the market for home microwave ovens in the United States, has, in fact, been developed by Japanese manufacturers. At least in this respect the record is clear. Back in 1967-68, a remarkable thing happened in Japan that was to have a powerful effect on the American market for microwave ovens. In the years from 1966-68, production of microwave ovens in Japan took off, rising from 15,000 in 1966 to more than 100,000 in 1968. Although the microwave oven had been invented in the United States, based upon Raytheon Corporation's development of microwave magnetrons and radar, it achieved mass acceptance much earlier in Japan than in the United States. As a result, in 1969 Japanese output of microwave ovens was roughly four times that of the U.S. industry.

By 1970, it was already clear, as Dan McConnell, then product planning manager for microwave ovens at Amana Refrigeration, pointed out, that whether American technical innovations could overcome Japanese ingenuity cost advantages would determine who acquired the major share of future sales in the U.S. market. And the industry's petition to the Treasury Department in the latest dumping action, while not admitting this necessary condition of competitive power, makes clear that it has not been attained.

While the petition does not fully explore the background and implications of the fact, it is common knowledge that although there have been some peripheral innovations by U.S. manufacturers in the 1971-79 period, the magnetron tubes that are the heart of the U.S.-made microwave ovens are in fact now made almost exclusively in Japan.

This loss of technological superiority may well be the root of the U.S. industry's apparent schizophrenia. It now seems most unlikely that the U.S. industry will recover the lead in innovation in the foreseeable future. But the evidence suggests that the outlook is not as dismal as the dumping petition claims. According to business researchers, the microwave oven market, an estimated US$1 billion-plus for 1979, is expected to reach US$2.5 billion in 1982. And at the pace with which Japanese manufacturers are rushing to set up production of microwave ovens in the United States, they clearly believe the market has good future prospects too.

Moreover, if there has been some shakeout in the industry, with marginal producers opting out, evidence concerning lower prices of both U.S. and Japanese microwave ovens shows that this is mainly the result of lower costs of production due to increases in worker productivity, technological advances, standardization of production and the rewards of greater economies of scale. Likewise, decreases in employment appear to reflect greater productivity of labor in the microwave oven industry as a whole.

Although it is true, as the Treasury Department points out, that the U.S. industry's capacity utilization rate has apparently declined, this decline has been accompanied by an expansion of capacity and heavy-capital investment — rather than the result of lower capital investment as claimed. Nor is the industry's structure nearly as static as the dumping petition contends. Not only have some of the marginal U.S. producers decided to throw in the towel, but most of the large Japanese manufacturers of microwave ovens have announced their plans to begin production in the United States in the near future. Some facilities will begin production as early as December 1979. Clearly, the stakes are worth the play and the industry is in a state of dynamic flux.

THE LONG TRAIL OF LITIGATION

From the end of the '50s to the end of the '70s, the U.S. electronics industry initiated some 35 investigatory and other legal proceedings against major Japanese electronic appliance and component manufacturers. The majority of these proceedings — 23 in all — were anti-dumping investigations initiated under the Anti-Dumping Act of 1921 and the Anti-Dumping Act of 1916.

YEAR	COMPLAINT	AGENCY	ADMINISTRATIVE AND JUDICIAL ACTION	RELATED DIPLOMATIC ACTION
		RADIOS		
1959	Transistor radio imports from Japan posed a threat to U.S. national security		Case dismissed, May 1962, after 2½ years of investigations	Japanese Government imposed voluntary export quotas on transistor radios to the U.S.
		COMPONENTS		
1961	TV tube dumping	Treasury Department	No dumping found	
1964	TV tube dumping	Treasury Department	No dumping found	
1970	Fixed resistor, transformer and capacitor dumping	Treasury Department	No dumping found	
1970	Tuner dumping	Treasury Department	Dumping found	
1972	TV tube dumping	Treasury Department	No injury to the U.S. industry found	
		TELEVISION RECEIVERS		
1968	Monochrome and color TV receiver dumping	Treasury Department	Dumping found, March 4, 1971. Duming duties imposed retroactively mid-March 1978.	

Year	Action	Agency/Court	Outcome
1971	Escape clause action, claiming injury to the U.S. television industry	U.S. Tariff Commission	Case dismissed
1970-72	Three separate complaints, claiming that exports of Japanese color TV and other consumer electronic products were subsidized resulting in injurious trade distortions	Treasury Department	Case dismissed, with negative determination
1974	Anti-trust violations by Matsushita Electric with the acquisition of Motorola's TV division	Justice Department	Acquisition not opposed
1974	Violation of anti-trust and anti-dumping laws	U.S. District Court, Philadelphia	Case pending
1976	Appeal against the Treasury Department's ruling on export-ing subsidies	U.S. Customs Court, New York	Unanimous decision that commodity tax rebate constitutes export subsidy calling for counter-vailing duties
1977	Appeal against U.S. Customs Court decision	Court of Customs and Patents Appeals: and U.S. Supreme Court	U.S. Customs Court decision reversed
1976	Unfair trade practices, including subsidies for color TV exports and dumping	International Trade Commission	Reversal of U.S. Custom Court upheld and Treasury Department's dismissal of the case sustained

YEAR	COMPLAINT	AGENCY	ADMINISTRATIVE AND JUDICIAL ACTION	RELATED DIPLOMATIC ACTION
1976	Allegations of 14 unfair trade practices coming within the scope of the anti-dumping, countervailing duty and anti-trust laws	International Trade Commission	"No contest" consent order, under which the ITC would monitor Japanese pricing practices in the U.S.	
1976	Escape clause action, seeking to impose import quotas on ground of injury to the U.S. television industry	International Trade Commission	Case dismissed. Recommendation to the President that higher import duties be imposed on color television to protect an injured U.S. industry. Recommendations were rejected by the President after conclusion of the OMA	Orderly Marketing Agreement (OMA) signed between Japan and the U.S., under which Japan agreed to limit color or TV exports to the U.S. to 1,750,000 sets annually

MICROWAVE OVENS

YEAR	COMPLAINT	AGENCY	ADMINISTRATIVE AND JUDICIAL ACTION	RELATED DIPLOMATIC ACTION
1979	Mictrowave oven dumping	Treasury Department, International Trade Commission	Treasury requested ITC to rule on the effects of imports of microwave ovens from Japan: ITC found injury. Dumping investigation by Treasury under way	

CHAPTER 10

Innovation and Internationalization

DURING THE FIRST HALF of 1979, Japanese exports of color television sets to the United States dropped precipitously, to half the rate of 1978 and approximately a quarter of the record 1976 level. In the process, Japan was replaced by Taiwan as the major exporter of color television receivers to the American market, and was running only slightly ahead of South Korea.

Yet despite this sudden shift in export trade patterns, there was no noticeable panic on the part of the Japanese consumer electronics industry. On the contrary, color television set manufacturers are operating new, highly-automated facilities at 80 percent capacity. Television picture tubes, of which 15 million can be produced annually by Japanese makers, are in short supply. Integrated circuit manufacturers are racing to meet rapidly expanding demand, and makers of "passive" components (as against "active" components such as valves and transistors) who were expected to suffer from the combined effects of declining expors to the United States and the increasing use of integrated circuits are feeling no pain.

In fact, the industry's total turnover and profit performance during the first half of 1979 exceeded that of the 1973 boom year for the first time.

By any standards, this was a most remarkable feat of industrial adjustment. In less than three years, the Japanese television industry had reduced its dependence on the U.S. market from 56 percent to 25 percent of total color TV exports, not only without a major upheaval in the industry but while maintaining record overall industry performance. At the same time, Japanese television manufacturers sharply reduced the labor content in assembly, and developed overseas production facilities with a total capacity of 3 million color television receivers annually, which is approximately half the industry's domestic output. And all this was achiev-

Reprinted from "Japan's New Electronic Revolution," *Far Eastern Economic Review*, 24 August 1979.

ed without massive industrial restructuring or laying off workers at home.

This adjustment was in part a positive response to the orderly marketing agreement signed with the United States in 1977, which limited Japanese exports of color television receivers to 1,560,000 complete and 190,000 incomplete sets per year. But neither that agreement nor the rising value of the yen during 1978 called for the kind of reduction in shipments witnessed in the first half of 1979.

The sharp cutback in exports of color television receivers to the United States was one of a set of radical and rapid changes in product strategy, production technology and marketing strategy made by manufacturers to ensure that they enjoy global competitive advantage and continued growth in the 1980s. In good judo (self-defensive) style, these measures are calculated to use the momentum of protectionism to strengthen the Japanese consumer electronic industry, and increase its share of world markets.

A shift to manufacturing in the United States has enabled Japanese makers to reduce exports to that market without prejudice to overall output. With the industry still operating at 80 percent capacity at home, the total new capacity of the seven Japanese manufacturers which have invested in the United States amounts to a substantial net addition to global output. Between them, the seven major Japanese TV makers now have an installed annual capacity of 2,310,000 sets in the United States. Once they are fully operational, the respective facilities can be expanded in line with market demand and with threatened restrictions on the import of other products, such as microwave ovens.

A similar, if more gradual, expansion of Japanese manufacturing of consumer electronic products in Europe is under way. Sony already has three production facilities in Europe, while Matsushita and Sanyo have two each. Hitachi and Toshiba are entering into major joint ventures with British partners, and specialized audio firms such as Pioneer and Trio-Kenwood have strategically located assembly facilities in Belgium.

These overseas investments by leading Japanese consumer electronics firms are more than ad hoc reactions to quantitative restrictions on imports, arbitrary imposition of anti-dumping duties, and administrative or legal harassment. They are the manifestation of a studied and deliberate internationalization of production required by market conditions in general, as well as by the economics of new technology.

Sanyo Electric, for example, has adopted a comprehensive global rationalization plan intended to balance domestic operations, exports and overseas manufacturing activity. In pursuit of this objective, Sanyo has moved production of radios, taperecorders, citizen's band (CB) radios and small stereos to its affiliates in South Korea, Taiwan, Hong Kong and

Singapore. Color TV production at its Warwick facilities in Arkansas is now running smoothly and is scheduled for expansion, while stereo manufacture has begun at the company's new plant in San Diego, California. Meanwhile, in Europe, Sanyo has formed a joint venture with Italian partners to produce a range of home entertainment equipment for EEC countries, adding to its already highly successful operations in Spain.

The group's production abroad in the year to November 30, 1978, is estimated at ¥220 billion (US$1.02 billion), accounting for 27 percent of total Sanyo sales. Although Matsushita produces a larger volume in its 29 overseas manufacturing facilities, the share of overseas production in Sanyo's total output is about twice that of its Kansai neighbor.

Global rationalization of production does not mean the increase of production at all overseas sites, however. In early 1979 a number of Japanese electronics manufacturers operating in South Korea made moves to withdraw from, or cut down, the scale of production there. At the same time, other firms shelved plans for expansion in South Korea.

These moves by Japanese producers are due in part to the high rate of inflation in South Korea, and the consequent 50 percent increase in wages. Moreover, South Korea is subject to restrictions under an orderly

Table 1.   Color Television: Company Volumes

(thousand units)

Company	1977 Production		Tubes
Matsushita	3,700	(2,700 Japan)	4,100
Philips	3,000		3,700
Sony	1,970	(1,400 Japan)	1,800
Hitachi	1,700	(950 Japan)	4,700
Sanyo	1,600	(1,250 Japan)	—
Toshiba	1,590	(1,400 Japan)	4,700
Grundig	1,170		—
Sharp	870	(estimate)	650
Telefunken	760		600
Mitsubishi	700	(estimate)	800
ITT	655		—
Blaupunkt	650		—
Thorn	450		—
Samsung	150		—
Tae Han	140		—
Decca	140		—
RCA	96		—
Gold Star	120		—

Source: Boston Consulting Group

marketing agreement limiting exports of color TV receivers to the U.S. market to 200,000 sets a year.

While costs of production are rising in South Korea, as well as in Taiwan and Hong Kong, they have been declining in Japan for some products. Through a combination of new technologies, Japanese consumer electronics manufacturers have been able to reduce man-hour requirements in assembly to such an extent that it is now more economic to concentrate production in large-scale highly-rationalized facilities at home or in other advanced industrial countries. Although employment costs are high and rising, average labor content for a 20-inch export color set, for example, had been reduced by 1978 to 30 percent of its level a decade earlier. The saving was particularly marked between 1974 and 1978, when labor content was reduced by half.

Japanese producers have adopted three basic approaches to the reduction of labor content in the assembly of electronic products: maximum application of solid-state technology to reduce the number of components used in the final product; design changes which reduce the number and complexity of assembly operations; and extensive automation, which reduces the number of workers required for those operations.

Large-scale integrated circuits now replace hundreds of discrete components, while improvements in tube design have greatly reduced calibration time in production. According to findings of the Boston Consulting Group (a U.S. management consultancy group), higher integration of semiconductor devices and pre-converged picture tubes have enabled Japanese manufacturers to reduce the component content of color television sets by as much as 40 percent in two years. And the industry expects this trend to continue well into the 1980s, with reductions in components of approximately 15 percent a year. This produces a double cost saving in labor content and lower material costs.

These economies obtained through the extensive application of solid-state technology have been enhanced through a shift in chassis design from modular construction to a single or double printed circuit board. This unified chassis is much simpler, with fewer components and fewer contact points, which, in turn, contributes to reliability by reducing the probability of failure and making testing easier. In a color television receiver, this basic design change may reduce the number of printed circuit boards from as many as 12 to one, a change which the Boston Consulting Group estimates could cut man hours by 18 percent.

In addition, the use of a unified chassis makes possible optimum advantages obtained through automated insertion of components in the printed circuit board assembly process.

By using a unified chassis, Japanese manufacturers are also able to obtain high volume output per chassis. This volume makes economies of scale possible and investment in pre-production design and testing economic. Through these pre-production steps, tighter specifications for component suppliers and heavy investment in automatic insertion and automated testing lines connected to the inserters, Japanese manufacturers are able to produce much higher quality sets at lower costs than those of manufacturers in the United States, Europe or other Asian countries, the U.S. consultants claim.

The result has been not only a remarkable increase in the competitiveness of consumer electronic products currently supplied from Japan, but the restoration of comparative advantage to Japanese production of monochrome television, taperecorders and other products which had previously been assembled most economically using labor-intensive techniques.

The Boston Consulting Group has quantified the materials cost advantage which Japanese producers enjoy on color-television set production relative to other major producers. Where the cost index, calculated in sterling, is 100 for a Japanese producer, it is 113 for South Korea, 119 for West Germany and 126 for Britain. The Japanese producer's advantage is based on his lower component requirements per set, the scale of his component suppliers' production and the producer's bargaining

Table 2. Japanese Color Television Production in the United States

Parent Company (local company)	Plant Location (new/acquired)	Annual Production (units)	Year of Operation
Sony (Sony America)	California (new)	450,000	1972
Matsushita (Quasar Electronics)	Illinois (acquired)	600,000	1974
Sanyo (Sanyo Manufacturing)	Arkansas (acquired)	700,000	1976
Mitsubishi (Melco Sales)	California (new)	120,000	1978
Toshiba (Toshiba America)	Tennessee (new)	200,000	1978
Sharp (Sharp Electronics)	Tennessee (new)	120,000	1979
Hitachi (Hitachi Consumer Products of America)	California (new)	100,000	1979

Source: International Business Information, Far Eastern Economic Review.

175

power resulting from his scale of purchase and backward integration into component manufacture.

With the opening of the China market, this regained comparative advantage could not have been more timely for the Japanese consumer electronics industry. In addition to the order of 100,000 color television sets the Chinese placed with Hitachi in January 1979, orders for monochrome receivers and taperecorders have been increasing steadily, making the China market a rather bright spot in an otherwise cloudy television export picture.

Imports of TV sets from Japan is strictly a stop-gap business, however, which will last until China has increased its output to meet growing demand. Japanese investments in the Chinese electronic industry would hasten higher production and shorten the period of exports.

Following recent improvements in television transmission in China — in both color and black and white — demand for receivers has been increasing much faster than local production can be expanded. Since it will take several years before production can meet demand, and Chinese authorities are now eager to achieve rapid diffusion of television receivers, Japanese manufacturers expect that exports to China will grow for a few years at least. And since the value-added content in monochrome television production is rising, Japanese manufacturers such as Tokyo Sanyo and Matsushita are currently expanding their production of these receivers by 20-30 percent.

This is only one of several major product strategy shifts to take advantage of new technology and new markets. Five important changes mark 1978-79 as a highly significant period of transition for the industry:

• For the first time since World War II, Japanese industrial electronics surpassed consumer electronics in total output.

• Video taperecorders replaced color televisions as the principal export item. Moreover, since exports of video taperecorders to Europe are growing much faster than those to the United States, for the first time Europe has emerged as the major growth market for Japanese consumer electronic products.

• Timely introduction of multiplex sound telecasting stimulated domestic demand for color television receivers and multiplex adaptors.

• The boom in "Invader" video games opens a vast new market for products which combine video and computer technology.

• The new generation of super hi-fidelity audio products, combining miniaturization, microcomputer controls, pulse code modulation sound reproduction and metal tape, set the stage for new developments in audio-visual synthesis once videodisc systems are perfected.

The ascent of industrial electronics — especially computers, telecommunications and automation equipment — is pregnant with implications for the consumer electronics industry. It is not so much that the consumer electronics industry has been eclipsed. Rather, the development of industrial electronics will spawn successive generations of consumer electronic products, blurring the distinction between the two sectors. Personal computers will use the television receiver as a visual terminal. Home facsimile equipment will reproduce images displayed on the TV receiver. And connected to the telephone, the TV set will bring visual communications to the private subscriber. These and other products, combining industrial and consumer electronics, are being developed in Japan.

For the immediate future, the hottest item in the industry is the video taperecorder. During the first half of 1979, exports exploded by a spectacular 78 percent over the corresponding period of 1978, even though exports to the United States remained stable during the first six months of the year, contrary to expectations that they would rise sharply from 400,000 units in 1978 to 750,000 in 1979. Despite the presence of

Table 3. A.   Impact of Component-Count Reduction on Labor Productivity of Japanese Color Television Manufacture

	Base	1974	1975	1976	1977	1978
Man-hours per Set	4.11	3.48	2.71	1.46	1.29	1.15
Index	100	85	66	36	31	28
Component Count	750	750	645	465	443	430
Impact*	4.11	4.11	3.70	2.97	2.89	2.84
Index	100	100	90	72	70	69

* man-hours per set
Source: The Boston Consulting Group.

Table 3. B.   Impact of Automated PCB Assembly on Labor Productivity of Japanese Color Television Manufacture

	Base	1974	1975	1976	1977	1978
Man-hours per Set	4.11	3.48	2.71	1.46	1.29	1.15
Index	100	85	66	36	31	28
Percent Automated						
PCB Assembly	0	16%	43%	55%	57%	75%
Automation Impact						
Impact*	4.11	3.68	3.12	3.19	3.20	2.95
Index	100	90	76	78	78	72

* man-hour per set
Source: The Boston Consulting Group

Philips and Grundig with their own jointly-developed systems, Japanese exports to Europe have soared. JVC alone will sell 300,000 systems in Europe in 1979: in addition to sales through its subsidiaries in France and West Germany, the leader of the VHS (Video Home System) group has contracted to supply its systems to leading French and German electronics manufacturers on an original equipment manufacturer basis. The VHS group is one of two Japanese video taperecorder groups using competing technologies: the group is headed by Matsushita and JVC, competing with the Betamax group led by Sony.

About 10 Japanese firms have entered the European video tape-recorder market in the past year, and are now locked in fierce competition for a share of the market. The race has only just begun, but the results so far have been startling. Accounting for virtually all the growth in video taperecorder exports during the first half of 1979, Europe has emerged as the most important growth market for Japanese electronic products.

With continued strong export demand by Japan, total video tape-recorder production is expected to increase by 40 percent in 1979 and expand at an annual rate of 30 percent over the next three years.

But this may all change again when videodisc systems are introduced for the consumer market. Although competition for the video tape-recorder market has just begun, major electronics manufacturers have already made plans for entry into the videodisc market.

In December 1978, Magnavox, an American subsidiary of Philips, launched its "Magnavision" optical videodisc players in the United States at prices considerably below those of video taperecorders. In January 1979, RCA announced that it would offer a system using two-hour discs with a selling price of US$400 or less. And just two days after that, the U.S. Pioneer Electronics group demonstrated to dealers and the press the industrial-institutional version that Universal Pioneer (a subsidiary of Pioneer Electronics) is producing for MCA (a large California-based entertainment concern) to sell in the United States.

During 1980 and 1981, industrial-use videodisc players are likely to be used widely for sales promotion, training and information storage by business and government. Indications are that Universal Pioneer will take the lead, as the company has already developed a prototype for mass production and began monthly production of 2,000 industrial-use systems in June 1979.

The timing of the introduction of multiplex sound telecasts in Japan in October 1978 was well calculated to spur domestic sales of color television sets and adapter units during the Christmas buying season, just as color television exports to the United States were turning down. Both television and audio manufacturers responded immediately with stereophonic multiplex adapters for existing TV receivers, and within a few weeks of the new telecasting system being introduced, new models of television receivers offered consumers built-in multiplex reception.

Japanese TV makers expect that market demand for multiplex sets and adapters will expand as telecasting improves and as its use becomes more extensive in Japan as well as being introduced abroad. American television networks are expected to begin multiplex broadcasting as soon as the Federal Communications Commission completes its inquiry, and bilingual countries are expected to follow Hong Kong's lead with simultaneous telecasts in two languages.

Further improvements in broadcasting are expected to follow with the introduction of teletext systems around the world, bringing renewed demand for new generations of television receivers. In addition, new features such as Sanyo's voice remote control system will make more varied use of microcomputers in successive generations of television receivers.

Just as multiplex telecasting is expected to be the forerunner of future systems requiring continuing refinements in television receivers, the present "Invader" boom will be followed by other TV games. Major suppliers, such as Hitachi and Fujitsu, have found a substantial new market for which they will develop a stream of new products combining improvements in television and microcomputers. Analysts believe the strength of market demand will lure new entrants.

Meanwhile, new developments in audio technology will bring a continuing flow of new products. However, this market is rather narrow, and the sophisticated nature of new systems defies mass market penetration. While pulse code modulation, metal tape and miniaturization are expected to sustain demand at present levels, audio makers have had to rely largely on an expanding export market for their growth. There are now clear indications that this is no longer possible.

Table 4. Industry Investment Trends

				(billions of yen / %)
	1977	1978*	1979**	1979/1978
Industrial Electronics	54.1	77.9	71.6	−8.1%
Consumer Electronics	65.8	62.3	59.9	−3.9%
Electronic Components	63.3	69.9	71.6	2.4%
Industry Total	290.6	330.3	311.2	−5.8%

* Estimated
** Projected
Source: MITI

The result is likely to be an industry shake-out. There are too many audio makers to share a virtually static market and all remain profitable. With the recent entry of full-line electronic/electrical appliance firms in the market for sophisticated hi-fidelity sound systems, specialized audio firms such as Pioneer, Trio-Kenwood, Sansui, TEAC and Nippon Columbia have come under increasingly severe competitive pressure. Their future growth, and survival, will depend on the success with which they diversify into new products, as Pioneer is doing by grasping the lead in development of the videodisc.

It is possible that the market for videodiscs will be larger and more profitable than the market for video taperecorders. In a sense, the disc is the next generation of the phonograph record, and will not compete directly with the video taperecorder. Moreover, entry into this market is not likely to be as expensive or as risky. For that reason, and because there will be at least five videodisc systems competing for the market, Pioneer may indeed be blazing the trail to future growth through a marriage of audio and video production and marketing.

A China Opening

Following discussions with China's Deputy Premier Deng Xiaoping in Beijing, Konosuke Matsushita, a doyen of Japan's electronics industry, has produced a grand design for Sino-Japanese cooperation in the modernization of China's fledgling electronics industry.

Matsushita's plan would mobilize Japanese capital, technology and management resources for a massive commitment to China's electronics revolution.

Initially, the emphasis is likely to be on consumer electronic products for use in China, though Beijing is anxious to acquire semiconductor technology which provides the key to modern "solid-state" development in both the consumer and industrial electronics fields.

From there, China could advance to finished-product exports, though the Japanese industry is not too eager to see this happen, and wants to write into any agreement a clause limiting exports to 10 percent of production. Even exports under that limit would have to be mutually agreed, with the Japanese partners having a veto.

The organization vehicle for participation in the fashioning of the most advanced stages of China's new technostructure would be, according to Matsushita's initial design, a single joint venture between the top 10 Japanese consumer electronics manufacturers and the entire Chinese electronics industry.

This remarkable notion was apparently conceived spontaneously by Matsushita, founder and now "director/adviser" of the company, during his discussions with Deng in Beijing in June 1979. According to Matsushita, Deng asked Matsushita Electric Industrial Company, the world's biggest electrical appliance maker, to provide technical assistance for the modernization and rationalization of the Chinese electronics industry.

Since such a mammoth undertaking could not be attempted by his company alone, Matsushita proposed that Chinese authorities consider forming a joint-venture company with the whole Japanese industry for the purposes of the project.

On his return from his initial 10-day visit to China, Matsushita disclosed that Chinese leaders showed keen interest in his proposal and urged him to take the initiative in establishing such a joint venture "regardless of the share of capital."

By early August 1979, Matsushita had prepared a plan which was spelled out in a working memorandum for discussion with the Japanese government and industry. Although the memorandum had not been made public, Japanese press reports indicated that the original Matsushita plan called for a single joint venture with the Chinese industry that would act as a kind of holding company with a 50-50 participation in equity capital.

Under this formula, 10 large Japanese electronics firms would divide half the equity among themselves, with a "main partner" taking a 25 percent shareholding. This main partner would provide most of the knowhow, machinery and other production facilities, with the residual requirements supplied by the other nine partners and about 40 other firms specializing in electronic parts or materials.

The main partner would provide the president of the joint venture, assuring virtual control of management by the Japanese participants. China's recently-published law on joint ventures stipulated that the Chinese side would appoint the "chairmen" to the boards of joint-venture companies, but China apparently agreed to bend this rule for the Japanese.

Although the outline of Matsushita's ambitious concept of Japanese participation in the upgrading of the Chinese electronics industry was apparently approved in principle when presented to industry leaders, and later to Prime Minister Masayoshi Ohira and Minister of International Trade and Industry Masumi Esaki, it became clear that there was no consensus within the industry or in government on details.

For a start, major Japanese electronic manufacturers were concerned about reactions in the United States and Europe, where the Matsushita plan could be seen as an attempt to establish a *de facto* Japanese hegemony over the Chinese market. Such a Sino-Japanese joint venture could also evoke ominous images of a major Asian combination intended to dominate the world electronics industry, and would almost certainly set off an explosion of anti-trust actions in both the United States and Europe, where rules of competition are interpreted relatively more strictly than in Japan.

The ambit of U.S. anti-trust law is very wide, and can extend to virtually any industrial arrangement or "conspiracy" anywhere which has an impact on U.S. trade and industry. U.S. interests have even been known to pursue successful actions along such lines against domestic industrial groupings in Japan.

Given the high stakes of the Japanese electronics industry in the American and European markets, it is most unlikely that manufacturers would seriously entertain any proposal which threatened to endanger those interests or trigger an intensification of friction with the electronics industries in the United States and Europe. A bird in the hand is still worth more than two in the bush, even where the huge Chinese market is concerned.

Industry objections apparently do not stop at this issue. Despite the capacity the Japanese have for intra-industry and business-government cooperation, competition among manufacturers in the marketplace is fierce, even excessive. Under these conditions, there is little likelihood of the industry accepting a plan which would reserve to a single manufacturer a predominant position in the Chinese market.

To meet these objections, a modified joint-venture scheme proposed the establishing of 10 regional joint ventures specializing in various product areas. The idea of organizing the existing plants in China under these joint ventures would be retained.

The main Japanese partner in each regional joint venture would be selected in accordance with the capabilities required, with each of the big 10 playing this role in one new enterprise. The main partner in each regional joint venture would provide technological knowhow.

CHAPTER 11

High-Definition Television

ON 14 JANUARY, 1984, the Japan Broadcasting Corporation (NHK) demonstrated to the press a newly-developed system which makes possible the transmission of TV images two to three times clearer than those of conventional broadcast, while using even less than the electro-magnetic-broadcast spectrum previously required by high-definition TV technologies. By squeezing the TV signal into a smaller bandwidth, NHK engineers achieved the breakthrough which may prove to be as important as the invention of color broadcasting.

The improvement of this new band-compression technology constitutes a major step forward in the long march of the TV industry, begun in 1884 with a patent of a complete TV system taken out in Germany. Stated simply, the progress in TV since that early beginning has been a function of the number of scanning lines used: the more lines a TV system has, the sharper and more detailed the picture it will produce. The greater the detail, in the language of the TV engineer, the higher the definition.

Since the beginning of TV, competition in the transmission and receiver-equipment business has begun with attempts to perfect higher definition pictures, which is a function of the number of scanning lines across the screen inside the receiver's cathode ray tube (CRT).

After more than a decade of research and development (R&D), RCA set the United States standard with the introduction of a 525-line system in 1940. In Europe, somewhat sharper images are obtained by the 625-line German PAL and the 819-line French SECAM systems after World War II. Then, in 1981, at the Society of Motion Picture and Television Engineers, representatives of NHK and a group of Japanese equipment manufacturers unveiled the working prototype of a high-definition TV (HDTV) system using 1,125 lines, more than twice the present U.S. standard. Using a somewhat larger screenwidth, the new HDTV

Reprinted from "Japan High Definition TV: The Sky Itself Is No Longer The Limit," *Far Eastern Economic Review*, 6 September 1984.

picture contains five times as much video information as a conventional one.

The main problem, however, was that to transmit this information, the orignial NHK system required 30 MHz of video baseband bandwidth, which is equivalent to the total radio-frequency broadcast spectrum of the five VHF or UHF TV channels in the United States. Put another way, a country that has five conventional TV channels could have only one HDTV channel if no additional channels were available under international conventions. Since the electromagnetic-broadcast spectrum is limited, bandwidths available for use in each country are limited by international considerations, a condition which virtually precluded the introduction of HDTV in its original form.

The NHK breakthrough with the development of its multiple sub-Nyquist* sampling encoding or Muse system overcomes this obstacle, clearing the way for the third generation of TV. This process squeezes TV signals into a channel less than half the normal size, using special equipment, not only at the TV transmitter but also in the TV receiver.

By overcoming the voracious appetite for bandwidth, NHK engineers capped an undertaking in TV systems re-design that began in 1968 with the fundamental question: given the physical limitations of visual perception and normal, comfortable viewing distances, what should the TV picture be like for optimal human comfort and information reception?

After a series of tests, it was determined that preferred viewing distance is two to four times the height of the screen and that our eyes resolve about one minute of arc, corresponding to 1,750 lines. But since even for a completely new system such bandwidths were impractical, NHK engineers finally settled on 1,125 lines, transmitted at 30 frames (or 60 fields) per second, and a 20-MHz luminance bandwidth: and for better, more comfortable viewing, a larger screen with a five-to-three aspect (width to height) rate was established as ideal.

Once these main specifications had been determined, everything had to be re-invented. A camera tube to obtain high resolution, the 1-in. diode-gun impregnated-cathode Saticon (DIS), was developed and made available to leading video manufacturers for production engineering. A three-tube color camera using DIS was produced with the necessary 1,125 scanning lines and frame rate, including circuitry to correct registration errors at 570 points in the picture. At present, given pick-up tube limita-

---

* Sub-Nyquist is part of proper noun nomenclature of MUSE bandwidth compression technology. It refers to the maximum separation in time which can be given to regularly spaced instantenous samples of a wave of given bandwidth. Numerically it equals (in seconds) less than one-half of the bandwidth.

tions, current camera technology is just barely able to produce a satisfactory HDTV image with studio-type cameras to deliver high-quality pictures.

The requirements of receiving tubes, which will eventually be used in computer and videotext terminals as well, were equally rigorous. A series of 30-55 inch high-definition picture tubes was developed to meet these demands. Experiments in flat screen were upstaged and remarkable progress made. A 70 mm laser telecine was perfected to convert film pictures into HDTV signals. Totally new transmission systems from field pickup equipment for program production to the terrestrial and satellite transmitters had to be designed and perfected. This, in turn, meant the re-design of videotape recorders used in broadcasting.

Two parallel developments in TV broadcasting became critical: direct satellite broadcasting and digitization. Since they at once expand the broadcast bandwidth availability and permit digitization of signals, two closely inter-related factors, direct satellite broadcasting is the key to HDTV. To best that key, in April 1978, NHK put an experimental medium-scale broadcast satellite, Yuri (or BSE), in orbit. By bouncing separate luminance, color and FM audio signals off the satellite in fixed geostationery orbit, major improvements in both picture and sound transmission were obtained.

Based on these experiments, the world's operational first direct-broadcasting satellite, the BS-2A, was put into orbit on 23 January 1984, within days after the announcement of the bandwidth compression breakthrough, adding powerful thrust to the TV revolution and the emergence of information systems of which TV is a critical component.

At present, due to the breakdown of two of the three transponders of the BS-2A satellite, NHK is operating only one channel which services remote mountainous areas, outlying islands and a million or so households located in congrested urban districts among high buildings. Direct-satellite broadcasting will expand in August 1985 with the scheduled launching of the three-channel BS-2B and is expected to be ready for HDTV broadcasts when the larger four-channel BS-3A goes into geostationary orbit in 1988. Although the final decision on the timing of the first HDTV transmissions has yet to be made, it is generally believed that the launching of the BS-3A will provide the appropriate occasion for the inauguaration of services.

A number of key problems remain to be resolved before the final decision is made, however. The problem of compatibility with existing TV receivers is especially difficult. Methods of chrominance multiplexing, luminance, bandwidths and line rates all differ from those currently in use in Japan, the United States and Europe. This means that a satisfactory scheme for phasing in the new system without immediately making the existing sets obsolete must be devised.

Or, put differently, since HDTV broadcasts cannot be launched until there are some receivers out there, the solution would seem to lie in the direction taken with color TV: that is, HDTV broadcast receivable by existing sets even though they cannot benefit from the full effects of the new system. It took 20 years after color TV was introduced to phase out monochrome sets. While the transition to HDTV is likely to be faster, this will largely depend on arrangements for gradual transition.

CBS, which has made a major commitment to HDTV in the United States, has devised a scheme for broadcast of 1,050-line transmission divided between two channels. Existing sets, according to this scheme, could receive the conventional 525-line broadcasts on a single channel.

But consensus on this approach is proving elusive. Some of the industry's wise men, especially in the United States, see no real need to shift to high-definition broadcasts soon. Pseudo-HDTV effects are obtainable with full digitization of the video circuits. Digital receiving systems (digital TV transmission systems have yet to be developed) have the advantage of higher fidelity in signal processing which can improve picture quality almost as much as HDTV, even with the existing broadcast signal.

By using large enough RAM (random access memory) IC (integrated circuitry) devices, it is possible to double or triple the number of scanning lines simply by storing them for display in the desired sequence. Digital receivers also can be designed for high-quality TV stereo, reception of video games, special effects (such as freeze-frame, split-screen and zoom), teletext storage and retrieval as well as flicker and ghost elimination.

Since neither bandwidth, obsolescence, nor problems of standards arise with digital TV, it is clear, as a Matsushita executive has affirmed, that the future of TV has to be digital.

There are other compelling reasons why this must be so. The number of components required in a digitized receiver is substantially lower, decreasing the adjustment steps in production while increasing reliability. Sony engineers reckon that through the use of digital circuitry, they can reduce the number of components from the 400 required at present to less than 100.

Unlike HDTV, which has been developed mainly by a broadcaster rather than a manufacturer, using entirely new technologies of Japanese origin, digital TV has found its champion in ITT Europe. So far, though the company is the only semiconductor manufacturer to produce digital TV chips, its West German subsidiary Intermetall has been selling the seven-chip Digivision circuits for as low as US$25 and in time the price will decline.

For leading Japanese makers, however, digital and HDTV are not mutually exclusive. Sony, Sanyo, Sharp, Matsushita and Toshiba have been among the early developers of digital signal processing using ITT chip assemblies. In August 1983, both Sony and Matsushita announced breakthroughs in digital TV circuitry, with Sony claiming more than 60 patents pending on a design that incorporated non-interlace scanning, which doubles the number of lines that make up a picture on the TV screen. To produce the 1,150-line picture, each line of the received broadcast is stored in a digital memory before being clocked out at double speed on to the TV screen.

Sony's new system also features complete separation of the luminance and color signals, reducing the coincidence of flickering, dots and color spilllover. A dynamic comb filter, developed by ITT and incorporated into one of the integrated circuits of the Digivision assembly, enhances picture resolution, providing reproduction of images free of interference.

At the same time, based on NHK's 1,125-line HDTV technology, Sony has developed a high-definition video system with video recording, time-base correction and other capabilities. Elements of the system, prototypes of which were demonstrated as early as 1981, include:

- A high-definition three-tube TV camera, which incorporates the NHK 1-in Saticon high-resolution pick-up tube.
- 1-in wide-band videotape recorder with a new high-density recording format.
- Wide-band digital time-base corrector, featuring a newly developed wide-band AD converter.
- 20-in. and 30-in. high-definition Trinitron monitors, with a fine-pitch picture tube.
- 100-in. HDTV projector with a wide-band picture tube.

Yet dramatic progress has been elusive, despite the best efforts of NHK and set makers. In May, both Sony and Matsushita announced that their first HDTV sets still will not be ready for market until 1986 and then at the handsome sum of ¥500-600,000 (US$2,050-2,461) per set, not including the parabolic antennae that might be needed to receive HDTV transmission by direct-satellite broadcast. At present prices, the dish antenna plus tuner is available for ¥200-300,000, all of which pushes the price for a total installation into the million-yen bracket.

At these price levels, industry analysts estimate, it is going to take seven to 10 years for the market to reach takeoff. Buyers in Japan can be expected to snap up almost any new product, but for anything such as a boomlet to develop, prices of the sets, themselves must be brought down to the ¥200,000 level and antennae prices lowered proportionately.

HDTV sets are expensive at the moment because the TV decoder needed for bandwidth compression will need memory chips totaling 10 Mb. However, as 256 K RAM devices decline in price with rising production, this memory cost will decline sharply. Makers have also indicated that they are taking other measures to trim costs. At the same time, antennae prices can be expected to decline with new technologies and rising scale economies with increased output.

Toshiba and Alcoa NEC Communications, an affiliate of NEC, will supply home receivers and parabolic antennae to Satellite Television Corp. (STC) of the United States, under agreements concluded at the end of 1983 as STC moved into the final stages of preparations for a major satellite-broadcasting blitz. Toshiba's system consists of a parabolic antenna for receiving super-high-frequency (SHF-12 GHz) signals and an outdoor converter which changes SHF signals to 1-GHz zone signals, plus a tuner connected to the TV set for reproducing pictures and digital sound signals — all for ¥240,000.

But new plastic antennae with built-in converters reportedly under development by half-a-dozen makers — including Yagi Antennae and Nitto Electric — could bring prices down to the ¥10,000 per set level, some financial analysts estimate. At this point, both direct satellite broadcasting and HDTV will become much more interesting propositions.

While TV digitization will not wait for HDTV transmissions, HDTV technology is finding its immediate applications on other fronts. Since it comes close to duplicating of 35 mm film when projected, HDTV systems are expected to be widely used in the movie-production process. Fully automatic HDTV cameras have the advantages of easy operation. Tape, unlike film, is not wasted in the editing process, but can be re-used more than once: and instead of shipping prints to theaters around the world, motion-picture companies will be able to distribute an electronic taped version by satellite.

With this use in mind, Sony now has an entire integrated line of HDTV motion-picture production equipment designed specifically to obtain economy and production efficiency in film making and distribution.

What is good enough for the film industry is good enough for the home. Since high-definition videotape recorders are an integral part of the NHK system, they can be marketed independently as stand-alone video systems appealing to the same people who purchase expensive audio equipment. High-definition tape can be reproduced directly from motion pictures, conserving the original quality.

For ultimate conversion of present systems of broadcasting to HDTV, several intermediary steps must be taken:

- Some agreement has to be reached among manufacturers and users worldwide on exact standards. Since HDTV systems are large-scale, they require large investments and all but public corporations such as NHK would likely hesitate to make investments on this scale, in the absence of universally accepted HDTV standards.
- There is the problem of the cost of band compression. Until this problem is resolved, either HDTV will be limited at the supply side with too few channels available to carry HDTV broadcasts or the sale of receivers will be restricted, causing hesitation among commercial and public broadcasters alike to undertake a major commitment to HDTV.
- Even if the chicken-and-egg syndrome is overcome, the transition from conventional to HDTV broadcasting is fraught with complexity. Some scheme, perhaps similar to the CBS formula, must be found which does not make all existing TV receivers obsolete. Digital TV could provide an answer through its rapid diffusion, which in turn would simplify the conversion process.

Two circumstantial factors suggest that these problems will be resolved or eventually vanish. All major Japanese electronic-equipment makers have developed a line of HDTV cameras, recorders and other equipment and the future growth of these companies depends to no small degree on the market for the next HDTV generation of TV receivers and video equipment. Likewise, NHK is fully committed to the development of HDTV to its ultimate potential and, given the preponderance of Japanese manufacturers in the TV industry, this alliance between NHK and the Japanese makers may well be sufficiently powerful to reduce the apparent obstacles to insignificance. Technology alone seems likely to mitigate the importance of standards, simply by programming sets to handle any number of lines, which would be tantamount to making Japanese receivers world-standard by dint of universal presence.

No one in the Japanese industry doubts that the powerful mix of HDTV, digital TV, direct satellite broadcasting and new media all add up to explosive and continuing growth. Among them, the three technologies provide a good 25 years of potential growth through product and production innovation to develop additional features and increasing value-added.

Add to this the consequences of the merging of TV and computer, TV and telephone and TV and printers in the new media. A foretaste of things to come is seen in Matsushita Electric Industrial's new digital set that can simultaneously process signals from regular TV stations, videotex services and home computers. In October, Mitsubishi Electric will begin marketing a US$1,062 TV set with built-in printer that turns out hard-color copies of images received on the screen, which means that TV is just a step away from becoming a facsimile terminal.

Mitsubishi Electric also joined the race for flat, full-color liquid displays with the announcement in 1983 of the world's first large system — with a picture field 1.2 x 1.8 m (the size of the *tatami* used in traditional Japanese flooring) — which can display full-color animation, patterns and characters input from a videocorder, video camera or computer. The system, initially priced at ¥10-30 million a m^2, will first be used for visual services at theaters, music halls and sports arenas, but also heralds the advent of the flat-screen wall TV.

Thus far Mitsubishi, Matsushita, Sony, Sanyo and Seiko — all working on active-matrix type liquid-crystal displays (LCDs) — have been notably silent about commercial-introduction dates for their new flat-screen color TV sets. But rapid-fire advances in technology have overcome many limitations of low-contrast, narrow-viewing angles and slow responses to electrical signals, thereby opening the door to such lucrative markets as color TV, computer displays, large-screen projectors and ultra-high definition graphics — functions which will all eventually be united in a common home terminal for reception of microwave, satellite and cable transmission.

Initially, such LCDs will be used in pocket color TVs, which are expected to appear on the market in 1986 or 1987. Japanese TV manufacturers are mobilizing resources to offer large-area matrix displays up to the size of the Mitsubishi wall unit for use in homes with HDTV, which is expected to be commercially operative in five years' time. Flat screen, many in the industry claim, could be the breakthrough that would solve the chicken-and-egg problem stymieing the industry today. The problem of obsolescence would be considerably mitigated for Japanese householders if that big box suddenly could be replaced with a thin wall-panel screen.

While many doubters remain, noting that large-size LCDs cost much more and perform less satisfactorily than the conventional CRT, few of them are in Japan. Even the conventional caveat that flat-screen technology still has some distance to go before it will make the 30-year-old CRT obsolete, is dismissed as relative nonsense. Distance, after all, is not an absolute but is relative to speed and the speed of technology change in ths field is approaching that of light.

Work on thin-film transistor materials, especially poly-crystalline silicon, is proceeding rapidly in Japanese laboratories. With the higher speed at which electrons move through these semi-conductor materials, responses to electrical signals is faster than other altenative materials, including amorphous silicon, which is a decided advantage for HDTV liquid displays.

Once this application is perfected by leaders such as Sharp Electronics — by far the largest producer of flat-screen displays at present — and major competitors such as Epson, Hitachi and Toshiba, the sky itself will not be the limit. The eventual market for full-size flat displays will grow at astronomical rates.

All things considered, TV is poised for explosive development that will make past booms look microscopic in comparison. HDTV, plus multifunctional digital TV, personal-computer TV, character-multiplex TV, direct satellite broadcasting TV, systems TV, inter-active CATV, teletext, videotex all add up to a mammoth and continuing tidal wave of new products which defies meaningful forecasts. Projections made by Nomura Research Institute are that, by 1990 at least 45 percent of the consumer electronic products sold in world markets will not have existed before 1980. This assumes that global demand for consumer electronics products will reach ¥45 trillion in 1990, or roughly three times that at the outset of this decade.

Combining steady improvements in production efficiency with a continuing flow of revolutionary changes in TV technology from their laboratories, Japanese consumer electronics firms are destined to play the predominant role — very much like the role they play in the continuing development of video casette recorders.

PART IV.

Semiconductors

CHAPTER 12

Brave New World of Microelectronics

THE JAPANESE electronics company Fujitsu introduced in 1978 a random access memory capable of storing 65 636 digits of information (the so-called 64K RAM) at about the same time as International Business Machines and Texas Instruments of the United States announced successful development of similar devices. Just four years earlier, U.S. semiconductor executives estimated that in technology terms they were still two years ahead of the Japanese. The gap had been closed at an astonishing pace.

This was no ordinary coincidence. Not only was it the photo finish of a major technological race between competitors on both sides of the Pacific, but the event marked the beginning of a new epoch in the dramatic development of modern electronics. At the same time as the semiconductor industry found itself catapulted into the age of very large scale integration (VLSI — the next stage in miniaturization of integrated circuits after the present LSI), promising a massive variety of new applications of electronic technology, a Japanese semiconductor manufacturer attained technological parity with leaders of the U.S. industry.

For the U.S. semiconductor industry, which had enjoyed virtually total pre-eminence in the world for over 25 years, technological parity with Japanese manufacturers is pregnant with peril. With the example of steel, automobiles and television firmly in mind, semiconductor manufacturers in California's "Silicon Valley" are convinced that they too will be faced with similar devastating competition by the early 1980s.

Industrialists in the United States began to feel concern as early as 1976 that the Japanese semiconductor industry was out to dominate what was previously a nearly exclusive American domain. At that time, exports of semiconductors from the United States to Japan reached a peak and Japanese competitors had begun to eat into the U.S. market.

But most worrisome of all was (and still is) Japan's cooperative government-industry R&D program established specifically to develop

Reprinted from "The March of the Japanese Micro," *New Scientist*, 11 October 1979.

very large scale integration processes and the technology for mass production of a range of devices.

The Specter of Japan Inc.

The threat of Japanese competition was enhanced by the specter of "Japan Inc." manifest in the joint VLSI research and development effort, which contributed substantially to closing the gap with American semiconductor technology. Spurred by that perception, a group of smaller California-based semiconductor manufacturers — plus larger firms like Motorola and Fairchild — formed the Semiconductor Industry Association (SIA) in 1977 to check the Japanese advance in the U.S. market and their declining exports of semiconductors to Japan by mobilizing political forces in Washington.

With the notable exception of Texas Instruments (the industry's leader, which did not join the association), U.S. semiconductor manufacturers openly say that competition with the Japanese industry is inherently a "zero-sums game." In anything but diplomatic language, vice-chairman Robert Noyce of Intel — the leading maker of microprocessors — led the attack, claiming that the Japanese were intent upon destroying the American industry and were using unfair competitive practices to do so. In this inflammatory tone, the U.S. semiconductor industry took its case to Washington and to the public with a barrage of speeches and campaigns to persuade industry to "Buy American."

In fact, the present state of the market provides little justification for the U.S. industry's reaction to the emergence of Japanese competition. As new applications continue to be developed, customers worldwide are demanding integrated circuits faster than the U.S. industry can produce them. Even with Japan's entry into world markets, supply is still not catching up with expanding demand. Even those "Buy American" campaigners have been forced to haul down the flag and turn to Japanese semiconductor makers for supplies. And U.S. semiconductor manufacturers still continue to hold well over 60 percent of the world market in 1979, compared with 24 percent for Japanese producers.

In 1978, U.S. production of MOS (metal oxide silicon) devices accounted for as much as 65 percent of total world output, while Japanese production amounted to only 20 percent, according to the market research firm Dataquest.

A Threatened Industry?

In the United States, Japanese suppliers had only about 2 percent of the market in 1979, while U.S. makers accounted for as much as 11 percent of all semiconductors sold in Japan, a Bank of America research affiliate estimates. American manufacturers supplied as many as 17 percent of the integrated circuits (ICs) used in Japan in 1978 (not including the output of Texas Instruments from its Japanese subsidiary) and accounted for a higher percentage still of MOS devices, which are the bulk of present and future sales. Clearly, this is not the performance record of a threatened industry.

But the increasingly rapid diffusion of semiconductor technology and production is certain to create vigorous competition in world markets, and consequently the share held by U.S. manufacturers and their subsidiaries will gradually be eroded. Most observers of the industry agree, however, that the American industry, with its high rate of R&D investment and the continuing support of the defense establishment, will probably never be seriously threatened by competition from abroad.

Smaller U.S. semiconductor manufacturers may fear the long-term implications of the joint industry-government VLSI R&D project in Japan, in comparison with U.S. efforts, but J. Fred Bucy, president of Texas Instruments, does not see the US$300 million spent by the Japanese as that overwhelming. American spending on semiconductor research and development, not including that of IBM and Bell Laboratories, was over US$200 million in 1977 alone. Knowledgeable industry sources estimate that IBM's investment in VLSI from 1977 to 1980 is about US$1 thousand million.

Then why all the fuss? What is there in the trade and investment flows between Japan and the United States in semiconductors that would provoke what is tantamount to a declaration of war by the cream of American industry? You can find some of the answers to these questions in the recent study of the Japanese semiconductor industry prepared by BA Asia Limited, the Bank of America's consulting firm based in Hong Kong.

The study shows that as the technological gap has closed and Japanese manufacturing expertise has increased, the Japanese semiconductor industry's ability to meet domestic demand for an ever-wider range of integrated circuits has improved. The output of ICs, in numbers of circuits, almost tripled during the four years from 1974 to 1978. The Japanese firms that build ICs into products have tended to buy fewer circuits from independent makers while concentrating on marketing those of their own products that they do not need.

As a result, the proportion of these items that Japan imports has steadily declined. And as the more sophisticated ICs have been those supplied by U.S. manufacturers or their subsidiaries in Asia, the Americans' share of the market has dropped sharply, although in absolute terms the total of their exports to Japan has continued to increase.

Japanese Quality Is the Key

Another factor is that the quality standards achieved by Japanese semiconductor manufacturers and end-users are substantially *higher* than those of their American counterparts. This lower reject rate for Japanese ICs is essential to the economics of highly automated assembly of the final consumer or industrial electronic products (the hallmarks of Japanese quality around the world). Naturally, companies there prefer locally produced ICs.

Nor is this a question of a "Buy Japanese" policy, as some American industry executives have claimed. The Japanese electronics industry decided that reliability was the key to improving its market share for consumer products, and it is now simply repeating that performance in microelectronics. And the reliability and cost of Japanese consumer electronics products, computers, telecommunications equipment and a wide range of other products depends on the quality of the semiconductors used in their production.

While the difficulties of testing large-scale integrated circuits have been the subject of a three-sided conflict in American industry — among the people who design and test systms, those who make integrated circuits, and those who use the circuits — Japanese semiconductor manufacturers have developed quality control into a new, highly sophisticated managerial art from. While it makes the most of advanced technology, this is not the only key to Japan's reputation for quality.

In the past the Japanese relied heavily on test equipment imported from the United States, including final test computers; they still do to a lesser extent. But Japanese IC makers do not simply buy a machine off the shelf and install it. They adapt it, and constantly improve it, often developing a substantially different machine and they then offer that to competing manufacturers around the world.

They take the same approach to production equipment. After a decade or more of heavy investment in automation, the Japanese semiconductor industry has better techniques and more advanced automated equipment than the United States, especially in the assembly stage. While leading U.S. makers have kept their labor-intensive assembly systems and moved them "off-shore" to low-cost labor areas, in Mexico or the Far East, the Japanese industry has sought the ultimate in automated production. As a result, several Japanese makers of semiconductor manufacturing equipment have emerged as suppliers to the world industry.

Anelva, a joint venture of NEC Corporation, and Varian Associates of the United States, has received orders from major American manufacturers for its automatic dry etching equipment — designed to etch the integrated circuit substrate by a planar electrode system at high rates and with low ion energy. A fully automatic mask aligning device developed by Canon to print circuit patterns with a minimum line width of 1 micrometer (the criterion for the "very" in VLSI) has also attracted export orders from leading semiconductor manufacturers in America and Europe.

Advanced equipment may indicate the higher level of production technology in Japan's semiconductor industry, but the keys to its efficiency and the reliability of its products are the management systems and the quality of the workmanship. Even in highly automated production, the human element is decisive. Testing in Japanese factories is not restricted simply to identifying good and bad products. It is a continuous approach that begins with design and permeates all stages of manufacture through to the product. Quality control is not one person's special job; everyone shares responsibility. All the staff are encouraged to offer suggestions for improving things. More important, their suggestions are carefully considered and often adopted.

This means that everyone must be informed of the whole production process. Everyone must know where he fits into the process and how his performance affects the end results. This calls for very intensive training and managers must avoid, or resolve, conflicts between people at various stages of production over quality control procedures. Intensive training is intended to assure total communication within the work group and thereby attain a high level of participation and cooperation.

Unfortunately, although the latest Japanese equipment can be acquired easily enough by American and European semiconductor makers, they tend to take a rigid approach to testing — an attitude whose roots are deep in the contract system and union demarcation rules — which tends to preclude the kind of team spirit that the Japanese display. The Japanese approach to quality control is concerned mainly with eliminating the causes of error, wherever they may occur in the production process, rather than simply with identifying and eliminating faulty ICs.

Test, Test and Test Again

American manufacturers have a rigid "acceptable quality level" (AQL), but they assume that a certain degree of failure is normal. The Japanese, on the other hand, begin with the premise that the only acceptable number of defects is none. At Fujitsu, for example, engineers meticulously inspect and test at each production phase and repeat this process from a variety of angles. To produce a semiconductor memory involves as many as 100 test steps. As its engineers identify and eliminate the causes of failure, they minimize the number of testing steps. And they use automated equipment wherever possible, to eliminate both manufacturing and testing errors. Failures are measured over time — the number of failures per thousand million hours of operation — not just at the time of testing. In addition to failure data from users, Japanese makers use test procedures which accelerate the rate of use and analyze the failures.

Life tests, environmental tests, function tests and failure origin tests are carried out in great number, over and over again, at all stages — from design and development through production. Using this so-called failures in time standard (FITS), Japanese semiconductor manufacturers consider 15 to 30 failures per billion hours of operation to be high quality, while 100 failures are rated good. (By comparison, the U.S. space industry sets 2 failures per billion hours of operation as the standard.)

Quest for Quality

This quest for quality is not motivated primarily by requirements of the space industry or military programs or for that matter by competitive market conditions outside Japan. Rather, it is dictated by the demands that the highly discriminating Japanese consumer makes and the fierce competition among Japanese manufacturers to meet those demands. Quality control is not a matter for concern only in the semiconductor industry. It is an all-pervasive philosophy of production that imposes rigid standards on quality of materials and workmanship at all stages of the manufacturing process of all products. Nothing could be more irksome for U.S. manufacturers, especially to the high-flyers of the microelectronics industry, than to be beaten on quality standards. And some have gone so far as to say, in public, that it is downright unfair. If they were to impose such quality standards in their production, they say, they would be forced to declare bankruptcy. Yet, they have had to impose just such quality standards to sell in Japan, even though this has meant establishing expensive post-production test facilities in Tokyo.

Texas Instruments has adopted an entirely different strategy, however. It decided that, rather than waste time and effort pushing water uphill, it would be far better to do what comes naturally: build a production plant in Japan, to meet the needs of users on a more-or-less equal footing with the local industry. "For Texas Instruments it is not a question of simply equaling Japanese quality," says an executive of TI-Japan, "we are determined to be better." As a result TI-Japan meets the higher standards of its Japanese customers at competitive prices and is steadily increasing its share of the market while other U.S. suppliers are losing ground. In fact, TI is doing so well in the Japanese market that it is now building two new plants to meet the growing demand for its products. It is no secret at TI that its Japanese subsidiary has the highest quality production in the company. Just as Japanese manufacturers have succeeded in America, so Texas Instruments is systematically using its base in Japan to upgrade quality control throughout its far-flung manufacturing operations.

Higher Productivity Too

Similarly, as the productivity of Japanese workers at TI-Japan is 20 to 40 percent better than in the United States (depending on the product manufactured) the company's top management in Dallas has adopted many of the Japanese management's techniques for use throughout Texas Instruments.

Another thing that concerns the U.S. industry, according to Thomas Kurlak (of stockbrokers Merrill Lynch, Pierce Fenner & Smith Inc.), is not the quantity of Japanese semiconductor imports, but rather the low Japanese prices that restrain American firms from raising their prices in the face of mounting demand. In brief, U.S. firms would like to set higher prices to realize higher profits which they could invest in new plants. Smaller Silicon Valley firms face a difficult task in making the technological transition in the next few years to VLSI, certainly one that will require extensive investment in machinery and facilities. Higher prices at a time when demand is at its peak would help enormously to make this transition less fraught.

The structural differences in supply and demand between Japan and the United States are now being extended beyond the Japanese market, into areas that have been the prime reserves of U.S. semiconductor makers. As Japanese electronic equipment manufacturers start setting up production plants abroad, and increase exports of ICs, it appears to be only a matter of time before Japanese firms begin producing semiconductors in the United States and Europe. During 1979, both NEC and Hitachi announced plans to manufacture in the United States.

These moves are clearly part of the global strategies of the two leading Japanese semiconductor makers. In early 1979 NEC also decided to assemble certain semiconductor devices in Brazil beginning in 1980 adding to its overseas production bases in Ireland, Southeast Asia and California. Hitachi plans to begin assembly of semiconductors in Europe, with Germany as the most likely location, giving a third dimension to its production operations abroad, supplementing similar plants in Penang, Malaysia, and Dallas, Texas.

It is not yet clear where the European industry will figure in this rapidly changing picture. Both production and market structures are fragmented, and governments, which taken together are spending even more than the Japanese on VLSI research and development, are still pulling in their own, often different, directions. As microelectronics will be the basis of the industrial system for the next 25 years, at least, the consensus seems to be that the European industry must do something. But the cost of catching up with the U.S. and Japanese industries promises to be staggering, and time is definitely not working for European firms in the field.

CHAPTER 13

The VLSI Revolution

DURING THE 1970s, the Japanese semiconductor industry emerged as a second pole of technological innovation in a global microelectronics revolution and won worldwide recognition for its prowess. In the next decade, microchip manufacture will be one of Japan's most rapidly growing sectors and a major force in the transformation of both Japanese and international industrial structures.

Propelled by rapid changes in microcircuit technology, the electronics industry will be the most dynamic sector of the Japanese economy during the next two decades. In 1979, the gross product (by value) of the electronics industry in Japan was less than half that of the steel industry and only about 40 percent that of the vehicle sector. By 1990, total output of electronic products will surpass US$100 million, not accounting for inflation, and exceed that of steel by a considerable margin. Only the vehicle industry is likely to remain larger in terms of toal output, and will itself have become a major consumer and producer of electronic devices.

If, as this forecast assumes, the output of electronic products grows at an average 10 percent per annum throughout the 1980s, industry planners expect semiconductor production to grow at an appreciably higher 16 percent a year. Most of this growth will be in integrated circuits (ICs), with output of discrete (separate) semiconductors (mainly transistors and diodes) remaining stagnant or declining.

During the four years from 1975 to 1979, the annual growth of IC production averaged almost 34 percent, exceeding the total production of discrete semiconductors for the first time in 1978. In August 1979, also for the first time, Japan became a net exporter of ICs, marking a significant strengthening of its position in world markets.

Prospects are good that, despite rapid growth in the home market, the Japanese industry will remain a net exporter in the future, substantially increasing its share in world marketplaces. Estimates by Nomura

Reprinted from "Semiconductors: Another Leap into the 1980s", *Far Eastern Economic Review*, 5 December 1980.

Securities, whose past performance would indicate are quite conservative, predict that domestic demand for ICs will expand more than two-and-a half times during the first half of the 1980s as a result of the combined growth of domestic demand and exports. According to the 1980 report on the industry by *BA Asia* — the Hong Kong-based consultancy subsidiary of the Bank of America — Japanese manufacturers could account for one-third of the US$20 billion world semiconductor market by 1985.

Such a substantial improvement over the present 26 percent share of world markets does not reflect simply (or mainly) the anticipated increase in Japanese exports of ICs, however. Rather, it is the expected result of a more rapid and pervasive diffusion of semiconductor technology throughout the broad spectrum of industrial production in Japan than will take place in North America or Western Europe.

Applications of ICs by Japanese industry have at once been more varied and taken place at a remarkably faster pace than in other advanced countries throughout the 1970s, while the computer industry has provided the single largest market for advanced large-scale integrated circuits (LSIs), and will continue to do so in the future. It accounted for only 25 percent of the ¥364 billion (US$1.72 billion) domestic consumption of ICs in 1979. Audio-video manufacture took as large a share of available ICs, assuring Japanese manufacturers a leading position in world markets for these products. Likewise, cameras, watches and caluculators have evolved through successive product design generations in pace with advances in semiconductor technology. And the diffusion has followed the same rapid pace in telecommunications, office machines, vehicles, home appliances, toys and a bewildering assortment of products.

Perhaps most outstanding of all, however, has been the rate with which semiconductors are applied to machinery in Japan. The expanding park of industrial robots which man Japanese factories round the clock exemplifies the special zeal with which semiconductor technology has been adopted in Japanese industry.

This zeal is not manifest haphazardly, however. As systmatic as semiconductor technology itself, the application of successive generations of ICs has been systematized into a separate technology — mechatronics. As the name suggests, mechatronics is simply the combination of mechanics and electronics, but done in such a way as to assure in each product the optimum combination of the two technologies.

This practice is not new, of course. But the development of the microcomputer and other semiconductor devices — all available at relatively low prices — has accelerated the process, giving rise to public policies and corporate strategies to encourage and direct the process of change.

Mechatronics in Japan has sped the process of semiconductor application through systematic and continued development, including:

- total replacement of conventional mechanical devices, whenever possible (calculators, watches);
- partial substitution of mechanical functions by electronic devices (sewing machines, cameras, copiers, vehicles);
- addition of electronic control devices to conventional machines (numerically-controlled machine tools, robots, electronic controls for engines).

As a result of phenomenal Japanese advances in application of semiconductor technology in calculators, watches, cameras, machine tools and robots, a high-growth market for ICs has been created. But the dynamics of the process does not stop here: the substitution of electronics for mechanical products also stimulates demand for the whole product.

The electronic watch well illustrates this dynamic interaction of semiconductor technology, market and production. From 1964 to 1972, unit shipments of watches in Japan increased at an average annual rate of just under 2 percent. After the introduction of the electronic watch in 1972, however, the growth rate has been over 7 percent a year. As a result, IC makers and watch manufactures have both benefited from this trend, not only from increased volume, but from higher value added per unit of production.

Semiconductor manufacturers have a special advantage, however. As systematic mechatronization increases domestic consumption of ICs, the resultant higher volume production enables them to improve their technology and efficency — making it possible to reduce production costs accordingly — with all the advantages which then accrue in highly competitive world markets.

This process is expedited, not so much by massive government subsidies or special tax advantages, but by an efficient financial system which makes possible the high and flexible rates of investment which the highly capital- and technology-intensive semiconductor industry requires. Japanese manufacturers in such high growth, high value-added sectors have access to capital on the sustained and rational basis which advanced and rapidly changing technologies require.

Equally important, the vertical and horizontal integration of large-sized, highly-diversified Japanese manufacturers of electrical and electronic products have a special advantage in the allocation and use of resources for rapid application of rapidly changing semiconductor technology. This explains in part why semiconductors were first applied to consumer electronics and calculators in Japan, rather than in the United States, where the basic technology was originally developed.

American IC manufacturers began as semiconductor manufacturers and merchants producing a broad range of devices to cover as large a market segment as possible.

This structually-induced efficiency in the application of new semiconductor technology by integrated Japanese manufacturers has been both cause and effect of a remarkably higher degree of specialization in semiconductor production than is found in leading American semiconductor makers. Specialization in Japan, to meet in-house product requirements, tends to assure greater economies of scale and learning so critical to semiconductor manufacture.

The next stage in the macrodynamics of this industry was entirely predictable. It is an immutable law of techno-economic behavior that basic technology flows to the point of most efficient application and production. During the first 30 years of the microelectronics revolution, basic technology was developed in the United States and transferred to Japan and other Asian countries where it was more readily, efficiently and economically applied. Now, in the second stage of the microelectronics revolution, the lead in development of basic technology itself is shifting to the point of application and production, where organizational, financial, and informational resources are optimally available in the necessary combinations for continuing innovation.

The 1980s have begun with the significant lead taken by Japanese manufacturers in the 16K Random Access Memory (RAM) marketplace. Fujitsu's lead in the fielding of 64K RAMs, ushering in the very large-scale integrated circuits (VLSI) era, was even more remarkable, signaling Japanese technological advantage in the crucial big-volume memory segment of the market. And, as a recent Daiwa Securities study pointedly notes, Nippon Telephone and Telegraph, Fujitsu and NEC succeeded collectively in developing the world's first 128K RAM, which has since been followed with the announcement by Japanese manufacturers of 256K bit VLSI chips.

In VLSI development, Japan has clearly taken the lead: and once more, Japanese industry is best equipped to get the greatest advantages of the new technology in new product development. But these spectacular changes in the industry, though momentous, are by no means the whole story. As *BA Asia's* 1980 report points out, Japanese applications for IC patents have been growing steadily in recent years. while foreign patent applications in Japan have stagnated. The dimensions of the trend are important: total semiconductor patent applications in Japan increased from 4,406 in 1974 to 6,397 in 1977, while foreign applications dropped from 10 to 7 percent of the total.

THE VLSI REVOLUTION

Most new technologies developed in Japan have been in assembly process and manufacturing rather than inventions of new devices. As a result, Japanese mass production and automation technology are generally agreed to be the highest in the world, assuring better production yields and greater product reliability.

Sony's recent perfection of a method to grow better silicon crystals under the influence of a magnetic field — increasing IC production yields by up to 20 percent — is a case in point. Reduction of imperfections in silicon wafers becomes especially critical in the production of VLSI chips, Sony claims.

Even less visible than such breakthroughs by major manufacturers has been the emergence of a number of smaller scale leaders in semiconductor materials and processing equipment technology:

- Shin-Etsu Semiconductos has grown to become one of the world's largest single-crystal silicon makers, with production facilities in Singapore and the United States.
- Kyoto Ceramic is the largest maker of ceramic packages for ICs and LSIs, accounting for 70-80 percent of the world market.
- Dai Nippon Printing, Toppan Printing, Sumitomo Metal Mining, Sumitomo Special Metals, Tamagawa Metal and Machinery and Mitsui Mfg. have developed technological strength in the production of IC lead frames of highly efficient conductors such as ferro-nickel, cobalt alloy, and silver-stripped phosphorous bronze.
- Dai Nippon and Toppan are also leaders in the production of photo masks, glass plates with circuit designs printed on them used to reprint the design on silicon wafers.
- Canon is the world's second largest mask-aligning equipment manufacturer.
- Kokusai Electric is a leading manufacturer of single-crystal silicon production apparatus, diffusion furnaces, ion implantation apparatus and other equipment used in IC production.
- The chief achievements of the government-sponsored VLSI Technology Development Union, which has already applied for over 1,000 patents, are the development of revolutionary electron beam exposure equipment and high speed electron beam drawing equipment which will enable the Japanese semiconductor industry to produce mega-class VLSIs.

As demand for ICs is expected to increase at approximately 22 percent annually through 1985, the outlook for production materials and apparatus is as bright if not brighter than the semiconductor industry itself. Demand for production apparatus will be supported not only by growth of the industry as a whole, but also in large part by an unusually fast replacement cycle which is accelerated by ever more frequent innovations.

At the same time internationalization of production, as the industry establishes plants in major markets of North America and Europe in response to protectionist pressures, will increase the demand for equipment and materials. Although this demand will be met in many instances by foreign suppliers, Japanese equipment and materials manufacturers will undoubtedly be important beneficiaries of the move to overseas production.

CHAPTER 14

Chip-Making Machinery:
The Birth of A Hi-Tech Industry

SIGNS OF A BASIC structural change in the global semiconductor industry are emerging with increased frequency and clarity. The share of Japanese makers in world markets for the most advanced microelectronic devices is climbing steadily with each new generation, mounting to an estimated 90 percent of the current market for 256K DRAM chips. Growth rates of semiconductor output in Japan have averaged 30.7 percent annually over the past decade, compared with the 20.7 percent annual growth of the U.S. industry. And in 1983 total capital spending of Japanese makers surpassed those of U.S. makers for the first time, assuring even more rapid growth of future production.

As significant as these signals are in themselves, a less obvious but even more fundamental change has taken place. While in 1975 fully 80 percent of all semiconductor manufacturing equipment purchased by Japanese chipmakers was of U.S. origin, just a decade later, in 1985, most new equipment installed by the Japanese industry — an estimated US$2.5 billion worth — is being supplied by domestic equipment manufacturers. Equipment imports now account for only about 20 percent of the market, and exports, expected to reach 10 percent of total Japanese chipmaking equipment production this year, are climbing steadily.

In the brief span of ten years, a thrusting new industry has emerged, driven by a mastery of state-of-the-art technology in microelectronics, optics, metallurgy and automated systems engineering. Although some large integrated manufacturers are included in their number, most semiconductor machinery manufacturers are small- to medium-sized firms specialized in particular fields of high technology.

In this respect the Japanese industrial structure resembles that of the United States where 70 percent of the 700-800 firms producing semiconductor equipment have an annual turnover of less than US$50 million. But here the similarity ends.

Japanese translation published in *Shukan Daiamondo* (Diamond Weekly), 7 September 1985.

Japanese semiconductor equipment builders tend to have closer ties with chipmakers, either as direct subsidiaries or in a client-supplier relationship that includes a two-way flow of resources — men, money and technology. Although most equipment manufacturers sell across group lines, competing fiercely for customers throughout the industry, each has developed more-or-less intimate relationships with principal customers as a result of continuing cooperation in the design and development of successive generations of more advanced equipment.

This structural feature is reinforced by the rapid growth of the semiconductor industry and equally radical changes in technology. To compete in the increasingly torrid race for market share, chipmakers have had to constantly up-grade practically all the 100-odd processes involved in integrated circuit manufacture to assure reliable mass production using higher precision required by design rules steadily approaching submicron line widths.

The immediate effects of this race to raise the density of integrated circuits has been to shorten the life cycles of each new generation of chips and increase the cost of more sophisticated higher precision production equipment. As a result, with entire production systems rendered obsolescent in 2-3 years after installation, frequent replacement with higher cost equipment requires sharply mounting capital outlays and continued close cooperation between equipment vendors and chipmakers.

Since capital has been relatively cheap and readily available to integrated electric/electronic machinery makers that produce most of the semiconductors manufactured in Japan, and since this vertical integration of production serves to reduce risks of heavy investments required by scrap-and-build manufacturing strategies, capital investments in the chipmaking industry have grown in cadence with the demands of technological change.

On average, therefore, the Japanese market for semiconductor equipment has grown in excess of 30 percent a year. As more expensive equipment is needed to produce the same volume of chips and the volume of IC production continues to rise at annual rates from 20 to 30 percent, even higher growth rates are assured for the equipment industry for the foreseeable future.

The same is not true, at least not to the same extent, for the U.S. industry, however. Constraints are multiple and multiplying for established U.S. semiconductor manufacturing-equipment builders, who are therefore steadily losing world market share. Not only has capital cost U.S. chipmakers at least 50 percent more than it has cost their Japanese counterparts over much of the past decade, by conservative estimates, but capital spending has also been forced to follow an erratic stop-go cyclical

Table 1.   Capital Spending By Japanese Semiconductor Companies

	Capital Expenditures ($ mil.)				% of Semi Sales			
	1981	1982	1982	1984E	1981	1982	1983	1984E
NEC	$180	$209	$281	$571	15%	18%	18%	23%
Hitachi	120	180	215	530	17	17	15	25
Toshiba	85	135	255	605	12	16	22	34
Fujitsu	148	174	231	408	34	36	33	27
Matsushita	60	40	126	448	18	9	23	37
Mitsubishi	58	98	153	216	18	26	28	24
Sharp	43	51	85	146	24	12	18	18
Sanyo	51	38	51	87	24	16	16	16
Oki	56	60	55	87	41	35	20	18
Total	$801	$985	$1,452	$3,098	17%	19%	21%	25%

Sources: Company annual reports; *Japan Economic Journal;* Hambrecht & Quist estimates.

Table 2.   Semiconductor Production and Capital Outlays

	(A) Semiconductor Production	Ratio of IC	(B) 9 Major Makers Capital Outlays	(B)/(A)
1985 (E)	2,880 (+23)	76	715 (+25)	25%
1984 (E)	2,340 (+50)	74	572 (+58)	24
1983	1,560 (+31)	73	362 (+55)	23
1982	1,195 (+12)	70	233 (+19)	20
1981	1,067 (+24)	65	196 (+26)	18
1980	864 (+36)	66	155 (+49)	18
1979	637 (+20)	60	105 (+71)	16

Notes: (1)   Production: C.Y., Capital Outlays; F.Y.
       (2)   Capital outlays are mostly on a parent company basis to maintain consistency of data. If subsidiaries' investments are included, 1984 investment should reach 650 billion yen instead of 572 billion.
Source: Nomura Research Institute.

pattern in response to gyrations of capital markets. To a much greater extent than for Japanese makers, which tend to be less reliant on capital markets for their investment needs, investments have been curtailed by U.S. semiconductor manufacturers at the critical stage of transition to production of each new generation of devices, resulting in underinvestment in new production capacity for 16K, 64K, 256K and now 1M DRAM chips at critical stages of their development.

Fig. 1 Capital Spending in the Semiconductor Industry: North America vs. Japan

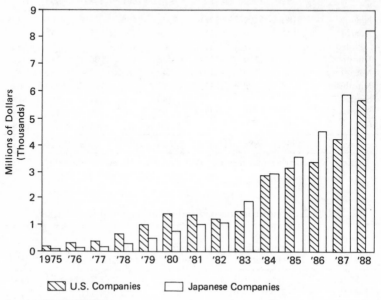

Source: DATAQUEST

During the successive downturns in the home market, U.S. equipment makers have thus become increasingly dependent on Japanese demand to sustain them. At the same time, however, they have steadily lost competitive advantage to Japanese competitors. Not only have Japanese equipment builders dramatically up-graded their technology for VLSI production, reducing and often eliminating the advantage of American makers, but they have also shown a greater readiness to make modifications in their equipment to meet customer needs. And since competition in the market for semiconductor devices is determined largely by superior manufacturing strategies and systems, Japanese chipmakers tend to be exceedingly exigent in their requirements for equipment modifications to their particular production system specifications.

U.S. equipment manufacturers, unaccustomed to making major changes in standard production models for American customers, have recently been subject to the added handicap of export controls that tend to reduce even further the flexibility in meeting Japanese market exigen-

Fig. 2 Capital to Revenue Ratios in the Semiconductor Industry: North America vs. Japan

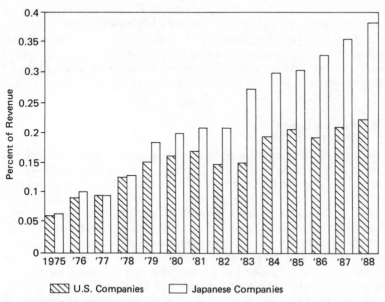

Source: DATAQUEST

cies. Exports of virtually every category of chipmaking equipment are subject to export licensing which requires U.S. Department of Defense approval, resulting in delays and uncertainties of delivery that few overseas buyers will accept. And these uncertainties quite naturally tend to act as added disincentive to making design changes in equipment to meet special customer requirements.

Ironically, this particular barrier to U.S. equipment export is based upon assumptions of technological superiority that is rapidly fading or no longer exists. In addition to their own substantial R&D efforts, Japanese semiconductor equipment builders have been major beneficiaries of a massive flow of new technologies from both public research laboratories and those of major chipmakers which are developing some of the world's most advanced technology.

Expenditures on IC research and development by Japanese semiconductor manufacturers alone has increased by approximately 30 percent annually over the past decade, with special emphasis given to VLSI pro-

213

duction technology. In addition to dramatic advances by private industry, seminal basic research at public laboratories such as MITI's Electrotechnical Laboratory (ETL), NTT''s Electrical Communications Laboratories (ECL), and the Institute of Chemical and Physical Research has contributed significantly to the improvement of production technology.

Profiting from this multifaceted effort, Japanese semiconductor equipment manufacturers now are in a position to produce and deliver to customer requirements a wide range of the world's most advanced standalone machines and complete production systems.

Microlithography

The heart of the integrated circuit production process, microlithography, and in particular electron-beam technology, was given highest priority in the VLSI Cooperative Research project conducted by the five leading semiconductor and computer makers under MITI auspices, from 1976 through 1980. As the density of integrated circuitry continues to increase by a factor of two every two years, and half of that increase is dependent on microfabrication, linear dimensions used decrease by half every 4 years. Predictably, then, lines of submicron width will soon have to be transferred from templates (masks) to silicon wafers, a task that taxes the limits of optical lithographic processes that have typically achieved this transfer through radiative exposure of a polymeric film. Eventually it becomes necessary to switch to more exacting lithographic processes using electron beams, ion beams, or X-ray systems employing synchrotrons.

Since electron-beam lithographic systems currently cost somewhere in the neighborhood of US$3 million each and X-ray systems about US$10 million, compared with the US$10,000 price of a conventional contact or proximity aligner used in the mid-1970s, microlithographic technology has become the critical parameter in the race to produce ever higher density ICs at progressively lower cost-per-bit. The task of the Japanese VLSI Cooperative Research project was sharply focused: develop new technologies to extend the reach of optical lithography to submicron dimensions and perfect entirely new systems that will be cost effective in the production of 16 and 64 megabit devices at the end of the 1980s and the first half of the following decade.

The results of this effort have radically changed the structure of world microlithographic equipment production. Canon, which in the latter half of the 1970s was on the way to becoming a leading supplier of first generation contact/proximity aligners, has now emerged as the sole supplier of a full range of optical lithographic equipment: proximity,

Table 3. Estimated FY1983 Sales and Exports of Major Semiconductor Production Equipment Manufacturers in Japan

(In Billions of Yen)

Companies	(A) Total Sales	(B) Equipment Sales	(B)/(A)	(C) Equipment Exports	(C)/(B)
Tokyo Electron	96.8	61.4	63.4%	0.0	0.0%
Takeda Riken	39.2	26.8	68.4	5.5	20.5
Nippon Kogaku	143.6	20.6	14.3	1.3	6.3
Ando Electric	28.3	14.8	52.3	1.5	10.1
Canon	374.1	12.2	3.3	3.8	31.1
Dainippon Screen	67.1	10.4	15.5	0.0	0.0
Tokyo Seimitsu	17.8	9.5	53.4	0.9	9.5
Kokusai Electric	66.6	9.3	14.0	0.0	0.0
JEOL	35.5	6.0	16.9	1.5	25.0
Nissei Sangyo	382.9	3.2	0.8	0.0	0.0
Nisshin Electric	50.5	1.5	3.0	0.0	0.0
Tabai Espec	10.6	0.8	7.5	0.0	0.0
Total	1,313.0	176.5	13.4	14.5	8.2

Notes: (1) Fiscal years of above companies end in March except for Canon and Tabai Espec, whose fiscal years end in December. Tokyo Electron has a fiscal year ending in September but figures shown here are adjusted to conform with other companies with March fiscal year.

(2) Semiconductor equipment sales and exports are Nomura's estimates.

mirror projection and step-and-repeat aligners. Nikon, a newcomer to the marketplace, has focused its resources on the latter stepper type of equipment to become the world's leading supplier. And by 1983, three Japanese equipment makers — Hitachi, Toshiba Machine and JEOL — had established themselves as major contenders for the growing electron-beam market.

In 1985, Canon will account for fully 95 percent of the world market for proximity aligners, having successfully extended their range into sub-micron geometries using deep-ultraviolet light. In addition, Canon now holds 50 percent of the rapidly expanding mirror projection market, which it shares with Perkin-Elmer of the United States, and is making a serious bid for a substantial share of the burgeoning stepper market. Already, in 1984, with stepper deliveries only begun, Canon's sales of microlithographic equipment soared upward 68.6 percent — well ahead of the overall market growth rate — to account for a lion's share of the ¥52.3 billion turnover recorded by the company's optical products division. If 1985 targets are met for stepper sales, ticketed at approximately ¥150 million a piece, which is twice the price of mirror projection aligners and five times that of proximity printing equipment, this new addition to the product line alone can be expected to boost turnover of the division by close to 30 percent.

The most immediate effect of Canon's entry into the wafer stepper market with a new production facility just opened in Utsunomiya will be to reduce Nikon's 70 percent share of the domestic market. Although Nikon's customer base is solid, demand for steppers is growing faster than supply as 256K DRAM production increases and chipmakers begin preparations for 1M DRAM deliveries later this year. Domestic demand, which is expected to rise to 550-650 machines this year, will outstrip Nikon's scheduled delivery of 320 units to Japanese makers. And, at best, GCA Corp., the leading American contender in the Japanese market, is expected to ship 170 systems.

Meanwhile, Nikon will make significant inroads on GCA's home turf, with export shipments estimated at 80 systems in 1985. And, since Canon already has a base in the U.S. and European markets for its proximity and mirror projection aligners, the world market for steppers is expected to become a three-way race among these leading contenders.

What will happen when technology shifts to more exotic electron-beam or X-ray equipment — perhaps at the end of the decade — is still anybody's guess. If past experience is prologue to the future, radical swings of market power can be anticipated. Hitachi, Toshiba and JEOL have each accumulated impressive strength in electron-beam technology and have the muscle to grasp the lead in world markets, given the necessary commitment of resources. And, if the world market for microlithographic equipment continues to grow at anything like the 40 percent annual rate of the past seven or eight years, the billion dollar volume of 1985 will reach the US$4 billion dollar level by the end of the decade, which is large enough to entice large-scale commitment by major firms.

Dry Etching

Once submicron patterns are printed on a silicon wafer, etching away unwanted materials to complete the circuit becomes the major problem of microfabrication. Prior to the VLSI era, layers of resist, metal and oxide were etched chemically, in acid baths. But these wet etching processes were not very precise. Acids have the bad habit of etching equally in all directions and are therefore not susceptible to the minute control

Fig. 3 Advances in Semiconductor Technology

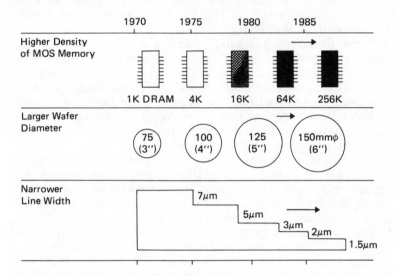

necessary for fine geometries. Development of dry-etching techniques was, therefore, one of the important objectives of the VLSI Cooperative Laboratories.

Plasma etching systems were developed using microwave for high efficiency in generating a stable plasma. To obtain 1 micron and below linear circuitry, the materials to be etched are bombarded with highly reactive ions of fluoride gas that cut cleanly, precisely and anisotropically. Proceeding from joint basic research, different systems were developed by Toshiba, Hitachi, Fujitsu, Mitsubishi-TEL and Anelva.

By 1979, Japanese equipment manufacturers, which already accounted for 100 percent of the domestic market for chemical etching systems, began adapting the results of the cooperative research effort. Production of the new systems, launched in 1981, has risen sharply in response to needs for 256K and 1M DRAM production. Sales in 1985 will reach an estimated ¥26 billion, an eight-fold increase in four years. And already in 1984, 63 percent of all domestic requirements for dry-etching systems were being met by Japanese suppliers.

Even more significant, however, was the share of Japanese makers worldwide. In a rigorous race among 38 dry-etching equipment manufacturers, eight Japanese firms ranked among the top 20 in 1984. And Anelva, with a 31.7 percent share of the Japanese market, had climbed to second position worldwide, ranking just behind front-running Applied Materials of the United States.

Anticipating strong demand in the U.S. market, Anelva opened a Silicon Valley branch in 1980. Although sales were slow at first, with the addition of sputtering equipment in 1983 and a lead in the development of 6-inch wafer processing in both dry etching and sputtering equipment, orders were received from Texas Instruments, Westinghouse and Advanced Micro Devices. And to hone more sharply the cutting edge of its market thrust, the innovative high vacuum equipment specialist established an engineering base in Dallas.

In response to the rapid growth of the domestic semiconductor industry and increased export orders, Anelva's sales of dry etching, sputtering and other advanced processing equipment rose from a modest US$13 million in 1979 to almost US$160 million in fiscal 1985. Driving this rapid growth has been strong competition from Kokusai Electric, ULVAC, Tokuda Seisakusho, Hitachi and Tokyo Ohka Kogyo — all beneficiaries of research work of the VLSI Cooperative Laboratories.

Tokuda, a Toshiba subsidiary whose products are sold in the United States by Tylan Corp., has developed automatic reactive ion etching systems with exceptionally high yields. Hitachi and Showa Denko have jointly perfected a dry etching system using two types of gases in plasma state — Fluorine 32 and Fluorine 41 — which allows selective etching of silicon nitride film at four to seven times the speed of conventional processes. Moreover, since the etching can be contained in the vertical direction, rather than horizontally, etching of submicron lines by this process is possible. Still another submicron etching system has been developed jointly by Tokyo Ohka Kogyo, the Faculty of Pharmacy of Chiba University and semiconductor makers in the Kawasaki region.

To meet this competition, aggressive foreign makers of dry etching equipment have formed joint ventures with Japanese partners. In July 1983, a leading U.S. manufacturer, LAM Research Corp. joined with Tokyo Electron (TEL) to produce automatic plasma etching systems following the successful entry of Applied Materials with a wholly-owned joint venture. And early in 1985 CIT-Alcatel of France announced that it would begin assembly of etching equipment next year in a joint venture with Canon Sales in an attempt to improve its share of the expanding Japanese market. As in much of the semiconductor equipment industry, in dry etching equipment a Japanese market share is becoming a *sine qua non* of survival for both U.S. and European manufacturers.

Film Deposition

If microlithography is the heart of semiconductor manufacturing, thin film technology provides the infrastructure. Even LSIs of earlier generations required up to 10 layers of ultra thin film coatings, and current VLSIs are built by packing 12 layers and more on a silicon or sapphire substrate. Different coatings of materials are used in the fabrication of various types of integrated circuits, and each has its appropriate deposition technique (electron-beam, sputtering, chemical vapor deposition, low pressure chemical deposition or plasma).

To provide interconnection lines between circuit elements, most silicon MOS (metal oxide semiconductor) and bipolar ICs are metalized with aluminum or one of its alloys since they can be vacuum-evaporated at comparatively low temperatures for depositing as a thin film on the wafer. Although chemical vapor deposition (CVD) and electron-beam evaporation deposition are suitable for metalization of some VLSI devices, when magnetron sputtering was perfected, automatic control of the process became feasible.

The lead in developing magnetron high rate sputtering technology for VLSI production has been taken in Japan by Anelva with a proprietary 4-chamber method of aluminum metalization that has found wide acceptance by Japanese and foreign chipmakers. In 1984, building on a 29.9 percent market share at home, Anelva began assembling sputtering equipment in San Jose, California, in a move to become a leading contender worldwide in six-inch systems. Before the end of the 1980s the company expects to sell half of its total output in the United States.

Other contenders for the expanding Japanese market for sputtering equipment include ULVAC, Tokuda, MRC and TEL-Varian. Like their counterparts in dry etching, MRC and Varian of the United States are producing in Japan not only as the most effective means of meeting Japanese competition at home, but to assure their position in world markets. Manufacturing in Japan is essential not only to attain the kind of flexibility in product design and reliability of delivery that Japanese semiconductor makers expect, but also to keep abreast of the most advanced production technology.

Ion Implantation

Simply stated, solid-state electronics is based on the ability to selectively produce, by "doping" with impurities, desired conducting properties in semiconductive crystals, most commonly of silicon. The doping process results in two different types of silicon. Where extra electrons have been added, by doping with tiny quantities of arsenic, it takes on a negative charge; where electrons have been vacated by doping with, say, boron, silicon takes on a positive charge. And by combining negative with positive-charged silicon in the appropriate arrangements, the semiconductor is made to serve as an amplifier or an extremely high-speed switch, and these arrangements can be repeated across the surface of a silicon wafer, producing the equivalent of hundreds of thousands, even millions, of discreet transistors.

A technique of pile-driving ionized projectile atoms into solid materials with enough energy to penetrate beyond the surface regions to selected depths and concentrations, ion implantation has become the pervasive method for doping silicon used in VLSI devices. The amounts of impurity introduced into the wafer can be controlled more precisely than with conventional doping in a diffusion furnace where silicon is exposed to impurity atoms in gaseous state at 1,000 degrees centigrade or so. Equally important, the entire ion implantation process can be performed at much lower temperatures.

Technological advance by Japanaese equipment makers has had remarkable effects. As recently as 1979, imported ion implanters built by Varian, Eaton, Applied Materials and Veeco accounted for 90 percent of the machines sold in the Japanese market. By 1984, the share of U.S. suppliers appeared to have dropped to 15 percent.

But, as in some other sectors of the semiconductor equipment market, foreign trade statistics in this instance do not reflect the subtle nuances of reality. National identification of producers is losing its significance. In the world of semiconductors, boundaries between nations are becoming increasingly fictitious.

Although statistics show that domestic production supplied 85 percent of the Japanese market requirements for ion implantation equipment in 1984, over half that amount was accounted for by a TEL-Varian joint venture. ULVAC, Nisshin High Voltage and Hitachi, along with the Nova division of Eaton Corp. of the United States, shared most of the remaining market. The success of TEL-Varian illustrates the effectiveness of combining U.S. equipment design technology with Japanese production, marketing and service in a joint venture with a competent partner. As long as TEL-Varian can continue to offer these advantages, chances of substantial erosion of its market share in the face of mounting domestic competition are not very great.

Test Equipment

The pattern in automatic test equipment is much more akin to that of microlithographic equipment, where corporate lines are more finely drawn along national lines. Japanese manufacturers began overtaking the U.S. front runners during the 1970s and, with the more rapid development of VLSI technology in the 1980s, have emerged in the lead. Imports, which once accounted for a major share of the market, shrank to about 18 percent in 1984, and in 1985 exports will surpass the level of imports.

This spectacular turnaround is the more remarkable since it has been brought about mainly by two companies: Takeda Riken and Ando Electric. Between them, they now combine 65 percent of the ¥48.7 billion Japanese market and account for most of the sales abroad. Both have close affiliations with leading semiconductor makers, an important factor in their technological success. Fujitsu holds 21.5 percent of Takeda Riken's outstanding shares, and Ando Electric is part of the NEC group. A third tester maker, Minato Electronics, plays a minor role in the market.

Takeda Riken, originally a manufacturer of measuring equipment for the electronics industry, began developing automatic testing equipment for sophisticated LSI production in the early 1970s and by 1975 had established itself among leading manufacturers with a 5.1 percent share of the world market. But the rapid advance came after 1979 with the advent of the VLSI era. Having pushed sales upward at an average annual rate of 44.4 percent over the past five years, Takeda Riken now commands 39 percent of the domestic market and more than 18 percent of worldwide sales of LSI/VLSI tester equipment. In the fiscal year ending March 1985, the company's export sales climbed 97 percent, reflecting increasing acceptance of Takeda Riken testers by chipmakers abroad, and especially in the United States.

The meteoric take-off of this once-obscure instrument maker to become the world leader in the most sophisticated automatic testing equipment used in IC production provides dramatic illustration of several representative traits of Japanese high technology management.

- Takeda's development of the world's fastest testing equipment was abetted by NTT's Electrical Communications Laboratories, which made available results of basic research on VLSI production under its own program that ran parallel to that of the MITI-sponsored VLSI Cooperative Laboratories. Although there were overlaps between the two programs, and the corporate participants in both programs were largely the same, development of advanced automatic testing systems was one of the aspects of VLSI production to which the ECL program gave particular emphasis in the basic division of labor between them.

• The process of diffusion of research results of the two projects differed, however, as did their basic organizational structures. While the VLSI Cooperative Laboratories operated under the umbrella of the VLSI Research Association, an ad hoc organization funded jointly by MITI advances and subscriptions of the various corporate members, the ECL project operated under the usual conditions of cooperative research and development between the NTT laboratories and selected members of the now defunct "Den Den Family." In this instance, Takeda Riken's participation in the baisc research for a new generation of automatic testers benefited from a *quid pro quo* understanding that ECL would buy two of the completed systems, which carry a US$3.5 million price tag, thereby helping to offset Takeda's estimated US$6 million development costs.

• Fujitsu's 1976 investment, which linked the resources of the largest computer manufacturer in Japan with those of a leader in measurement-related technologies at a critical point in the development of tester technology, has brought benefits to Takeda far more important than the capital contribution itself. Added to the technological synergy of this relationship is the flow of production technology that is inherent to closely affiliated companies in related fields of semiconductor R&D and production.

• Building on these sound technological foundations, Takeda's own expenditures on R&D have added the necessary forward thrust of systems development. Since 1980, R&D outlays by the company have risen five-fold from ¥1.2 billion to ¥6.1 billion in 1985, resulting in a product line-up headed by testers with speeds up to 100 megahertz (MHz) for both logic and memory chips. This is considerably higher than the mainstream test speeds which most test equipment manufacturers are now upgrading from 20 to 40 MHz.

Takeda Riken has taken a clear lead in the market for 40 MHz production testers, and has been the only firm capable of delivering 100 MHz testers for use in research laboratories of major world IC manufacturers. Now the company is preparing the launch of production models of 100 MHz testers that will allow its customers to up-grade their present systems to test chips of higher density at sustained high speeds without major modifications of software. Since the costs associated with development of application/operation software by users is high, the ability to upgrade systems using the existing software base is a critical factor in the quotient of competitive advantage.

Two further changes in Takeda's technological repertoire will add to its market strength. Greater emphasis on logic testers for microprocessors and gate array chips has given breadth to its product line and new impetus to sales. In fiscal 1983, logic testers exceeded those of memory testers for the first time, reflecting a significant change in semiconductor production patterns in the Japanese industry. And in 1984 the company introduced analog IC testers and charged coupled device (CCD optical sensors) testers. Fueled by rapid technological advance and a broadening product line, Takeda's sales of test systems rose a stunning 7.5 fold in the five years 1979-1984, for an average annual growth of 47.6 percent.

Although the appearance of innovative products by other makers could alter competitive positions rather quickly, since Takeda Riken already has products (100 MHz testers) in hand for testing chips up to the equivalent of 4 megabits, the chances of its yielding its leading position to others in the product development race seem small. And if the company continues to pour more than 10 percent of sales into R&D, as it is at present, the likelihood of continued Takeda leadership of the technological competition must be rated high.

Such technological prowess has transformed the structure of the entire world industry since 1978. By 1984, three of the top ten chipmaking equipment manufacturers in the world were Japanese: TEL (Tokyo Electron) was third, Nikon sixth and Canon tenth. At their current growth rates, all three are destined to move up rapidly.

Not all of this structural change has been a direct result of Japanese technological advance, however. Tokyo Electron's rapid climb from ninth to third in the worldwide industry was achieved mainly by managing the mix of U.S. technology and Japanese production, marketing and service. With joint ventures that have assured TEL dominant shares in a wide cross section of domestic semiconductor equipment markets, the trading-company-turned-manufacturer has an approximate 40 percent share of the Japanese wafer-processing equipment market.

With this momentum, as a diversified semiconductor equipment supplier, TEL has also developed its own technology in critical product sectors. To capture the momentum of factory automation in Japan, TEL engineers have perfected fully automatic probers with a pattern recognition capability that replaces manpower in adjusting wafers for probing. Although there has been strong demand for this equipment abroad, no export shipments have yet been made due to the growing demand in Japan, where TEL holds 80 percent of the market.

This situation is bound to change, however. All indications are that TEL and other Japanese equipment makers will gain market share worldwide for the remainder of the 1980s and well into the following decade, for long-term structural and technological reasons.

- The Japanese semiconductor industry will continue to grow at higher rates than in the United States, manifesting a higher rate of investment in R&D and plant and equipment.

• As successive generations of new devices are introduced first in Japan, a higher proportion of Japanese production will be devoted to the most advanced integrated circuits. The inevitable result will be higher demand for the latest and the most efficient equipment in Japan.

• The imperative of higher automated wafer production to eliminate dust and error-prone production by humans will continue to find more ready response in Japan than elsewhere. This means, among other things, that equipment makers can no longer rely on stand-alone machinery sales. Rather they must become systems engineers, adapting their equipment for compatibility with the user's system. Proximity to, and close relations with, customers will become even more important than before, as will the ability and willingness to design or adapt equipment to specific customers' requirements.

Any surprise-free predictions for the future state and structure of the industry then must envisage a continuing decline of market share held by U.S. manufacturers and a proportionate rise of Japanese makers. Before the end of the 1980s, the Japanese will most likely overtake the American industry in global market share and move out in front, the most advanced ICs being produced on Japanese equipment throughout the world.

Not all this growth of the Japanese industry will come from existing equipment manufacturers. Rapid technological change and high growth rates, which are characteristic of the equipment industry and tend to alter competitive positions rather quickly, are attracting a wide variety of new entrants. Komatsu, the world's second largest construction machinery maker, now offers equipment to produce epitaxial wafers. Minebea, the leading miniature bearing manufacturer, is producing a new line of semiconductor inspection systems. Other machinery makers, such as precision machine tool builders (Tsugami) and textile machinery makers (Fujido Seisakusho), are turning their expertise to development of automated semiconductor production systems.

These new entrants, like the current leaders of the Japanese equipment industry, are large diversified and integrated firms with resources to sustain the high level of investments in technology required to stay in the running. Smaller and more specialized U.S. equipment makers will not only be handicapped by erratic market behavior at home, but also structurally less capable of effectively managing rapid technology development.

The recent formation of the Semiconductor Equipment Association of Japan with 158 regular and associate members, marks the coming of age of a key industry, one as important to the era of information technology as the machine tool industry has been to the age of mechanical technology.

PART V.

Computers

CHAPTER 15

Big Blue Beseiged

FOR THE PAST quarter of a century, few things in the business world have been more certain than IBM's mastery over the computer industry. IBM set the standards for the industry, developed the technology and had by far the largest customer base.

But there are signs that this image of invincibility is a surrogate for reality. Although IBM has continued to grow at impressive annual rates, its share of the world market has declined from 80 percent in 1964 to under 60 percent in 1979. In recent years, new entrants into the industry have been steadily chipping away at IBM's market position, and the technological lead which sustained that position has now virtually vanished.

The industry is poised for an era of pluralism in which IBM will share the computer firmament with a galaxy of global alliances combining varying mixes of Japanese, American and European production, design, software, marketing and financial strength. Given the nature of the industry, the contest, perforce, has assumed global dimensions — although, as in other fields of advanced electronics, the focal point of the action is Japan.

When Fujitsu and Hitachi overtook IBM in 1974 with the introduction of more advanced general purpose computers, technological leadership of the industry was placed in question. IBM has since shifted its strategy to direct the full force of its market power against the Japanese industry, introducing a plethora of new products, shortening life-cycles, abandoning its usual practice of announcing large models before smaller versions, "unbundling" software and cutting prices on hardware to the bone.

The new models themselves offered few surprising technological advances. But their proliferation, the depth of the price cuts and the rapidity with which unbundling (selling software separately from hardware) has been pursued have sent successive shock waves through the industry.

Reprinted from "Japan's Computers Get One Step Ahead," *Far Eastern Economic Review*, 27 July 1979.

Even Lloyds of London, with nearly 300 years of experience in insuring risks, was a major casualty of IBM's product strategy and price-cutting, and liable to claims estimated in excess of US$220 million on underwriting leasing companies whose equipment became outdated within the insurance period.

Japanese computer-makers, and their new allies among vendors of machines with compatible plugs, stayed with IBM all the way, and by the end of 1978 Fujitsu and Hitachi could boast of the world's largest general purpose computer, offering superlative advantages.

The leading Japanese manufacturers were allowed little time to savor their success, however. IBM's long-awaited fourth generation of computers, the 4300 processors, announced in most markets in January 1979, was unveiled in Japan just over a month later. Although Japanese manufacturers were once more visibly stunned with IBM's offer of 64K bit LSI (large-scale integration) in machines priced 30 percent below those of Systems 370 equipment, in rapid-fire succession NEC and Mitsubishi Electric introduced competitive systems in early February 1979; Fujitsu followed with its riposte in its announcement of the first four Model F systems; and by mid-year Hitachi had unveiled three H Model computers claimed to be up to 1.9 times more powerful than the latest IBM offerings.

Fujitsu also made public its intention to mount 64K bit LSI in its network computers to be delivered to Nippon Telephone and Telegraph in October, and to apply high-density LSI chips to its general purpose computers in December. Both NEC and Mitsubishi will follow suit in January 1980, and Hitachi will introduce single-phase power supply 64K bit random access memories mounted on multi-layered ceramic boards to enhance performance and reduce processor size of the three H Model computers to be ready for delivery in January 1980.

Closing the Gap

The closure of the technological gap between the Japanese industry and IBM is not likely to be the decisive factor in determining the future contours of the global computer industry. It is widely understood in the industry, and by Japanese industrial policy makers, that it takes much more than technology to compete effectively with IBM.

Economies of scale, organizational power and financial resources are likely to be critical. To achieve these prerequisites, the Ministry of International Trade and Industry (MITI) has been promoting pooling and amalgamation of resources into increasingly larger groups.

In the early 1970s, MITI sought to merge the existing six mainframe manufacturers into three groups. Then, capitalizing on its role in the very large-scale integration research and development project, it organized the five leading computer makers into two groups for technical cooperation, leaving Oki Electric to withdraw from the general purpose computer race and concentrate on peripherals and special purpose computers. Indications are that MITI would like to see the entire industry merged eventually into a single computer company, though the chances are not great.

There is little likelihood that Fujitsu and Hitachi, the two leaders of the industry, will ever be willing or able to combine their computer operations effectively. Nor is there any clear sign that either NEC or Mitsubishi Electric are going to withdraw from the general purpose computer field, notwithstanding recurring rumors that both are contemplating specialization in medium-sized and small business systems.

However, apart from the fierce competition among the four remaining mainframe makers (Toshiba has withdrawn in favor of NEC), optimum advantages of scale and organizational power cannot be attained on the national level. IBM has the distinctive characteristic of being one of the world's truly global corporations, with strategies to optimize resource allocation worldwide. And as late-comers to world markets, it is most unlikely that Japanese computer manufacturers, singly or collectively, will be able to develop effective global structures on their own.

Domestic and international competition are intricately linked, and leading Japanese computer firms clearly see that the key to their success in the raging competition at home is in the effectiveness and speed with which they develop world markets. Indications are that each of the four mainframe makers has set out to balance their domestic and foreign sales, at least by 1985, to achieve global spread and commitment similar to that of U.S. firms.

Beginning with Fujitsu's initial investment in Amdahl of the United States in 1972, Japanese manufacturers have been gradually but steadily forging worldwide networks linking with computer manufacturers or vendors abroad through acquisitions, joint-ventures, sales agreements, supply arrangements with manufacturers of original equipment, cross-licensing and technical assistance contracts. It is this network, and particularly the ties between Japanese makers and U.S. plug-compatible machine vendors, that has provoked IBM's multi-directional counter-offensive.

The main thrust of IBM's fourth generation strategy is aimed directly at the Japanese industry, with the intention, at the same time, of maximizing pressures on the plug-compatible machine vendors — particularly Amdahl, Itel, Magnuson and Control Data.

IBM's primary objective is to check the Japanese thrust into North American and European markets by putting the major Japanese mainframe manufacturers on the defensive in their own markets.

To do this, IBM reversed its customary strategy of introducing larger systems first, a policy which prevents users wanting the latest technology from switching to smaller, cheaper systems. IBM strategists decided to acept the risks of users moving downward or sideways with the new "bottom-up" strategy. By opening with the 4331 and 4341 small- and medium-sized systems, IBM focused on precisely those segments of the Japanese market where domestic makers are strongest, holding over 70 percent of sales.

In the late 1950s, when the Japanese electronic data processing market was confined to large users — almost exclusively corporate giants ranking among Japan's top 500 companies — IBM held an 80 percent market share in Japan, as in other markets. But as the Japanese market broadened to include national government organizations and medium-sized firms, and then local governments and business enterprises, IBM steadily lost its market share to Japanese computer manufacturers, which concentrated on medium- and small-sized computer systems rather than confront IBM head-on.

In order to strengthen their competitive power in these more rapidly growing segments of the market, Japanese makers offered more extensive computer services than IBM had been accustomed to supplying to larger users maintaining their own in-house electronic data processing service organizations. As a result, IBM's overall share of the total market dropped below 30 percent, and its share of the small- and medium-sized systems market is even lower.

If IBM is going to inflict any significant damage on the Japanese electronic data processing industry at home it must make substantial gains in this segment of the market. Indeed, IBM's aggressive price-performance strategy is calculated to attract medium-sized and small computer users and to force Japanese suppliers to strain their already tight resources in an effort to follow IBM in both performance and pricing.

IBM clearly intends to add to those pressures by unbundling and relying on a pricing policy which takes full advantage of its lead in software. Parallel to the unbundling process, IBM Japan established a software support center in Tokyo, on an experimental basis, to provide services covering approximately 60 different programs. The formal service starts on a nationwide basis in January 1980.

It is not clear that IBM's elaborate new strategy will achieve its intended objectives. Leading Wall Street computer industry watcher Ulric Weil believes IBM's profitability could be cut by users in the U.S. switching to smaller models or renting rather than purchasing (taking advantage of IBM's new 15 percent discounted two-year lease arrangements).

In Japan there are no signs that IBM will succeed in its effort to increase its share in the medium- and small-sized segments of the computer market. Since Japanese mainframe manufactures have been quick to match or improve on IBM offers in terms of price and performance, there will be no compelling reason for users to convert to IBM. Moreover, computer users in this market segment are much more dependent on outside services which IBM, with its high cost levels, is not prepared to offer on terms equal to those of Japanese computer makers, who subcontract to lower-cost specialized service bureaus.

Economies of scale do not apply to the service sector, where IBM's rigid cost/expense structure constitutes a handicap in the Japanese market.

More important, the rapidity with which Japanese manufacturers have been able to leapfrog IBM's latest technology has dissipated much of the punch of the fourth generation strategy, in vivid contrast to the effects of the spectacular advance IBM made with its third generation 15 years earlier.

The Illusions of Invincibility

Japan's computer companies delivered a swift and substantial reply to IBM's new fourth generation system. NEC brought out two models of the ACOS System 250 to compete with IBM Systems 38 and 4331, while Mitsubishi Electric unveiled COSMO 700III and 700S. These represent a breakthrough: they are the first medium-sized general purpose computers which can form "multi-processing systems" in which modular processors are connected. Mitsubishi claims that the first system beats the IBM 4341 for performance by 100 percent, while the second is four times as powerful as the 4331.

Fujitsu was not far behind with its answer to IBM's new generation, putting on the market four new computer systems which the company judged are better value in terms of price and performance than the latest from the United States. The new Fujitsu Model F systems, which reputedly outpower by 60 percent the IBM 4300 models, will beat the 4341 on delivery date by several months.

Fujitsu has chalked up a lead in the 64K bit memory applications and will mount the 64K bit LSI on its network computers for Nippon Telephone and Telegraph, while applying high-density chips to its Model F general purpose computers.

IBM and other competitors will find it costly to match some of the major functional innovations that Fujitsu has come up with. The model F systems have great potential in processing Asian languages. They will be able to process Japanese *kanji* (Chinese characters) and *kana* (Japanese characters) as well as alphanumeric characters. The system also promises to do the same with such languages as Korean, Chinese and Thai.

A major strength of the Japanese challenge to IBM has been the ability to anticipate accurately the American company's main technological advances, and then to match them, virtually eradicating IBM's initial advantage. Hitachi's latest innovations prove this point once again.

The new Hitachi Model H, available from January 1980, incorporates fourth generation high density LSI, multi-chip modular structures and multi-layer ceramic printed circuit boards. Running time of control programs is cut by 15-20 percent by using firmware in which much of the operating systems is micro-coded on LSI chips.

Imitating IBM's software strategy, Hitachi offers some 124 different software programs designed for the new system. With the closing of the gap in this field, it seems that IBM's lead in software may be as illusory as was its once apparent overall invincibility in the whole industry. The computer scientists of Japan are hot on the heels of the U.S. giant and have proved they can respond quickly and effectively to its advance.

Since Japanese computer makers are able to offer better performance at prices roughly equal to or lower than those of IBM, the world industry leader has not substantially improved its competitive position with the new generation of machines. In fact, IBM has still to match Fujitsu and Hitachi M-200 very large processors with their fifth generation 550 gate LSIs, though it is expected to do so with its forthcoming announcement of a new H series of large-scale computers.

IBM's main advantages are still its vast base in a market that is conditioned to IBM's leadership in both hardware and software technology. As a result, IBM's programed leaks of advance information about its 303X series of large processors in 1977 and the 4300 at the beginning of 1979 virtually froze the market. Potential buyers preferred to wait for the new improved price/performance machines, thus neutralizing much of the competitive advantage of the plug-compatible machine vendors and their Japanese suppliers.

By its very nature, however, this advantage is largely lost as soon as the new models are announced. Once it becomes clear that IBM offers no substantial advantages, and customers become unhappy with the long waiting line for delivery of new IBM models, plug-compatible machine vendors should recover much of their former competitive power and elan.

Indeed, the main effect of the IBM strategy may well be the strengthening of ties between vendors of plug-compatible machines and their principal Japanese suppliers, Fujitsu and Hitachi.

With technological development dissipating the apocalyptic vision of IBM's fourth generation announcement, it remains problematical how much the preoccupation with the IBM threat to the home market has detracted from the overseas thrust of Japanese computer manufacturers.

The IBM strategy may have dampened the fervor for a direct overseas marketing drive, with the heavy initial investment this would entail in marketing and servicing organizations, though it is doubtful that this was ever the main option open to Japanese general purpose computer or peripheral manufacturers. If anything, IBM strategy would seem to have confirmed leading Japanese makers in the wisdom of the strategy they had already adopted — that of forging international networks of links with manufacturers and vendors in major markets through various forms of technological cooperation, production sharing, software development and marketing.

This clearly does not portend a diminished flow to world markets of Japanese electronic data processing hardware and software. IBM's chairman Frank Carey was not exaggerating when he stated that he expected to see Japanese computers "everywhere in the world." Market forces combined with IBM's fourth generation stratgegy make exports imperative for Japanese computer makers.

Since the failure of Mitsubishi Electric in its first attempt to develop the European market for Melcom minicomputers in the early 1970s, Japanese computer manufacturers have developed a refined appreciation of the costs and pitfalls involved in direct marketing. It has also been clear for some time that computer markets are highly sensitive politically. These considerations, added to the inherent problems Japanese manufacturers are likely to encounter in direct marketing of hardware, as well as software and computer services, are compelling reasons for a cooperative rather than a competitive approach to world markets.

In addition to the advantages of market entry and development in supply and production sharing with original-equipment manufacturers, there are unmistakable signs that Japanese corporate and public policy makers have decided that these forms of cooperation are the best long-range solution to some of Japan's pressing problems in the field of international economic relations.

In a remarkable pronouncement early in May 1979, International Trade and Industry Minister Masumi Esaki revealed that Japan would promote joint international technological ventures in such areas as electronic computers, aircraft and cars. Accordingly advocates of further amalgamation of the computer industry are reported to have seized this pretext to seek the merger of the two computer research consortia, established to develop very large-scale integration technology, into a single Japanese group to work with prospective overseas partners.

Whether or not such an industry-wide consortium is formed, however, the wisdom of enhanced international cooperation in the computer industry is unquestionable. Fujitsu demonstrated the advantages of this approach, first with Amdahl and more recently with Secoinsa in Spain and Siemens in West Germany. Hitachi, also, concluded an initial research and development tie-up with Britain's ICL, opening a wide spectrum of possibilities for further cooperation in the future.

Significantly, the Fujitsu pattern of international cooperation began with just such a research and development arrangement with Amdahl, the difference being that it was Amdahl which provided the technological assistance. Since then, Fujitsu has extended its linkages abroad to include effective production-sharing arrangements, as well as marketing and sales agreements which have generated a substantial volume of export orders more effectively and at less cost than would have been possible by direct marketing strategies.

Synergistic relationships benefit all parties involved. It is becoming increasingly apparent that links with the Japanese industry provide optimum prospects for survival and growth for several key computer manufacturers and vendors in Europe and North America which would otherwise be likely to succumb in the torrid heat of IBM market power.

In this brave new computer world the emerging global alliances are likely to strengthen their combined market share, eventually reducing IBM's position in a pluralistic global industry to something approaching the role it now plays in the Japanese market.

This does not mean that IBM will cease to grow as it has in the past. With the explosive growth of the world electronic data processing market, there will be more room for new global competitors. And the alliances which the Japanese industry is forging are clearly the most logical candidates for an increased share of this vast new market.

CHAPTER 16

Turbulent Transition

WHEN IBM INTRODUCED the 4300 series of fourth generation computers in 1979, it was clear that a primary objective was to check the mounting Japanese thrust into North American and European markets. By opening its new series with small- and medium-sized systems and an aggressive price-performance strategy, IBM sought to cut into those segments of the Japanese market where domestic makers were strongest and at the same time pull the rug from under the plug-compatible machine (PCM) sellers through which Japanese manufacturers had entered the US and European markets.

With the sudden demise of Itel's computer division, which was selling Hitachi-made PCM equipment in world markets, the effectiveness of IBM's strategy as well as the conventional wisdom about the industry leader's invicibility seemed to be confirmed. Wall Street analysts, industry experts and computer-users who had foreseen the end of the PCM industry were momentarily heralded as prophets.

But that is not the way things have turned out. In fact, the most severe difficulties of PCM sellers were suffered before the IBM announcement when customers were delaying buying new computers until it was clear what IBM had to offer. Once the word was out, however, and it became clear that IBM was not in a position to meet the pent-up demand for the more powerful and inexpensive 4300 machines, even those PCM sellers competing directly with the 4300 series became the unintended beneficiaries of IBM's seemingly well-designed strategy.

By the second quarter of 1980 the PCM industry had recovered from the shock treatment and unit shipments climbed to a record. According to International Data Corp. (IDC) estimates, deliveries of medium-sized PCM machines in the U.S. market alone were expected to surpass the 520 level in 1980, which would amount to a 45 percent growth rate over the previous year.

Reprinted from "IBM's Strategy Misfires: Down It Goes to No. 2," *Far Eastern Economic Review,* 5 December 1980.

Even for those experts mesmerized by IBM's size and past performance, this was dramatic evidence that there have been some profound changes in the industry which they had failed to perceive. To be sure, IBM's strategy misfired in part because of its overly-successful pre-announcement promotions. The flood of orders for the new series pushed back delivery dates by as much as two years and many customers simply could not afford to wait.

But even more important, for many there was little advantage to be had in waiting for the IBM machines. Fujitsu and Hitachi, main suppliers of equipment to PCM sellers, more than matched IBM's new performance standards with their own fourth generation models and they were able to meet the stiff price competition.

Significantly, Amdahl Corp., in which Fujitsu has a major stake, did not follow Itel's path to ruin, as some analysts had anticipated. And when National Semiconductor took over the sales and service network of Itel to form National Advanced Systems (NAS), Hitachi was assured an even more effective marketing channel for its giant PCM equipment than it had previously. While these two firms were consolidating their hold on 12 percent of the IBM-type equipment market, other PCM suppliers such as Magnuson Computer Systems Inc., Two Pi Corp. (a Philips subsidiary) and IPL Systems Inc. moved rapidly to take advantage of opportunities which in large part are of IBM's making.

For IBM this is not the worst of the bad news, however. If the 4300 strategy failed to eliminate PCM sellers in the U.S. and other major markets, in Japan it had a near disastrous boomerang effect. Since the announcement of the new series, IBM has continued to lose Japanese market share, dropping from its long-held position of market leadership to second place behind Fujitsu.

Fujitsu's sales have risen steadily in the face of the best IBM can offer, increasing 8 percent in fiscal 1979 to ¥326.8 billion (US$1.53 billion), while IBM Japan's computer sales in the Japanese market declined for the first time since its founding. Total sales of IBM Japan, including exports to the U.S. parent and affiliates in other markets, were ¥324.2 billion in 1979, but domestic computer deliveries accounted for only ¥267 billion, down slightly from ¥270 billion the previous year.

Exports of medium- and small-sized machines, magnetic tape devices and electronic components rose during the period by 26.3% to ¥57.2 billion to bring 1980 total sales up just 2.8% over the 1979 level.

Here again, IBM was the victim of its own strategy. Japanese computer users who might have bought IBM machines were hesitant to make buying decisions because sufficient numbers of the new 4300 series were not available for delivery.

After January 1981, when new production facilities will enable the increased delivery of 4300 models, IBM Japan could possibly recover some of its lost market share. Meanwhile, in an interim attempt to boost declining sales, IBM announced in November 1979 that two new models of the 3033 would be available to Japanese users, and the purchase price of other equipment would be reduced by 15-30 percent.

At the same time, IBM Japan finally entered the *Kanji* (Chinese character) processing market with two systems announced by its data processing and general systems division.

And in November 1980 the parent company in New York announced the long-awaited top-of-the-line Series H central processing unit, called Model 3081. Ever since news of this new line of giant processors was first leaked in 1978, customers and competitors alike had expected its unveiling to give another shock to the industry, with even more profound effects than the 4300 announcement.

But the combined force of the announcement of the new series and price cuts ranging from 3 to 22 percent on older models in IBM's 3000 series has been much milder than anticipated, and remains largely defensive in character. Rather than a major advance over Japanese competitors, the 3081 is apparently little more than a match for Hitachi's high-powered machine. Although more details about the huge US$3.7 million-and-up processor are awaited with interest, initial reactions of competitors in the United States and Japan have ranged from relief to ecstasy. IBM has not regained the technological leadership in the performance race, its prices can be met, and much longer delivery schedules than expected provide comepetitors sufficient time to respond with new and improved machines.

Significantly, IBM Japan decided not to recruit new staff in fiscal 1980, signalling a major restructuring of the company's business organization and the introduction of more automated systems in clerical and managerial operations to trim rising manpower costs. During the last three years of the 1970s, IBM Japan's manpower costs reportedly exceeded the growth in revenue and had to be curtailed without breaching the company's life-long employment commitments to its staff.

In part, IBM Japan's financial problems were a reflection of changes in cash flow resulting from a customer shift from purchasing to leasing computers, a practice which IBM's strategy has itself tended to encourage. But the basic problem runs far deeper than this.

At least since 1976, IBM has been losing its market share not only to Fujitsu but also to Hitachi and NEC. From 1976 to 1979, Hitachi's computer sales increased by 50 percent and NEC boosted its data processing revenues by more than 75 percent, compared with a 25 percent increase in total IBM turnover. And there is nothing on the horizon that suggests a reversal of this trend. Quite to the contrary, all signs suggest that IBM can expect things to get worse and, possibly, *never* better.

In November last year, NEC unveiled three new ACOS systems reputed to average 30-40 percent better performance than comparable IBM 4300 series equipment. And, if NEC's experience with the ACOS 250 is any measure of the reception these machines can expect, they spell trouble for IBM. In the seven months following the announcement of the ACOS 250, fully 60 percent of the 360 systems orders were from users of other makers' machines or those who had not previously been using computers. At least some of this gain was at IBM's expense.

It is by no means an accident, of course, that the most rapid growth in the Japanese computer market has been recorded by Japan's leading manufacturer of semiconductors and communications equipment. NEC's worldwide lead in memory technology has served it particularly well in computer development in recent years, and its combined strength in integrated circuitry, communications and computers is likely to have important synergistic advantages in the future.

Indications are that NEC will continue to build market share, not only in Japan but also in foreign computer markets where it is mounting an aggressive marketing effort. If all goes according to plan, NEC will increase its computer exports six-fold from an estimated ¥11.2 billion in fiscal 1979 to around ¥60 billion by fiscal 1983. Should this target be met, NEC's exports in that year will almost equal total Japanese computer exports in 1978.

At the same time, Fujitsu is mounting an even more ambitious bid for world markets. From fiscal 1978 computer exports of ¥34.3 billion — amounting to 11.3 percent of the company's total computer sales — Fujitsu expects to increase overseas deliveries to approximately ¥100 billion or 30 percent of sales for fiscal 1980. If these targets of Japanese makers are achieved, export growth rates, which have averaged 30-31 percent during the past five years, could easily top 50 percent during the next several years.

Until now, Fujitsu's major export drive has been in the form of' original equipment manufacturers (OEM) supply to computer equipment firms in North America and Europe. Amdahl, in which Fujitsu has a 26.7 percent shareholding, boosted its purchases of Fujitsu M series computers from 41 units in fiscal 1976 to 110 units in fiscal 1978, taking the leadership in the PCM market. During the same period, Fujitsu supplied Memorex in the U.S. with magnetic tape drives which were marketed under the buyer's brand.

Although Amdahl remains Fujitsu's primary business partner in the

American market, and also sells Fujitsu-made PCM computers in Europe through Amdahl International Ltd. (a 50-50 joint venture between Amdahl and Fujitsu), in May of this year Fujitsu announced that agreement had been reached with TRW to establish a separate marketing joint venture in Los Angeles. The new company, TRW-Fujitsu, is 51 percent-owned by Fujitsu with the remaining 49 percent equity held by the American partners. The new company has begun marketing a wide variety of peripheral and terminal equipment in additon to small general-purpose computers, becoming the first overseas sales organization to offer Fujitsu computers under the manufacturer's own brand name.

The logic of the TRW-Fujitsu joint venture is compelling. TRW is able to use its existing resources and facilities for marketing and maintenance of aerospace and other electronic systems to sell a complementary line of computer equipment. Fujitsu gains a ready-made marketing organization for a broad spectrum of its most competitive data-processing equipment, which is not sold by Amdahl. This at once provides a hedge against risks inherent in the PCM business with Amdahl and makes Fujitsu's position in the U.S. market less vulnerable to IBM attack.

Sales by the new joint venture got off to an auspicious start soon after its doors were open for business with an order for a computer system consisting of POS terminals and modular store processors from a leading Detroit department store.

Although the main thrust of Fujitsu's present North American strategy is towards building marketing structures, the leading Japanese EDP equipment maker has already established some solid foundations for a major industrial presence in the market when the time is ripe.

Fujitsu America, which distributes semiconductors and other electronic components, already assembles Amdahl computers in San Jose. In Canada, Fujitsu holds a 23.6 percent share of Consolidated Computer Inc., which it has licensed to produce its terminal equipment. And in autumn of 1980, Fujitsu began semiconductor production in a new ¥2 billion wholly-owned facility in San Diego, California.

In contrast to this impressive beginning in North America, Fujitsu's presence in the European market is less well-developed. Total EDP equipment sales in Europe have been only a third of those in the United States. As in North America, Fujitsu's approach to the more fragmented European market has been through organizational links with key partners. To take on IBM and struggling European manufacturers at once in a direct frontal attack on the market would be financially imprudent and politically impossible.

In Spain, Fujitsu formed a joint venture with INI, the public-sector industrial giant, to manufacture small general-purpose computers under license. Computers and non-impact printers are supplied to Siemens of Germany on an OEM basis, an arrangement which could well be the beginning of a much broader-based industrial cooperation.

Siemens and Fujitsu have a long-standing close relationship which dates back to 1923 when Siemens took a major shareholding in Fujitsu's parent company, Fuji Denki. Much of Fujitsu's original communications technology came from Siemens, and more recently — through Fujitsu Fanuc — the two firms have been closely linked in selling numerical control equipment in European markets.

As in the United States, while Fujitsu has opted for cooperative strategies in entering the European numerical control and computer markets, it has decided to go it alone in semiconductors, with a new wholly-owned manufacturing facility to begin operation in Ireland some time in 1981. Both European firms and Fujitsu have found common ground for cooperative ventures designed to chisel away at the predominant position of IBM in continental markets. But the competitive environment of the semiconductor industry in Europe has been more conducive to Japanese investment in wholly-owned manufacturing ventures, just as it has been in North America.

Fujitsu not only differentiates its strategy from product to product in major industrial countries; it has also adopted special strategies designed for the business environments of other markets. In Korea, the Philippines, Australia and Brazil, Fujitsu has established wholly-owned marketing subsidiaries. Beginning in January 1981, however, a Brazilian electronic equipment manufacturer, EDISA, will begin assembling Fujitsu small general purpose computers under license. Until production is under way, using completely knocked down (CKD) computers supplied by Fujitsu, EDISA will market Fujitsu computers purchased on an OEM basis.

Meanwhile, Hitachi is following the route taken by Fujitsu in the U.S. and European markets. Large computers are supplied to NAS, which has non-exclusive marketing rights in world markets. Basically the same equipment, an advanced version of the powerful Hitachi M200H, has also been offered to Olivetti and BASF for marketing under their respective trade marks.

Since the Hitachi machine is IBM-compatible, these arrangements enable all three firms to compete with IBM in this segment of the market. And since the new Hitachi computer is probably the most powerful of its kind now available, each of the three firms is confident that they can generate substantial sales even though they are competing with essentially the same machine in the same marketplace.

Olivetti, the first to sell a giant Hitachi machine in Europe, expects to install 30 by the end of 1982. NAS is even more sanguine: with 170 computers already installed in Europe, a base acquired from Itel, NAS Europe estimates that the market for machines of this size could well be 150-200 by 1983. And through its multichannelled approach to the European market, Hitachi stands a good chance of supplying a significant share of that demand.

As yet, Hitachi has not made a concerted bid for other segments of the U.S. and European computer markets, however. Chances are that when it does, it will opt for joint ventures or further OEM supply arrangements in both markets.

All Japanese computer makers, including about 20 manufacturers of small business computers, recognize that survival in this industry ultimately depends on strength in world markets. Still, at present, exports amount to only about 8 percent of total sales by the Japanese industry, compared to approximately 50 percent for the automobile industry and 42 percent for color TV.

In some respects, the computer industry today stands at the same take-off point reached by the motor industry in 1965. But in some important ways the computer industry is in a much stronger position for global expansion than the motor industry was at that time.

Japanese computer manufacturers are organizationally and financially stronger than Japanese motorcar manufacturers were in 1965. And even more important, compared with the competition, Japanese computer makers are also relatively stronger technologically than were the car manufacturers in that earlier period.

It is reasonable to expect, therefore, that the impact of the Japanese computer industry on world markets might well be as great in the 1980s and 1990s as that of the Japanese motor industry in the past two decades.

The main difference is likely to be in the strategies of Japanese computer makers. Ministry of International Trade and Industry (MITI) policymakers, as well as corporate stategists, have clearly opted for cooperative modes of global expansion, developing major markets jointly with leading foreign manufacturers and sellers.

To be sure, in a real sense, IBM's pre-eminence in world markets makes this approach imperative. IBM's strength lies in its global structure and Japanese computer makers recognize that they cannot challenge that structure alone.

But it is also true that there is a growing realization in Japan that the nature of the global village in an information age calls for cooperative arrangements to reduce tensions and assure optimal advantages of advanced technologies.

CHAPTER 17

New Strategies and Structures

THROUGHOUT THE 1970s, as Japanese computer manufacturers steadily gained market share at home and slowly established beachheads in North American and European markets, industry observers continued to discount their ability to mount a serious threat to IBM, long No. 1 in the computer business. With its 60 percent world market share and a massive cash flow that raised the capital costs of challenging its leadership to seemingly prohibitive levels, IBM appeared invincible. In 1980, R&D expenditures of US$1.5 billion, which exceeded the combined outlays of the entire Japanese computer industry, seemed to assure the global giant of technological leadership throughout the decade. However great the manufacturing prowess of Japanese computer makers, their lower expenditures on technology and a seemingly inherent handicap in operating software design precluded their emergence as a major force in markets outside Japan.

IBM itself responded to Japanese advances in the U.S. and European markets in alliance with national PCM vendors in a predictable way, befitting a firm with long experience in the exercise of superior market power. In the words of the computer giant's former chief executive officer, T. Vincent Learson, the threat was easily disposed of. IBM would "murder them."

But that proved to be more easily said than done. Beginning with the introduction of the 303X series in 1977, IBM launched a counteroffensive discernibly designed with just that intent. In 1979, with the announcement of a new medium-scale 4300 series, the company that had set the *de facto* standards for the EDP industry since its inception unleashed the full fury of its aggressive pricing power. After Fujitsu succeeded in moving ahead of IBM 370 architecture and technology in the mid-1970s, IBM strategists apparently concluded that the only way to prevent the imminent ascendancy of Japanese computer makers was to

---

Reprinted from "The Shape of Future Japanese Computers Industry," *Electronics in Japan '83-84*, March 1984.

nip it in the bud by drastically lowering prices on medium-sized mainframes in which their strength was centered.

When the high-flying Itel, which was marketing Hitachi mainframes worldwide, was forced to withdraw from the business in 1979, observers took it as a sign that the IBM juggernaut was as invincible as ever. Fujitsu also felt the full impact of the broadsides as customers began to lease rather than buy, in anticipation of cheaper computers from IBM, slowing cash flow from its sales through Amdahl Corp., in which it had a 28 percent interest at the time.

But both Fujitsu and Hitachi, as well as NEC, met the IBM moves head on. Within weeks of the IBM announcements, all three leading Japanese mainframe makers introduced new series designed to outperform the IBM 4300, and at prices competitive with those of the industry leader. As a result, IBM's strategy boomeranged. Not only were their new product announcements and pricing moves unsuccessful in forestalling the Japanese thrust into foreign markets, but the price-cutting inflicted serious penalties on IBM itself. In 1979, for the first time in two decades, the firm's net income declined from that of the previous year, and in Japan, for the first time ever Fujitsu outstripped IBM in computer-related revenues.

Then, in late 1980, IBM fired its ultimate weapon, the long-awaited 3081 large-scale computer, intended to check Fujitsu and Hitachi in their OEM sales to PCM vendors in the United States and Europe and to regain market share for IBM Japan. Once again, leading Japanese mainframers met the IBM riposte with larger and faster machines: Fujitsu's M-380, Hitachi's M-280H, and NEC's ACOS 1000.

Since 1977, Japanese computer manufacturers have consistently matched every one of IBM's price reductions and introduced new, plug-compatible as well as non-plug-compatible, central processors delivering computing power higher than each new IBM model and with better cost/performance ratings. IBM got the message: the Japanese industry is determined and able to match and exceed the hardware performance of each new offering in the market to meet every price reduction regardless of short-term impairment of income, something no American mainframer has been able to do.

As a result, PCM vendors supplied mainly (as much as 90 percent in 1980) by Fujitsu and Hitachi had succeeded in taking 19 percent of the U.S. market at the turn of the decade, hammering down IBM's market share to a modest 58 percent and quickly eroding the price/performance advantages that the non-IBM-compatible mainframe manufacturers have traditionally maintained over the computer giant.

With very little investment in sales and service organizations and minimal software development costs, by supplying PCM vendors in the North American and European markets, Fujitsu and Hitachi have become a powerful force in the international industry. By mastering production technology of both mainframes and their microelectronic building blocks, Japanese makers managed in a very short time to change both the structure and the style of the global computer business.

Not only have those firms supplied by Fujitsu and Hitachi on an OEM basis (Amdahl and National Advanced Systems in the United States; Siemens, ICL, Olivetti and BASF in Europe) strengthened their positions in their respective markets, but IBM has broken out in a rash of technological change, shifting resources into new growth areas. In contrast, what has come to be known as the BUNCH (Burroughs, Sperry Univac, NCR, Control Data and Honeywell) of non-plug compatible makers have seen their total U.S. market share drop sharply from 29 percent in 1977 to 11 percent in 1981.

The changes brought about by this new competition have been of a major order of magnitude. After the unhappy financial experience following the aggressive pricing of the 4300 family, coinciding with price cuts on peripheral equipment and on the relatively new large-scale 303X series, IBM began to change its pricing, product announcement and marketing strategies. To improve its cash flow, leasing and maintenance prices as well as license fees were substantially increased, while purchase prices on many products were cut. As a result, in 1980 outright sales of IBM mainframes rose to record highs.

Ironically, the effect of this new purchase-oriented market is to remove the barrier to entry constituted by IBM's previous leasing policy, easing the financial constraints on the Japanese mainframes and their PCM customers. As the PCM vendors took an increasing market share, IBM's account control began to erode, further opening the door to compatible mainframe vendors from Japan. And, at the same time, IBM has added to the flow of Japanese EDP products, breaking established practice to subcontract, after careful "buy-or-make analysis," the production of copiers and personal computers with Japanese makers.

Combined, the push of better prices and quality, along with the pull of changing market forces, boosted Japanese exports of computers and related equipment from 1976 to 1980 to an annual increase of 41.5 percent. Exports as a share of total production rose from 6.8 to 10.7 percent over the five-year interval, and are expected to continue their upward thrust to 30 percent of output in 1985. By 1990, industry sources anticipate, Japanese exports will account for as much as 30 percent of the total world EDP market. These projections assume a conservative average increase of 20-25 percent annually over the remainder of the decade.

COMPUTERS

Apart from the mainframe market, Japanese makers are expanding their efforts to develop overseas markets for high-volume hardware where superior manufacturing technology assures distinct cost advantages. Japanese-made hand-held computers already hold a pre-eminent place in world markets, and Japanese firms are beginning to take significant shares of overseas markets for matrix printers, floppy disk drives. CRT-based teminals, as well as copiers, facsimile equipment and PABXs. In the months ahead, exports will be fueled by the entry into new and relatively untouched markets for intelligent robots, personal computers, supercomputers, word processors, laser printers, graphic devices and voice recognition equipment. At the same time, Japanese computer makers are expected to make substantial progress in the development of exportable application systems — especially flexible manufacturing systems, computer-integrated manufacturing (CIM) systems, computer-aided design and manufacturing (CAD/CAM), computer-generated graphics and electronic banking.

Significantly, leading Japanese computer makers have adopted export strategies similar in some respects to those of the automotive industry. In both instances, the main markets targeted for development have been those of the United States and Western Europe. But quite unlike automotive manufacturers, Japanese computer makers began by exporting the high end of the product line rather than beginning with lower-priced models and trading up as market position improves and costs increase. This high-value-added export strategy was possible, of course, because mainframes have been sold mainly to PCM vendors rather than through direct sales channels. Where exports have been direct, through their own overseas sales organization, Japanese makers (especially NEC and Mitsubishi Electric) have been careful to avoid a direct confrontation with IBM. Their exports of EDP equipment have therefore begun with small business or personal computers, sold under their own label. Once established in the market, with appropriate sales organizations, NEC expects to introduce its medium- and large-scale mainframes in direct competition with other Japanese and American makers.

Contrary to perceived wisdom from the Japscam affair, the PCM strategy of Fujitsu and Hitachi is not a sign of technological weakness and dependence. To produce computers which can be used interchangeably with IBM requires a high level of technical capability. Makers must proceed with the understanding that they will receive no assistance from IBM. Nor can they rely on technology of other mainframe makers. Moreover, they must be able to anticipate IBM's future strategy and be prepared to launch a competitive product with price/performance and delivery advantages at about the same time IBM announces a new computer series. Thus, a successful PCM strategy depends upon a parity of technology with IBM.

It is not by accident, therefore, that Fujitsu was first to take this route in Japan. Fujitsu was the only Japanese mainframer which developed its own technology from the outset, producing its first in-house designed computer in 1954. In 1966, solid-state international-standard large-scale computers developed under the direction of Drs. Hanzo Omi and Toshio Ikeda were exported for the first time. And by 1972, Fujitsu had reached a technological parity with IBM. What was needed was a continual flow of information regarding IBM's marketing plans. Most important, it has been necessary to know which in a series of potential new machines IBM will introduce at a given point in time, with what kind of software, and at what price. While answering these questions has been no easy task, a continuing flow of information regarding IBM's strategies and plans can be gleaned from the growing legion of IBM-watchers, information services, consultants and recruiting organizations.

Fujitsu and Hitachi have consistently been able to manage this information flow, however, successively introducing new products which are sufficiently better and lower-priced than those of IBM to assure an adequate flow of orders. No attempt has been made to leap-frog IBM introducing an entirely new machine, for this would defeat the purpose of the IBM plug-compatible strategy. The main tenet of that strategy is to let IBM create the demand through their own development, announcement and promotion, tapping into effective demand by offering better machines at lower prices and earlier delivery. In fact, the PCM strategy, to the extent that it renders unnecessary heavy investment in independent software development and international distribution networks, has enabled Fujitsu and Hitachi to invest more heavily in hardware design and development technologies.

NEC has adopted a totally different strategy, however, undertaking the development of both its own operating systems software and international marketing organization. This strategy, NEC insists, enables greater flexibility (since it is not necessary to follow IBM slavishly) and stability (which is not directly affected by aggressive product and marketing strategies of IBM). It has, however, meant a heavier drain on NEC capital resources, keeping profitability of computer operations at a lower level than those of Hitachi and Fujitsu.

The rewards of NEC's strategy of heavy investment are now being reaped. These investments, carefully calculated to achieve a synergistic combination of the company's computer, communications and integrated circuit technologies, have assured domestic market leadership in communications, semiconductors, office computers, personal computers, and intelligent terminals. And in mainframes and minicomputers, NEC is a strong second with a 15 percent market share in each sector.

There is a pervasive conviction at NEC that, although the PCM strategy assures early and easy entry into foreign markets, this is at best a short-term expediency. In the long run, success will depend upon marketing branded products through rational distribution and production organizations. And this, in turn, requires independent software capability.

Management at both Fujitsu and Hitachi agree with this general proposition. The difference is mainly a matter of timing. If they are to become full computer systems vendors, some lingering software development problems must be solved. But this presents no major problems.

Contrary to common wisdom on the subject, the Japanese industry suffers from no inherent handicap in software development. Indeed, over the past twenty years Japanese computer firms have developed some highly sophisticated applications software. The facts are, the Japanese computer industry has solved its software requirements quite differently from other industries. Most software development is under the aegis of the mainframe makers. Hitachi, for example, has 15 subsidiaries producing software, and its software factory has been operating since the late 1960s. Fujitsu has a total of 10 software subsidiaries, six of them having been established in the short period of two years, 1979-81. In addition, these mainframe makers employ large numbers of software subcontractors — numbering about 1,000 — to meet their growing needs.

By 1987, Morgan Stanley's Ulric Weil believes, the Japanese computer industry will have closed any remaining gap in software capability. And once the software gap is finally closed, Japanese computer makers will compete in world markets on a systems basis and worldwide software prices will come down. Already a growing number of independent software houses are offering applications software packages, responding to the booming demand for personal computers. By 1985 an estimated 10,000 to 15,000 applications software packages will be available for Japanese personal computer owners, and by the end of the decade Japanese software houses fully expect to be doing a booming export business. Within the next five years, Japanese software may well be functionally superior and significantly lower priced than that of the competition, and by the 1990s the economics will in all probability be right for a boom in Japanese exports. Software, by then, will have become a commodity just as computer hardware has become in the 1980s. Once the market is there, a proliferation of software desginers can reasonably be ex-

248

Table 1. Comparison of Very Large-scale Computers

	Fujitsu M-380	Hitachi M-280	NEC ACOS 1000	IBM 3081
Processing speed (relative)	1.5 (1)	1.1 (1)	—	(2)
Maximum storage capacity	64 megabytes	32 megabytes	64 megabytes	32 megabytes
Power consumption	27.9 kilovolt-ampere	40KVA	—	31.8 KVA
Maximum integration of logic circuits	1,300 gates	1,500 gates	1,200 gates	704 gates
Efficiency of logic circuits	0.35 nanosecond	0.45 ns	0.5 ns	1.2 ns
Memory integration	64 kilobits	16K-bits	64K-bits	16K-bits

Note: Figures in parentheses show the number of central processing units. Relative processing speed is calculated by the Nihon Keizai Shimbun from data available. Uncalculable for ACOS 1000 because of the difference in architecure.

COMPUTERS

Table 2. Sales of Eight Major Computer Builders

(Computer division; in billions of yen)

	FY 1976	'77	'78	'79	'80	'81	'82
Fujitsu	239.6	274.5	303.0	326.8	382.0	448.4	535.0
Hitachi	142.0	160.0	190.0	216.0	250.0	288.0	330.0
NEC	114.0	137.5	166.8	200.7	240.3	332.5	389.5
Toshiba	59.2	59.1	43.0	50.4	80.3	95.0	115.0
Oki Electric Industry	48.3	44.4	47.9	62.8	78.8	109.1	131.0
Mitsubishi Electric	32.0	38.0	45.0	53.0	62.0	73.0	88.0
IBM Japan	275.4	293.8	315.3	324.2	338.3	428.9	n.a.
Nippon Univac	70.6	67.8	71.6	73.6	78.6	90.9	100.0

Notes: (1) The term ended December for IBM Japan. The accounting term ends in March for other companies.

(2) Products included in the computer division were expanded in fiscal 1980 for NEC.

pected to master the software creation process, offering original packages at very low prices. The importance given to software development in government support for the next computer generation will provide added impetus to this trend. Support for basic software technologies through the Computer Basic Technology Research Association, emphasizing the development of operating systems for network control, data base management and Japanese language processing will go far towards bridging the present gap with the U.S. industry.

Meanwhile, major changes in electronic data processing markets and technologies tend to favor the existing advantages of the Japanese computer industry. The continuing trend toward distributed data processing broadens the market for office computers, professional personal computers, terminals, and telecommunications systems. Superior manufacturing technology and the integration of computer and communications technologies give Japanese industry a decided cost and systems design advantage. Miniaturization of computers and terminals is rapidly transforming them into consumer commodities, subject to price and income elasticities of demand which make scale and learning economies decisive in the calculus of competitive advantage. Here again broadening market demand justifies volume production at which Japanese industry excels.

Both of these trends entail diminished unit prices, which, when combined with rapidly escalating selling costs, mitigate against providing direct sales and support services to a myriad of small data-processing users. Computer makers are now turning to new channels of distribution — including captive retail stores and mass merchandisers such as Sears,

250

Roebuck; J.C. Penney and K-Mart — which have in the past been the vital links with the market which Japanese manufacturers manage with particular skill. So long as direct sales and support activity is needed to market computers, Japanese makers are at a decided disadvantage. So the switch of distribution to independent dealer networks or mass merchandisers not only eliminates a major disadvantage but replaces it with the advantages Japanese manufacturers enjoy in the management of disciplined dealer networks and mass merchandiser supply.

The combined effect of these radical changes in the electronic data processing industry is to reduce sharply the relative importance of mainframes. While in 1975 mainframes accounted for as much as 65 percent of the EDP market, with the remaining 35 percent spread over a range of terminals and new products such as word processors, by 1985 the pattern of the market will be the reverse. New products for the automated office and home will be at once the largest and fastest growing segment of the market. And, once again, this shift is especially auspicious for those things Japanese industry does best.

Finally, changing geographical patterns of demand tend to favor Japanese manufacturers. While in the past the U.S. market was by far the largest for EDP products, giving IBM and other American mainframers a decided advantage, the greatest potential growth in the future will be in Europe, East Asia and newly industrialized countries in other regions where the business environment is receptive to the demonstrated Japanese willingness to adapt to local preferences for joint ventures and other forms of industrial cooperation.

In this rapidly changing global market, the growth strategy of the Japanese computer industry for the 1980s is built around five key advantages.

Structure: Unlike the U.S. computer industry, which is characterized by specialized equipment manufacturers, Japanese computer makers are highly-diversified vertically-integrated firms producing either a wide range of electrical/electronic products or telecommunications/electronic products. Signficantly, all six of Japan's computer makers are leading manufacturers of communications equipment. And all of them, with the exception of Oki, have a base in consumer electronics production. The effect of this structural feature of the Japanese industry is to make possible higher sustained investments in R&D, plant and equipment and marketing.

As a result, all six major Japanese computer makers are fully integrated vertically from semiconductor production to end product; and the four mainframers — Fujitsu, Hitachi, NEC and Mitsubishi Electric —

have diversified product lines including personal and office computers, peripherals and teminals, word processors and robots. In July, Fujitsu added a further dimension of diversification with the introduction of a supercomputer for scientific use, claimed to have the world's fastest processing speed, a niche which even IBM has opted to leave to specialized computer makers such as Cray Research and Control Data Corp. Both NEC and Hitachi are also developing their own supercomputers as well, making the three leading computer makers the world's most diversified.

System and synergy: Since the late 1950s, these diversified and vertically integrated firms have been systematically developing information technologies to obtain the maximum synergy from the related technologies in which they are especially competent. With the present trend towards distributed data processing, the synergy between communications and computers, which Japanese makers such as NEC have long recognized and sought to develop systematically, is becoming a principal determinant of competitive strength in the industry. Similarly, the combination of consumer and industrial electronics manufacturing and merketing capabilities is increasingly important as miniaturized computers become consumer commodities. The world's most advanced consumer manufacturing technology and most extensive consumer electronics marketing networks are increasingly being employed in the service of data processing products.

Advanced semiconductor technology: The prowess of Japanese computer manufacturers is founded mainly on their great strength in advanced microelectronics. Not by accident, the leaders of semiconductor technology and production in Japan are also the leading computer manufacturers. Advances in 64K RAM production give to Japanese computer systems production a versatility and efficiency which made possible their rapid response to IBM's latest bid to strengthen its position in large-scale computers. The vast storehouse of VLSI technology developed since 1976 positions Japanese mainframers to take the lead in future developments of fourth generation computers, if they choose to do so. And a growing string of innovations in semiconductors — the high electron mobility transistor (HEMT) pioneered and developed at Fujitsu, Josephson and galium arsenide devices perfected by NEC — provide the necessary building blocks for the much-touted fifth generation computer to be developed cooperatively over the remainder of the 1980s.

Accent on production: Advances in electronic technology are being turned systematically to the improvement of both factory and office output. Just as manufacturing technology has served to gain for the Japanese computer industry its main advantages in the world marketplace in the past, new automated equipment, robots, CIM, CAD/CAM, and flexible manufacturing systems will serve as the sinews of competitive power for successive generations of computers.

Internationalization: Japanese computer makers will further rationalize production and sales in the pursuit of global strategies. Because of their technological and production prowess, Japanese mainframers are in a position to build international linkage with American and European firms into an effective global response to IBM. In the future, these networks are likely to be strengthened as BUNCH companies turn to form partnerships and other ties with Japanese makers in their bid for survival in an increasingly competitve environment. The technical tie-up between Sperry Univac and Mitsubisihi Electric is a case in point. For its part Mitsubishi abandoned plans to take the PCM route in favor of a broad-based two-way cooperative relationship with a leading mainframe computer maker which desperately needs to diversify its product range to stay in the field. As it becomes more widely recognized that international cooperation of this kind is the only remedy for fallout from economic warfare, attempts at legislated partitioning of the world's industry will be abandoned as more harmful than beneficial.

With a growth strategy firmly grounded on these elements, the Japanese computer industry is destined to become a major factor in the global electronic data processing field, attaining significant market shares in North America and Europe by 1985. By the turn of the decade, the leading Japanese makers can be expected to manufacture Japanese systems based upon their own architectural concepts, utilizing state-of-the-art technology, and offering proprietary non-IBM-compatible software support. Industrial policies, which elevate this objective to the level of highest national priority, assure the necessary resources and cooperative effort required for its attainment. Finances will be available on terms and in quantity appropriate to the undertaking. And the pervasive cooperative mode of management in Japanese enterprises provides an industrial environment most conducive to the rapid diffusion and application of computer technology in all aspects of industrial activity.

CHAPTER 18

The Great Supercomputer Sweepstakes

WHEN THE JAPANESE National Superspeed Computer Project got underway in January 1982, few observers would have believed that just three years later it would be among the highest priority items on the U.S.-Japan diplomatic agenda discussed by President Ronald Reagan and Prime Minister Yasuhiro Nakasone at their California summit. At the time of their meeting in early January 1985, there were still scarcely more than 130 supercomputers in operation worldwide. Less than ten of these had been built by Japanese computer makers, and virtually all the rest were products of two American firms: Cray Research, Inc., and Control Data Corporation (CDC), a duopoly which had dominated world markets since the first ultra-high-speed number crunchers were built at the beginning of the 1970s. And about 70 percent of the supercomputers then in operation had been supplied by Cray Research.

The immediate problem which had been raised at earlier technical meetings between representatives of the two countries was access to Japanese markets — particularly universities. But in the background loomed a larger issue. The Japanese national supercomputer project, not to be confused with the Fifth Generation Computer Project intended to develop and incorporate concepts of artificial intelligence in the next generation of computers, had set for its specific ojectives the design of machines with a sustained execution rate of about 10 billion floating-point operations per second (gigaflops) — the standard measure of computing speed — or 100 times faster than supercomputers available from leading American makers and ten times faster than the projected peak speed of the next generation of American machines.

This speed would be attained by using entirely new semiconductor devices, distributed parallel-processing architecture, one billion bytes (one gigabyte) of memory and a memory bandwidth of 1.5 gigabytes per second. To meet these specifications, the Scientific Computer Research

Reprinted from ''Japan on the Fast Track: The Rush is on to Develop Ultra-High-Speed Machines'' *Far Eastern Economic Review, 31* January 1985.

Association was established under MITI auspices, combining six leading integrated computer-semiconductor manufacturers under the coordination of the national Electrotechnical Laboratory in Tsukuba Science City.

The immediate U.S. concern over possible Japanese inroads in supercomputers, until then an exclusive American domain, was evidenced in earlier reports of teams of scientists from Los Alamos and Lawrence Livermore National Laboratories after their visits to Japanese computer centers in the Spring of 1982. But when their conclusions that the Japanese industry could eventually surpass the American companies was dramatically substantiated in the following fifteen months by successive announcements of progressively superior machines to be placed on the market by Japanese computer manufacturers by 1985, supercomputers

SUPERCOMPUTER PRIMER

Although the concept of supercomputing has been traced back to Charles Babbage in the 1830s, and even to the Greeks of Salamis in 100 B.C., the terminology of today is calculated to confuse, especially since it is badly tangled with vocabulary of fifth generation computers.

Supercomputers — Currently computers which process data at 20 megaflops or higher speed, but the fastest supercomputers today operate at sustained speeds of 400 megaflops and peak speeds of 1 gigaflop or higher. Speeds on supercomputers vary depending on how much of the data flows through parallel scalar or through vector processors. In the future, supercomputers may operate simultaneously in multidimensional arrays with many parallel operations being performed in a pipelined way. Over the last 30 years the raw speed of the fastest computers has approximately doubled each year.

Scalar Processor — Operates on individual data elements with instructions that yield one result for each instruction. To perform a single operation on all elements of an ar-

ray (such as calculating a 5 percent turnover tax), the scalar processor must loop through the table of numbers constituting the array, repeating the same instruction on each element to achieve the desired results. Scalar processing therefore requires fast circuitry for high performance.

Vector Processor — Uses only one instruction to perform a single calculation on an array of data and achieve an array of results, all in one operation. In supercomputers, the more data vectorized, the faster the operation. Vector processing is usually not available on general purpose mainframe or minicomputers. However, vector processing is a feasible alternative to scalar processing only when repetitive operations must be performed.

Array Processors — Do not include scalar processing. Array processors are usually configured as peripheral devices on which both mainframe and minicomputer users can run vectorizable portions of programs. Synonymous with vector processors, but operating as independent units, array processors can, in some applications, serve as an inexpensive alternative to supercomputers. Unlike the more elaborate parallel became the most sensitive US$200 million market in the history of the information processing industry. After years of being virtually a one-company and — in the personage of Seymour Cray, its founder — almost a one-man industry, the market was teeming with new players, projects and funding. A supersensitive, security-related technology which the United States had gone so far as to specifically deny to France for nuclear development in the late 1960s, was clearly out of control.

That this should have come as a surprise to U.S. industry leaders and government policymakers is in itself suprising to Japanese. For at least a decade, and in fact since 1957 when the electronics industry was identified as a key industry, it has been plainly evident that information industries and information technology are considered essential to the economic development, and survival, of Japan. In that techno-economic

---

processing supercomputers, array processors are produced and sold in volumes akin to minicomputers.

Parallel Processing — Concurrent application of two or more processors to perform simultaneous operations in a single task. Superspeed computers must use high speed vector processors in tandem with scalar processors, dividing each task for optimal use of the multiple processors under a central operating system control. The processors may or may not be pipelined.

Pipelining — Speeding computer operations by breaking down instructions into discrete steps for processing in an assembly-line system, with different steps in the preparation and execution of an instruction performed simultaneously. In a strictly sequential processor, in contrast, all operations on one instruction are completed before processing of the next begins. An essential adjunct to pipelining is the use of vector registers — high speed memories to store temporarily and then feed instructions to the pipeline at a speed that is greater than possible when instructions must be called up from the computer's main memory.

Floating Point Operations — Refers to the binary version of representing numbers called scientific notations. In ordinary scientific notation a number is represented as a product in which one factor (the mantissa) has a magnitude between 0.1 and 1, and the other factor (the characteristic) is an exact power of ten. Thus 7,700 is represented as $.77 \times 10^4$ and 77 is represented as $.77 \times 10^2$. The computer relies on a binary version of floating-point notation, making possible a standard representation of a very wide range of magnitudes. A single floating point operation is the addition, subtraction, multiplication or division of two floating point operands to get a floating point result. Computing speeds are, therefore, expressed in terms of floating-point operations per second (flops).

Megaflop — One million floating-point operations per second.

Gigaflop — One billion floating-point operations per second.

Operand — A quantity that is to be the subject of a mathematical calculation.

Fifth Generation Computers — The proposed next generation of computers, to be produced in the 1990s, which will have radically new inferential architectures to exploit the potential of developments of artificial intelligence.

context, computers were obviously of central import, a seminal technology which would have a major impact on other leading technologies affecting the way research and development systems will function in all industrial and scientific fields.

Japanese policymakers had unequivocally acted on the premise that the wealth of nations will be largely determined in the future by information technology, and that, in the absence of other than human resources, Japan is more dependent on that technology for its economic future than are most other advanced industrial countries. New generations of computers will therefore serve as the prime movers in an increasing range of industries, helping to improve efficiency where productivity has in the past been little influenced by automation. But computers are also much more than critical tools that will determine the future comparative advantage of Japanese industry; they will also be essential for the management of the ecosystem and dealing with a myriad of problems of an aging society.

Supercomputers answer special needs of a growing community of governmental, scientific and industrial users for capability to execute complex parallel processing of massive amounts of data to aid in the mathematical simulations of multidimensional physical phenomena. They therefore constitute an essential element of any comprehensive information technology strategy. Here is the rub. What in a techno-economic context is vital to the security of Japan is seen in the United States as a threat to national supremacy in critical military technology.

It is these apparently conflicting vital interests, then, that have resulted in the current U.S.-Japan confrontation, not an inadvertent trespassing of U.S. national security interests by Japanese industrial policymakers or an attempt to replace U.S. domination of this sector of the computer industry. Japanese information policy specialists, scientists and industry leaders have made clear their intentions to push forward the state-of-the-art in supercomputing, as well as Fifth Generation technology, and their readiness to develop their own approaches to creative research and development in this field, no longer confining their efforts to copying the U.S. leaders. By setting goals of creativity and leadership in research and development, pioneering new projects in high technologies such as supercomputing, the intention is not only to assure the competitive strength of Japanese industry and the wealth of the nation, but to contribute to the forward march of human progress.

None of this suggests that the Japanese view their action as a move in a zero-sums game, in which their success necessarily entails the loss of present industry leadership by American firms or a threat to U.S. security. The end of the U.S. duopoly that has enjoyed world exclusivity in the supercomputer sector for the past decade, with Japanese makers gaining market share at home and abroad, is not tantamount in their minds to a replacement of U.S. domination by Japanese. Given the high technological level of the Japanese electronics industry, the expanding market for supercomputers, and their vital importance to Japanese industry, the development of supercomputer production capability is seen rather as a natural course of open worldwide competition, likely to lead to the entry of other players as well.

Japanese computer makers take seriously the expressed confidence of Cray Research executives that they will be able to match Japanese advances in supercomputer technology. They know how important it is to be first in the field, and that the existing park of Cray supercomputers constitutes a formidable advantage in the competitive marketplace. There are also few illusions among Japanese industry leaders that they will be able to penetrate readily the U.S. government market, which has accounted for the bulk of supercomputer sales in the past and will undoubtedly continue to be an important market in the future. And, if there is some dismay at the tone and thrust of the rhetoric in the United States over a market that is currently running at barely more than US$200 million a year, the riposte of the U.S. government with increased outlays for R&D and other assistance for American industry comes as no particular surprise to Japanese informed by experience in worldwide computer and semiconductor competition.

Rather, the Japanese computer industry has formulated its strategies in anticipation of this reaction and the understanding that the world supercomputer market, just as other segments of the information industry, will be shared among an increasing number of suppliers — not only Japanese and American, but also European, and possibly other manufacturers. The notion that any country, or firm, can dominate major sectors of the information industry, or control the course of the information revolution, in any real sense, is not entertained or advanced as a serious proposition by Japanese industry leaders. From Tokyo and Osaka, the industry appears as a pluralistic, dynamic and constantly shifting system in which the free flow of information — if not always products — is the critical determinant.

For these, as well as other reasons, Japanese supercomputer strategies differ significantly from those of the U.S. pioneers. Supercomputers were not built by the leading U.S. computer manufacturer, IBM, but by "dwarf" companies looking for niches in the market where they can survive unmolested by the Big Blue. When Seymour Cray left Control Data, which he had helped create but later found to be unable or unwilling to sustain a sufficiently high commitment to the advancement of superspeed computers, he originally intended to build only one of his projected giant number crunchers a year for scientific purposes. He saw no great market for these machines, and even after the favorable initial response from the Los Alamos National Laboratory, which took delivery of the first Cray 1 in 1976, the new company that began producing these advanced computers in an abandoned shoe factory in Chippewa Falls, Wisconsin, increased production only gradually to four machines a year in 1978 and 13 in 1984.

In 1982, when Japanese computer makers announced their imminent entry into the field, there were only 50 supercomputers in operation worldwide. Thirty-five of those had been built by Cray Research, 14 by Control Data and 1 by Denelcor, a small Colorado builder of multiprogramming machines. Fully 38 of those superspeed computers were operating in the United States, 10 at Los Alamos and most of the others at various government laboratories and agencies. It was a restricted market, for which the U.S. builders had developed special machines that were programed in a completely different manner from standard general-purpose mainframes.

The new Japanese entrants into the market — Fujitsu, Hitachi and NEC — are, in stark contrast with U.S. firms in this field, major highly-integrated electronic companies producing a full line of computers, from personal to the largest and fastest mainframes, and are at the same time among the world's top ten semiconductor manufacturers. Not only did these makers build their new superspeed machines to operate on the same software used by their standard mainframe models, but this meant that both Fujitsu and Hitachi would supply machines compatible with mainframe models built by IBM, which had not so far given any sign that it would add supercomputers to its product line.

This does not mean that IBM will not defend its accounts in all markets from inroads by Japanese supercomputers. IBM has always priced the two-processor version of its mainframes very aggressively, and a two-header top-of-the-line scalar Sierra model will be enough to deter many IBM users from upgrading to supercomputers. But the PCM (plug compatible machine) market itself is sufficiently fertile ground for Fujitsu and Hitachi to reach their objectives without any major victories in IBM territory, and, in addition, they can reasonably expect to take some share of the IBM market during the remainder of the decade.

The prospects are good enough, at least, so that from the outset Japanese manufacturers adopted a total, global strategy designed to take maximum advantage of the absence of IBM from the world supercomputer marketplace. By producing IBM-compatible machines, Fujitsu and Hitachi offer the widest number of computer users at home and abroad the option of moving to more powerful machines without heavy expenditures of time and money for adapting new special software.

But the global strategies of Japanese makers — including NEC — are based upon other elements of a perceived economic reality that have not weighed as heavily in plans and operations of American leaders in the field:

• Firms in a widening range of high-technology industries, and virtually all of the key industries to which Japan's economic future is linked, are beginning to feel the need of ultrafast computer processing capabilities. Advanced computer graphics, computer-aided design, structural design of buildings and complex machinery, as well as distribution systems all require massive volumes of data, necessitating processing improvement of two to three orders of magnitude during the current decade.

• Development of supercomputers contributes to the development of other computer systems, including, of course, fifth generation intelligent computers, and is expected to be a vital factor in the development of data processing in the 1990s.

• Supercomputer development is also linked synergistically with progress in semiconductor technology. Not only is the computing power which it offers vital to very large scale integrated (VLSI) circuit design, making the availability of supercomputers a critical factor in assuring a competitive edge in future generations of devices. Supercomputers themselves require microelectronic devices with increasingly higher switching speeds, speeds which can be attained only by developing entirely new integrated-circuit technologies, thus providing an important forward thrust to the semiconductor industry.

• By effectively responding to potential worldwide demand for supercomputers, Japanese makers see the possibility of cutting prices sharply. Although the limited market served by U.S. supercomputer builders has a relatively low price elasticity of demand, which has tended to limit their search for broader strategic options, industrial users are likely to be much more sensitive to lower prices.

With the 1980s at mid-passage, supercomputers still occupied the niche that first generation computers had occupied in the early 1950s. Those earlier machines cost even more, if inflation is factored into the pricing calculus, and were so powerful that only a very special group of users could find applications for them or afford them. Yet in the United States alone there were thirteen computer manufacturers. Remington Rand was the leader, IBM was only beginning to break with its traditional punch card technology, and the two accounted for only nine electronic computer installations between them.

By comparison, the worldwide supercomputers park at the beginning of 1985 had reached 130, with only nine installations in Japan: four Fujitsu VP 100/200s, three Hitachi S-810/20s, and two Cray machines. By 1990, it is estimated that at least 100 supercomputers will be operating in Japan, about the same number installed worldwide in 1984.

World market estimates for 1990 vary widely from 400-500 machines to as many as 1000. Not surprisingly, the higher estimates of demand are confidently forecast by Japanese makers who are aiming at a broader range of users. And each of them have mounted their global marketing effort that is calculated to make it all happen.

Fujitsu has already concluded arrangements with the Amdahl Corporation, in which it holds a 49 percent equity, to market its supercomputers in North America and Europe. Initial benchmark tests in the United States indicate that the sustained throughput — which measures both central processing unit speed and input/output speed, reflecting a machine's overall performance — of the Amdahl 1100 and 1200 (which are identical to the Fujitsu VP-100 and VP-200) compares favorably with the competition. The 1100 sustains speeds of 175 megaflops and the 1200 at 300 megaflops. The 1100, in sustained throughput, is 1-2 times faster than the Cray X-MP uniprocessor and twice as fast as the two-pipeline Control Data Cyber 205, while the 1200 is about as fast as the Cray top-of-the-line X-MP two-way processor.

While some computer scientists discount the advantage of IBM-compatibility, pointing out that if a customer is prepared to vectorize the scalar software code, the same results can be obtained on a Cray with an IBM up front; others assert that programs developed from the start for vector processing on supercomputers will run 20 to 30 times better than scalar programs converted to vector processing.

Wall Street computer industry analysts have been consistently bullish on Amdahl's ability to reach a new class of users. Based upon an estimated 10 letters of intent to purchase machines at the outset, Amdahl supercomputer shipments are expected to reach between US$50 and 75 million during 1985, rising to at least US$200 million in 1988. But this respectable figure, which appeared to some to be on the conservative side before Fujitsu's January 1985 announcement of its new VP-400 with 1 gigaflop processing speed, will most likely be revised upward in future projections. The earlier ambitions of the Amdahl Corp. to take on 20 percent share of the rapidly expanding U.S. market now appear feasible.

In Europe, Siemens, which announced the introduction of vector processing systems in September 1984, will market Fujitsu-made supercomputers under its own brand name in competition with Amdahl, from already well-established market positions for Fujitsu-made mainframes.

Table 1. Genealogy of the Supercomputer

Maker	Model	Availability	Maximum Rated Performance (Megaflops)	Maximum Main Memory (Megabytes)
Cray	1M	September 1983	250	32
	X-MP1	June 1983	630	32
	X-MP48	1985	1,000	n.a.
	2	1985	1,000	n.a.
	3	1985/1986	n.a.	n.a.
CDC/ETA	205	February 1982	400	64
	2XX	1986/1987	1,200	256
	GF10	1986/1987	10,000	n.a.
Denelcor	HEP-2	1985/1986	1,000	2,000
Fujitsu	100	November 1983	250	128
	200	November 1983	500	256
	400	1985	1,000	n.a.
Hitachi	S810/10	November 1983	315	128
	S810/20	November 1983	630	256
NEC	SX-1		570	256
	SX-2	March 1985	1,300	256

Amdahl and Siemens, both offering Fujitsu-built machines, are expected to cut into Cray and CDC European market share even more deeply than Amdahl can on its own in the United States. Although major European governments all have supercomputer projects and will spend heavily over the coming decade to develop their own technology in ultrafast computer processing, in the interim they may opt for Japanese machines to limit dependence on U.S. suppliers.

Hitachi has followed the Fujitsu strategy in the United States supplying National Semiconductor's subsidiary, National Advanced Systems (NAS), on an OEM (original equipment manufacturer) basis, building on its existing market position in superspeed machines established with its earlier array processor capable of a peak throughput of 28 megaflops. In 1982, Hitachi responded to the Fujitsu entry into the ultrafast market with its two-model S-810 series — the S-810/20 with speeds up to 630 megaflops, and a second model S-810/10, rated at 315 megalfops. In Europe, Hitachi-made supercomputers are being offered by NAS, Olivet-

Fig. 1 Supercomputer Development Trends

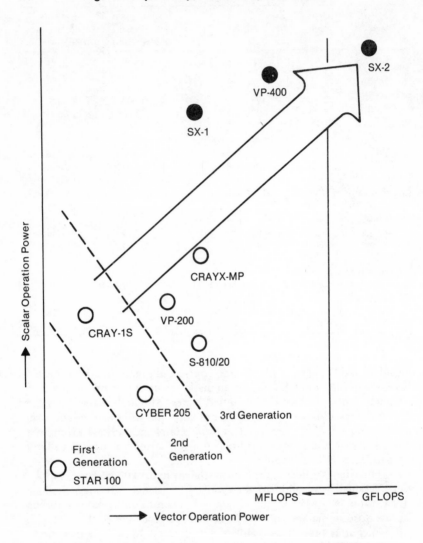

ti and BASF as vector processing extensions of the Hitachi-built line of mainframe computers the three vendors have been marketing for the past five years or so.

The third Japanese supercomputer builder, NEC, is following quite an independent strategy. While its new SX-series has software compatibility with NEC's own mainframe models, neither are IBM-compatible. "IBM architecture is 20 years old," NEC supercomputer mrketing manager Akihiro Iwaya points out. "To assure IBM-compatibility, some performance compromises are necessary, which NEC chooses to avoid." As a result, NEC succeeded in surpassing both Fujitsu and Hitachi, unveiling machines with operating speeds up to 1.3 gigaflops, or ten times the rates capacity of the Cray 1S and twice as fast as the X-MP two-way processor, which has a peak 630 megaflop throughput.

NEC has led with advanced semiconductor technology which Iwaya believes to be the ultimate arbiter of success. Both NEC machines, the SX-1 and SX-2, feature highspeed logic circuit LSIs with a density of 1000 gates per chip and delay times of only 250 picoseconds. The SX system utilizes unique, high-density packaging with 36 LSI chips mounted on a ceramic substrate 10 cm square, and minimal wiring length between packages, achieving higher speeds as a result of the shorter distances signals travel. Cooling, which is a critical factor in supercomputer design, is a direct liquid system within the LSI packages, but the cooling unit itself is air cooled and requires no special equipment. High speed main memory elements with a maximum data supply rate of 11 gigabytes per second is supplemented by an extended memory offering, significantly higher performance than obtainable with a magnetic disk unit, improving input/output and greatly reducing the run time for large programs.

The architecture of the world's first giga-level machine is based on multiple parallel processing using four sets of four vector pipelines, enabling a maximum of 16 parallel vector operations to be performed simultaneously. Although for the immediate future NEC expects this powerful system to be most useful for customers handling complex scientific calculations, weather forecasting and architectural design, a broader customer base is sought in energy-related, manufacturing and distribution companies, which are expected to find the increased cost-performance of the new superspeed machines attractive. "While our SX-2 is ten times as fast as the Cray 1S," Iwaya notes smiling, "the price is substantially the same."

Although Hitachi encountered some embarrassing difficulties with the operating systems software for the S-80 series, which was found to infringe IBM software, both Fujitsu and NEC have demonstrated originality and competency in their software systems design. Fujitsu's software has been given high marks from users, offering what it calls a highly advanced compiler (Fortran 77/VP) that can vectorize standard Fortran 77 and Fortran IV programs. Although such compilers are not new, Fujitsu achieved something of a breakthrough in its system's ability to significantly speed up IF statements (in computer programing, an instruction to jump to another specified statement if a specific condition is satisfied) which has proved impossible with conventional methods. And yet this software is an extension of its current system, operating like a mainframe and requiring no special operator training or cooling.

NEC also features Fortran 77/SX, a powerful compiler which automatically generates vector instructions for complicated IF statements, intrinsic functions and list vectors, improving systems performance by substantially increasing the vectorizing ratio for programs.

If, as past experience suggests, supercomputer sales tend to follow good software, since somewhere between 20 and 80 percent of all Fortran code is theoretically vectorizable, these approaches in the main software issues are likely to prove advantageous. But, more important still, such software design achievements lay to rest a long-outdated myth that the Japanese industry is somehow inherently limited in its software capabilities.

And this is just the beginning. The six major Japanese semiconductor manufacturers, which include the three supercomputer makers, have joined with MITI's Electrotechnical Laboratory (ETL) in the Scientific Computer Research Association (SCRA) formed to implement the National Superspeed Computer project. This project targets the development of a new breed of supercomputer with sustained data processing speeds of at least 10 gigaflops. Begun in 1981 with a ¥23 billion (US$100 million) nine-year budget funded by MITI, to be supplemented by additional money from each of the six participating companies, the project got underway in January 1982 and was scheduled for completion by 1989, a year before the completion of the parallel but separate Fifth Generation Computer Project. After the completion of the project, all results — device technology, architecture and software — realized by the SCRA will be available to member firms, as well as for licensing to others.

Specific reasons given for the project significantly confirm and fortify the techno-economic considerations which have informed Fujitsu, Hitachi, and NEC supercomputer strategies:

• Supercomputers are required for upgrading national industrial standards, promoting efficient utilization of scarce natural resources, and improving the environment, and

• Supercomputers are also necessary to ensure the progress of extractive and manufacturing industries, as well as the development of efficient distribution systems.

Mention of military and space applications is notably absent, as are the obvious implications of the project for exports. These considerations, which have hardly been overlooked, are apparently not sufficient reason for its undertaking.

The first three years of the project were devoted to the development of new device technology for high-speed logic and memory. By 1983, significant advances had been made in Josephson junction devices, high-electron mobility transistors (HEMT) and galium arsenide. Evaluation of the new devices is expected to be completed in 1985, and then preparations can be made for their production.

While the Electrotechnical Laboratory (ETL) has been conducting research on all these devices, its development in 1983 of the world's fastest Josephson junction logic gate with a delay time of merely 7 picoseconds per gate, the first in the world to break through the 10 picosecond barrier, has been its main contribution. But in 1984, the ETL announced the perfection of another world premier, the first development of a niobium nitride Josephson junction, which in laboratory tests performed four digit multiplications in less than one nanosecond — four to five times faster than that of a galium arsenide semiconductor device.

Despite this remarkable progress in Josephson junction research, this technology has the least likelihood of selection for use in the supercomputers to be developed in the context of the present national project; since the devices themselves require development of new methods of fabrication, packaging, testing and maintenance and a computer using the device would have to be immersed in liquid helium to attain the minus 273 degrees Celsius, or absolute zero, temperatures at which Josephson junctions function optimally.

Much more likely is the choice of galium arsenide (GaAs) devices developed by Toshiba, NEC, Hitachi, Mitsubishi Electric, Fujitsu, and Oki. All six of the participants in the project already have various GaAs devices in mass production. Even before the national supercomputer project began, Japanese makers had focused their efforts on GaAs devices with low power consumption and a large number of gates, compared to most work in the United States and Europe done on depletion mode transistors. The choices reflect the techno-political priorities of Western countries where the emphasis is on military and space applications, and the techno-economic priorities of Japanese firms that targeted the computing market from the outset. Now, this emphasis is reaping dividends in the form of supercomputer applications.

The most dramatic development has been the high electron mobility transistor (HEMT), a superlattice aluminum-doped GaAs device which springs originally from an idea proposed in 1969 by Leo Esaki and Raphael Tsu at IBM's Yorktown Heights research laboratory. But it was Fujitsu that pushed forward to perfect super-high-speed HEMT technology in 1980. Using Fujitsu's HEMT devices, a supercomputer can be built for slow operations at room temperatures to facilitate checking and debugging and then be cooled to the temperature of liquid nitrogen (77K) for maximum performance. But despite the very important advantages of HEMT, Fujitsu computer designers themselves believe that less complicated GaAs devices have the best prospects for selection to be used in the National Superspeed Computer.

From 1985 to 1989 the focus will be on development of the actual supercomputer, stepping up work on distributed parallel processing architecture and related software. At this juncture, no constraint on the hardware seems likely to limit the success of the superspeed computer project, as the hardware engineers have already proven their ability to deliver computer speed as it is needed. And, while software probems of parallel processing pose more difficulties, there seems little likelihood that they will deter the SCRA member from attaining their goals.

Ultimately, the industrial impact of the project will be largely determined by the integration of semiconductor and supercomputer production. Supercomputer circuitry innovation is the key element in the undertaking, and its parameters are economically determined. While U.S. supercomputer makers do not have integrated chip-making capability, specialized U.S. semiconductor firms are not particularly interested in working with supercomputer manufacturers because the orders are too small, especially at a time when computer aided design costs are soaring. Integrated Japanese electronic companies making both semiconductors and supercomputers are not faced with this problem.

What will happen after 1990 is still anybody's guess, however. Chances are good that by 1990 the three Japanese supercomputer makers will have as high as 50 percent of the world market, which is likely to be running at US$2 billion a year and climbing. After 1990, much depends on the level of commitment of resources to research and development during the remainder of the 1980s. That level is clearly very high for Fujitsu, Hitachi and NEC, as well as for the other three members of SCRA which at present are focusing their resources on the development of critical device technology. But there are other actors involved in Japan, not the least of which are MITI's Electrotechnical Laboratory, Nippon Telegraph and Telephone Electrocommunications Laboratory and the National Research and Development Corporation. Nor should the announcement by the Institute of Physical and Chemical Research of the world's first supercomputer system designed exclusively for processing differential and integral equations at high speed be left from the picture. This machine, developed in cooperation with Mitsui Shipbuilding Engineering Co., Fujitsu, and Cambridge University of Britain at a cost of ¥400 million will be used in making calculations for high energy physics, nuclear fusion and the design of ULSIs (ultra large-scale integrated circuits). Taken together, this amounts to a mighty thrust that will doubtless render Japanese supercomputers highly competitive, if not leaders, in the 1990s.

But the U.S. industry most certainly will not be standing still. Quite to the contrary. If CDC and its subsidiary ETA are doubtful elements, Cray Research will remain a major power in the marketplace for the foreseeable future. Limited mainly by its lack of semiconductor production capability, the likelihood that this structural weakness will be overcome by a strategic acquisition must be considered great. Another generally expected development is IBM's eventual entry into the market when it reaches volumes approaching US$2 billion annually, or even before. And, quite clearly, the U.S. government and cooperative research projects will mobilize substantial resources to protect vital national security and the industry's competitive interests.

National projects in Great Britain, France and Germany will, in all probability, bring new actors onto the scene. Special thrusts can also be expected from other countries, including possibly the People's Republic of China and most certainly the Soviet Union and some of the East European countries. Although a few firms are likely to lead the field, most probably American and Japanese, the suprise-free scenario would seem to be pluralistic rather than a market dominated by any one firm or group of firms from a single country. It remains to be seen whether any future inventions will emerge to bring about a significant departure from that projection. In the end the advantage will rest with the firms that are best organized to optimize the critical synergy between semiconductor and supercomputer technology.

PART VI.

Computer Services

CHAPTER 19

Hard Facts About Japanese Software

EVER SINCE Japanese computer makers demonstrated that they could match IBM's best offerings early in the 1970s it has been generally observed that, while the Japanese excel at hardware design and manufacture, they are less accomplished in the production of software: and much evidence seemed to sustain this observation.

The development of the software industry, as a distinct sector with viable independent producers, until recently has been less than remarkable. Software houses in Japan are still small in size with the top 150 having, on average, fewer than 100 employees and an annual turnover of little more than US$3.8 million, while the production of packaged software has been almost non-existent. Moreover, the Fujitsu decision, in 1973, to adopt a plug-compatible machine (PCM) strategy producing computers for world markets that use IBM software, seemed to endorse the common view that the Japanese industry was lagging in this critical area of data processing by a gap of at least 10 years. The Japscam affair in the United States in 1982 — involving Hitachi men spying on the U.S. computer industry — tended to confirm this construction.

But as is often the case in the hazardous game of gapology, this perception was always at best a gross oversimplification obscuring the actual state of affairs. At worst, it was calculated to console those for whom the advance of the Japanese computer industry appears as more of a threat than an opportunity.

In fact, past assessments of Japanese software technology have fallen foul of three common fallacies. The focus of attention has been on the form of industrial organization and the packaging of the product, rather than on the substance and application of the technology. Since the Western form — taken as standard for the industry — was notably absent, it was assumed that therefore Japanese technology was retarded and — perhaps the most serious error — it has been further assumed that

Reprinted from "Hard Facts About Japanese Software," *Far Eastern Economic Review*, 3 December 1982.

what is not readily susceptible to mathematical measure must not exist. Now, suddenly, in a short span of two to three years — as in so many other industries — the 10-year software gap seems to have all but vanished.

In retrospect, it has become quite clear that the remarkable Japanese achievements in computer applications, as well as the rapid advance of Japanese computer manufacturers, are as much the result of mastering software engineering as hardware production.

Japanese superiority in industry after industry — steel, ship-building, motor industry production, consumer electronics, fine optics — has been due largely to the world's most advanced computer systems designs, mainly by Japanese software engineers. Nor, of course, is this a recent phenomenon. The *shinkansen* bullet-train that has been the epitome of railway systems since 1964 would have been impossible but for ingenuity in software engineering. Likewise, since 1964 Japanese banking has led the world in the development of on-line computer systems which now link all major commercial banks in a single national computer network, a formidable achievement in large-scale, computer software engineering that may not be as dramatic as a moon-landing but is at least as significant for the prowess and progress of the economy.

In other aspects of industry, the lag theory is no more helpful in explaining the Japanese reality. If Japanese industry today employs approximately 70 percent of the world's robot population, it is most certainly not due to superiority in the production of robot hardware: some of the world's most advanced robots are still designed and manufactured in the United States. The unexcelled ability to use robots entails, among other things, a keen capacity for software systems engineering. It is also patently obvious, on cursory observation, that the remarkable achievements of Japanese industry in quality control are essentially the result of prowess in software applications.

To say that Japanese industry lags in software engineering but leads the world in production efficiency, robotics and quality control is a bald contradiction where bold perspicacity is most important.

The reality is that, far from suffering any inherent cultural or social handicap in software engineering, as is often suggested by those who hold to the gap theory, Japanese industry has a proven record of innovation in the design and application of complex software systems. What *is* different about the software industry in Japan is its organizational structure, which tends to obscure the visibility of software production and renders it more difficult to define and to measure.

Partly for historical reasons, but chiefly due to the nature of Japanese industrial organization and management systems, software

Fig. 1   Typical Order Dispatching Procedure for Software Products

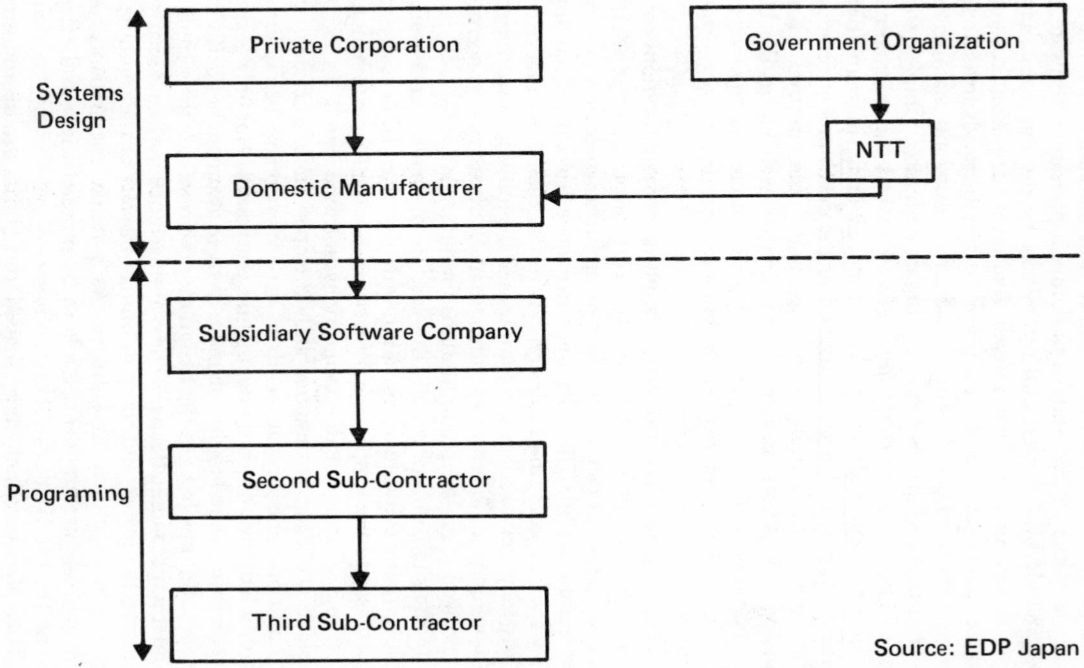

Government Organization

NTT

Private Corporation

Domestic Manufacturer

Subsidiary Software Company

Second Sub-Contractor

Third Sub-Contractor

Systems Design

Programing

Source: EDP Japan

engineering differs from its U.S. and European counter-parts in both its locus and content. Although in Japan basic software for computer operating systems is usually designed and developed by mainframe makers, as in the United States and Europe, applications software is produced mainly by the users and secondarily by the computer manufacturers. Rather than rely on software packages of outside consultants, Japanese users prefer to tailor software to specific needs involving from the outset those who will be responsible for its use and maintenance.

Historically, Japanese industry began designing complex software systems before Japanese computer-hardware makers had reached maturity. The first large-scale computers employed in Japan were IBM and Univac machines, which were engaged in the development of advanced steelmaking, shipbuilding and on-line computerized banking technology. Applications software was neither available in the United States nor applicable to the needs of Japanese users, making it necessary to develop in-house software engineering capabilities and organizational structures.

The preference for in-house, custom-tailored applications software was not purely circumstantial, however. In general, Japanese are not inclined to pursue universal solutions to problems: specific situations require specific action. Thus, Japanese managers reject packaged applications software, preferring to design systems for their particular needs. In this way they seek to adapt the computer to the job, the organization and the humans who will use it, rather than fit these vital concerns to a preconceived ready-made applications formula.

In the short run, of course, this costs more. It means maintaining a sizable in-house software organization with high fixed labor costs entailed by Japanese lifetime employment practices. But the long-run benefits are compelling. Custom-made applications software tends to be more reliable. Since it can be designed and developed under strict quality control conditions, the software tends to have fewer bugs, thus reducing or obviating lengthy and costly debugging procedures. Not only is applications software tailored to the job more likely to assure higher performance efficiency, but involving in the design those who will use the system tends to render its use more effective. The man-machine interface is so arranged to reinsure the reality and sense of human mastery of the computer and other machines it may manipulate. As a result, maintenance of the system by in-house engineers tends to be more efficient and effective, reducing the life-cycle cost of the software.

Three factors which have shaped the U.S. software industry are notably absent in Japan. The largest customer for software engineering services in the United States is the government, especially the defense and aerospace establishments which, for a variety of reasons, fill much of their software needs by relying on private contractors. Private users, operating under short-term performance constraints, have tended to opt for software packages with the best immediate cost-benefit ratios and, since IBM has a predominant position in the hardware market accounting for 70-80 percent of the total U.S. computer park, IBM software engineers often find they can do better by resigning to establish their independent software operations to design packages or provide consulting services for users of IBM equipment.

Since Japanese management tends to give more importance to long-run performance considerations and neither government expenditures for external software services nor IBM's market position provide the kind of inducement they do in the United States, the software industry has developed along quite different structural lines in Japan.

Fig. 2   Distribution of Software Houses by Annual Revenue

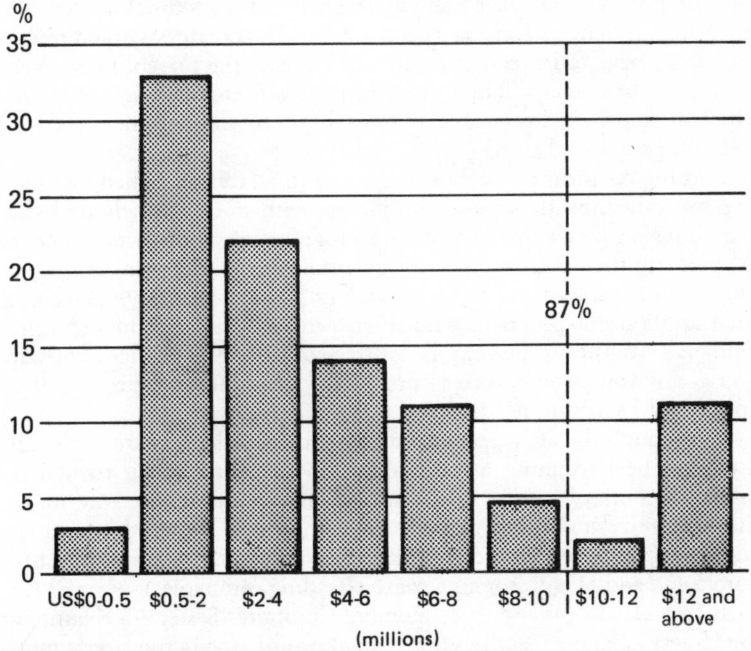

Source: SIA Survey, March 1979.

COMPUTER SERVICES

By far the largest number of software engineers and programers are employed by computer users and manufacturers and are therefore difficult to identify statistically. According to the latest Ministry of International Trade and Industry (MITI) survey of data processing in Japan, the 5,272 users covered in its sample employed 135,009 people in software activities or an average of 25-26 per company. But this is only a fraction of the total employment in software by computer users. A recent Fujitsu customer survey revealed that the average number of data processing personnel required for its medium- and large-scale M series computers in 78 companies was 134 systems analysts and programmers. If this is taken as an average only for the 3,324 large-scale computers in operation in Japan as of September 1981, users would require more than 445,000 data-processing personnel; and this would not include those software engineers and programers required by the remaining 10,319 medium-sized and 29,380 minicomputers in operation at the time of the survey.

No one seems to know just how many software engineers and programers are employed by computer users in Japan. What is clear, however, is that the shortage of such personnel is acute and has reached crisis proportions. As Sakae Shimizu, manager of the technology development department at Toshiba's central research laboratories noted recently: "If we have to increase software engineers at the present rate, all the people on the planet will be turned into software engineers in a few years. This of course is impossible, so we have to develop new software technology to avoid this."

In the meantime, users have resorted to establishing their own captive software subsidiaries and contracting with so-called dedicated software houses which supply them with programers and other personnel to supplement their permanent software employees. This organizational arrangement enables users to create and implement nearly 90 percent of their applications programs with their own software personnel and programers. Another 7 percent is contracted with independent software houses and computer makers to provide the remaining 6 percent of their applications software needs.

Although the aggregate expenditure for outside software services by users has been running at only about 15 percent of their total data-processing outlays, captive subsidiaries of large computer users are among the 40 largest data-processing firms in Japan. Mitsui Knowledge Industry, Sumisho Computer Service and MSK Systems.— subsidiaries of Mitsui, Sumitomo and Mitsubishi *sogo shosha* (trading companies) — rank 14th, 17th and 21st in the industry. Nomura Computer System, a creature of the largest Japanese securities house, ranks third among the largest information service companies and Daiwa Computer Service, a subsidiary of

278

Daiwa Securities, is 23rd. Major banks, steelmakers and private railway groups have also established separate computer service subsidiaries that are now counted among the top 40 in the industry.

But this practice is not confined to large companies alone. Medium-sized enterprises such as machine-tool makers, now riding the wave of factory automation, are compelled to establish special software subsidiaries and use outside subcontractors to meet their expanding software requirements. One such toolmaker, Toyota Machinery Works, now employs 50-60 software engineers but needs 100 more. To obtain them, the company is contemplating establishment of its own software house.

User preference for custom software has shaped the industry in still another important way. In order to sell the hardware, fiercely competing mainframers offer not only systems and utility software but also provide assistance in developing applications software — all free of charge. In order to provide these services, there has been a rapid development of software activities within the major computer makers, each of which employs between 3,000 and 6,000 software engineers and programers in-house. In addition, to meet the burgeoning demand, they have established a bevy of software houses and employ large numbers of subcontractors throughout Japan.

NEC currently employs 3,000 software specialists in the company and another 3,000 in 11 dedicated software subsidiaries working solely for the parent company. Three of these subsidiaries — NEC Software, NEC Information Services and Nippon System Development — rank among the 40 top data-processing companies in the country. The others, distributed geographically in various domestic sales regions, provide software support to NEC computer users.

Supplementing this permanent organization are 4,000 more software engineers and programers supplied by 250-odd subcontractors. Most of these 4,000 outsiders work inside NEC, supplementing the permanent software staff. For the most part, these contract workers are programers working under the supervision of permanent personnel, performing more labor-intensive tasks.

NEC, now No. 3 computer company in Japan after Fujitsu and IBM Japan, has not followed Fujitsu and Hitachi in their PCM strategy. Instead, NEC relies wholly on software of its own design and gains market share precisely because it is impervious to IBM's strategies of technological change aimed at the plug-compatible manufacturers. NEC firmware-based operating system used in the current ACOS computer models is equal to IBM's latest technology and the large variety of general purpose software systems which the company offers is finding a growing market in Japan and abroad. Numerical programs — such as process control, econometric modeling, sicentific calculations and operations research — are quite advanced, while text processing and general business programs tend to be less developed because of difficulties of the Japanese language and the lack of standardization of procedures among Japanese companies.

At Fujitsu, of 36,000 employees fully 6,000 are engaged in software design and development: 2,000 of these, working at the Numazu computer center, are employed in operating systems development and implementation: another 1,000 at the company's systems laboratory in Kamata are developing applications software while 3,000 more systems engineers provide customer service support throughout Japan. In addition, to supplement and support in-house software activities, Fujitsu has been creating an average of five new software subsidiaries annually since 1980. The largest, Fujitsu Facom Information Processing Co. (FIP), ranks among the top 10 information processing firms with a turnover for the fiscal year ending March 1982 of ¥9.5 billion (US$34.93 million). Not all of this represents work performed for Fujitsu, however; nor is FIP restricted to software activities: it maintains its own computer center providing computation services to clients and since mid-1982 has incorporated Fujitsu's time-sharing division. Approximately 70 percent of the time-sharing services are rendered to firms outside the Fujitsu group.

Although Hitachi has been overtaken in the domestic computer market by NEC, it was not from lack of commitment to software development. Hitachi has 15 subsidiaries providing software services, in addition to its large in-house data-processing contingent. Two of the top 10 data-processing firms in Japan are numbered among these subsidiaries: Nippon Business Consultants with 1980 revenues from diversified information services exceeding ¥15 billion and Hitachi Software Engineering (HSE) with a turnover of ¥8 billion, the largest dedicated software house in the country.

Founded in 1969, HSE began primarily as a supplier to Hitachi of programers who did not work merely as temporary manpower support however, but were assigned specific tasks within the Hitachi organization to augment technical strength of the parent company while minimizing business risks. Now the 1900-strong organization not only has as one of its major activities the mainframer's operating-system development and maintenance, but earns 40 percent of its revenues from services to computer users. According to Tomoo Matsubara, who supervises product planning at HSE, the company will soon extend its activities to the United States and Europe where it already works closely with software firms and universities.

All three mainframers have embarked on a concerted program to im-

Fig. 3 Annual Revenue of Hitachi Software Engineering

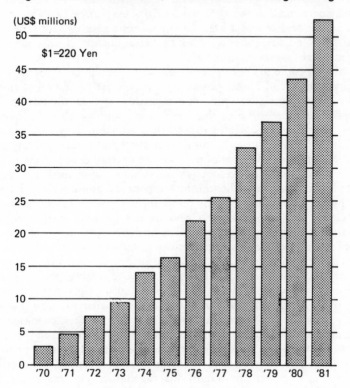

Fig. 4 Number of Employees at Hitachi Software Engineering

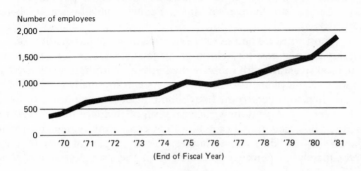

prove techniques for software production. At NEC, Dr. Yukio Mizuno, associate senior vice-president, heads a company-wide effort of quality management in all software-related activities. As he puts it: "Our objectives are the realization of high quality software and high software productivity. We believe that if we ask for better quality, at the same time we can obtain better productivity."

Since people are the heart of software work, NEC's approach to quality control and higher productivity entails the perfection of teamwork. "Team members who work together are, in their collective wisdom, far more effective than individuals working alone," Mizuno notes, "so we are encouraging the formation of voluntary software quality control [SWQC] circles to improve personal and team capability and build quality into all stages of the software development process." The NEC SWQC information center publishes a variety of manuals and handbooks for guidance in quality control methods and group activity. Education programs conducted by the center include group leaders' teach-ins and classes for junior managers who are not normally invited to participate in SWQC group activities. The objective is to involve the entire organization in quality and productivity improvement.

HSE has an independent quality assurance division which aims to achieve the same objectives. To assure software quality, the division's engineers conduct extensive statistical testing during the software debugging stage. Using empirical data and statistical analysis, they provide feedback for the implementation group to enrich their test conditions to elminate error. HSE also encourages employee participation in this process through SWQC circles as an important element in increasing productivity and improving quality. Realizing that productivity is generally preceded by improvement in individual capabilities, HSE makes a major educational investment in its employees — an investment that pays off in employee loyalty. The annual 3.2 percent personnel turnover at HSE, significantly below the Japanese industrial average of 5.8 percent, is itself rated highly as an element in the company's overall effort to boost productivity.

For most software houses such extensive training and quality-control programs are not possible. However, generally software houses form a hierarchical structure with software divisions of mainframers and users at the top. Users implement their software systems internally with the help of programers supplied by the body shop type of software house — those which make it their business to supply supplementary data-processing personnel. But users account for only 40 percent of the aggregate revenue of Japan's 2,000-odd software houses: the larger part of their business (approximately 60 percent) is derived from the six mainframers. Most software houses work primarily for one of the major computer makers: of their total number, 90 percent are small firms with an average of 30 programers. The larger software houses, with revenues exceeding US$10 million, are — except for a few independents — subsidiaries of the mainframers or large users.

Entry by new firms is relatively easy as capital requirements are negligible. "All you need to establish a software house these days," quips Mizuno, "is a desk, a chair and four or five programers. It's a process of biological reproduction as explosive as a colony of amoeba. "Given what seems to be an insatiable demand — with simultaneous booms in office automation, personal computers, intelligent terminals and factory automation, all fueled by the development of new Japanese-language processing systems — Mizuno expects there will be a steady stream of new entrants into the software service business throughout the 1980s. Personal computer, small business computer and word processor users will require and use packaged software more readily than larger computer users, opening a whole new field for innovative software engineers and programers.

Judging from the recent growth in software house revenues, the impact is already being felt. In 1981, turnover of software houses from software development activities was more than five times that of 1975, with annual growth rates mounting as high as 60 percent during this same period. However, employment in the industry scarcely doubled to 105,891, which means that productivity rose a remarkable 2.5 times due, at least in part, to quality control and productivity-improvement programs of the mainframers.

The outlook now is for a 10-fold increase in Japanese software production in the 1980s, rising to at least ¥2 trillion at the end of the decade: and if productivity continues to improve at present rates, output in 1990 could be considerably higher still.

Much depends on how the criticial manpower problem now confronting the entire industry is resolved. The shortage of software engineers has reached crisis proportions and is steadily growing worse. According to the Industrial Structures Council, a MITI advisory organ, there will be a shortage of as many as 179,000 computer specialists by 1985 and the Japan Software Industry Association (JSIA) is predicting an even greater shortfall.

Both MITI and JSIA, through its software technology committee, have programs to improve the training and professional competence of software systems engineers. The Institute of Information Technology, established by MITI in 1971, has focused its resources on the training of high-level engineers and senior programers with a solid background in on-the-job experience. These trainees return to their original companies — users, mainframers or software houses — where they participate in the development of junior engineers and programers. The larger Japanese companies seldom employ experienced people. Since computer science departments, of course, are rare — and not very practical in high schools and colleges — new recruits generally receive two months' training in programing before assignment to relevant departments for four months of on-the-job training. Engineers and programers are required to attend periodic refresher courses to keep abreast of new technologies and broaden their knowledge and they may even be given the opportunity to leave their jobs to study abroad for a year or two.

But development of human skills in this rapidly changing field of technology is a lengthy and costly process. As software development costs rise, one obvious solution which software managers are forced to consider is maximizing the use of existing software, avoiding the development of redundant new systems. An increasing share of the total cost of users' data-processing divisions is for software maintenance and less for new software systems development. More computer users are now depending on independent software houses for the development of high-cost, custom-applications software and those packages which enable them to save electronic data-processing manpower are gaining popularity. Computer users are coming to realize that the cost as well as time saving in purchasing a software product rather than developing a custom system can range from 50 to 90 percent. In the words of NEC's software guru, Mizuno, ''The 1982s will see the dawn of the golden age of packaged software in Japan.''

Another major growth opportunity is in the sale of hardware/software systems, commonly known in the business as turnkey system or system package supply. As it becomes easier and cheaper to buy the components of a computer system to assemble the hardware and package it with software, more software houses are entering this field. Growth, especially in the sales of systems using personal computers, is expected to exceed 20 percent annually for the remainder of the 1980s.

Quite apart from measures to improve the efficiency of the software industry at home, mainframers and software houses have embarked on some novel forms of international cooperation which will be harbingers of things to come.

• Last year, a JSIA delegation to mainland China, headed by association chairman Tadashi Hattori, submitted a plan to Vice-Premier Fang Yi for the training of Chinese students by Japanese software engineers. Graduates of these special training programs would be engaged to make computer programs for Japanese computer makers and software houses. Accordingly, both Fujitsu and NEC plan to open software training centers in several Chinese cities and Fujitsu has begun negotiations on a joint venture with the Chinese Science Commission of Tianjin for the development of a computer hardware and software system for the Chinese language. Anticipating this development, Sord Computer Systems, the feisty personal computer maker, has already trained five Chinese software engineers and engaged the services of a Chinese software specialist to design software at the company's Tokyo headquarters.

• Computer Applications Corp. (CAC), a leading independent software house with a staff of 240 computer specialists, shifted the coding of software programs to CAC Taiwan, a joint venture with Systex Corp. of Taiwan formed in 1980. Having found the Chinese quick to learn skills required for software production, Shigeru Ohkubo, CAC president and co-founder, determined that wage differentials between Tokyo and Taipei would make possible as much as a 30-40 percent reduction of costs. Since coding requires little or no contact with the customer and therefore need not be done in Japan, the entire operation could be shifted to Taiwan: and with the coding group located just three hours away, no changes were necessary in the method of operation at CAC.

• Serendipitiously Fujitsu has discovered a new source of software production support in its South Korean subsidiary. Facom Korea was originally established simply to supply sales and technical support for Fujitsu data-processing equipment. But one of the conditions imposed by the South Korean government on its authorization for a sales subsidiary was that Fujitsu should establish a center to train South Koreans in software production. To turn this cost center activity into a profit center, Facom Korea's president Masayoshi Kuribayashi set out to develop a software production capability for the parent company. Presently, of 150 people employed at Facom Korea, almost a third are engaged exclusively in producing software contracted by Fujitsu from Tokyo. South Korean programers do all the coding and debugging just as programers dispatched from software houses in Japan do. The group began in 1977 by producing systems software for a batch environment and developed a fast one-pass Fortran compiler now being used at a number of Japanese universities. Since then work has progressed to more advanced software and, if the Koreans at Facom have their way, will lead to development of the nucleus of operations sytems software.

Possibilities of shifting programing software for applications in Japan are limited to those countries where work can be done in Japanese, for instructions come from the computer maker or user in that form. It is also advantageous to have the programing done in the geographical proximity of Tokyo. But there are also possibilities in other countries, such as Brazil, where there is a large enough Japanese community.

More interesting is the potential for joint ventures abroad to develop software applications for Japanese computers exported to overseas markets. Just as U.S. software houses seeking to enter the Japanese market would do well to join forces with a Japanese partner, Japanese are likely to find this to be the most effective route to overseas expansion. This is the view of Haruyasu Nakayama, president of DPC America, one of the first Japanese software houses to open a branch in California to offer software services to Japanese computer-users operating in the United States.

But, as Japanese software output develops, mainframers and software companies are moving to establish overseas marketing organizations. There are already Japanese software operations in Singapore, Malaysia, Britain and the United States. Fujitsu has decided to take the route of direct marketing of basic software to be used for technological research and development on IBM machines. The Fortran 77 software, a high-level programing language developed by the leading Japanese computer manufacturer, is highly regarded in both the United States and Europe where it has already been used. Fujitsu America Inc., of Santa Clara (California) is distributing the software in the U.S. market. This introduction of Fujitsu's programing language to the United States will serve as the occasion for at least a few competitors to undertake an agonizing reappraisal of the relative strengths and weaknesses of Japanese software technology.

A Case in Point

Few new Japanese companies have created as much sensation, and few have enjoyed such instant success, as Cosmo 80. Within months after Yutaka Usui, 47, resigned his position as a top executive of Ishikawajima-Harima's (IHI) computer business development center to establish the new data-processing venture, it had caught the imagination of a bearish investment community sending successive speculative waves through the Tokyo Stock Exchange. Even though Cosmo 80 is still not listed on any Japanese exchange and its shares are not traded over the counter, investors looking for new opportunities rushed to buy shares of any listed company that had any ties at all with the new venture.

Founded in May 1981 by Usui and two other dissident IHI managers, Cosmo 80 sent shock waves through the business community as a steady stream of IHI computer experts left in groups of 10-20 at a time to join the new company. Within five months, 84 of the 230 computer specialists employed by the large diversified shipbuilder and heavy machinery maker had made the move, giving Cosmo 80 an inside track in the high-growth software market. In the following six months the company's sales were already in excess of ¥1 billion (US$3.85 million) with profits of ¥30 million.

There is much about this sudden exodus of software engineers and computer programers from one of the industrial giants which led Japan's high growth in the 1960s that is symptomatic of the radical transformation of the country's industrial structure in the 1980s. Shipbuilding, once the pride of Japan's heavy industries, was one of the sectors hardest hit by the oil crises of the past decade. Capacity was curtailed throughout the industry by 35-40 percent and, to avoid massive lay-offs of personnel, the larger shipbuilders diversified rapidly into new fields of activity.

As part of this process, the IHI computer department was transformed into a profit center. A computer business development center was established to contract computer services to outside clients and in 1977 Usui became its manager. Having joined Kure Shipbuilding Co. fresh out of high school in 1954, Usui was assigned to the company's computer department in 1964, four years before its merger with IHI in March 1968. From then on he advanced rapidly as computer applications became central to ship design, productivity and quality control. By 1974 he had become chief of the computer room of IHI and a year later he was promoted manager of the computer-planning department.

Under his management, by 1980 the computer business development center had boosted software sales to ¥1.6 billion, with profits of ¥200 million, which placed IHI among the top 50 information service companies in the country. Among its clients were leading manufacturers such as Sony, JVC and Bridgestone as well as quasi-public enterprises such as Japan Travel Bureau.

Then, for reasons which Usui says he still does not fully understand, IHI top management decided to abandon the computer services business in February 1981, making the announcement out of the blue at a labor-management negotiations session. The company's data-processing organization was thrown into disarray and its customers were utterly confused.

At underground meetings, the data-processing staff reassessed the situation, equipped with simulation models on the future of software and other data-services markets. The decision was made to continue service to clients whose confidence they had won on their own. Once assured of support by their major clients, Usui led the exodus with his resignation in April 1981.

With a third of the new company's ¥200 million capital subscribed by the 84 IHI dissidents and the remainder taken by client companies at the time of establishment, the success of Cosmo was all but guaranteed from the outset. With a substantial cash flow almost instantaneously generated, as quickly as the organization could be put in place the new company could envisage sales of ¥20 billion and a work force of 2,000 in just five years.

In the first year, the staff was increased to 120 professionals and made important acquisitions in anticipation of a strategic diversification into hardware. "We cannot live on software alone," Usui maintains. "We must take advantage of rapid changes in telecommunications technology. Nippon Telegraph and Telephone's new digital information network system will require a vast array of new terminals for home and office. This will be a very good market and we expect to have a piece of it."

In May 1982 word was out that Cosmo 80 had acquired a third of the equity of Olympic Fishing Tackle Co., wresting it free from the influence of Mitsubishi to enter into an arrangement for the production of computer terminals.

In fact, it was Mitsubishi which introduced Olympic to Cosmo to open the way for diversification of its long-time supplier's precision machining activities and negotiations were finalized in August 1982. Olympic, which had been over-extended and had paid no dividend since 1978, suddenly attracted wave after wave of buyers pushing up its stock which had been languishing below ¥100 a share for several years.

Two months later, in October, Cosmo 80 made its second acquisition, this time Hokushin Keiso, the lackluster instrumentation subsidiary of Hokushin Electric. Then, when rumors swept through the market that Cosmo 80 had concluded ties with Towa Sankiden Co., a leading electronic cash register manufacturer, for the distribution of hardware, Towa's shares soared 90 percent in three weeks from ¥395 to ¥739.

Usui's logic is clear and compelling. "Cosmo 80 is a software house with a difference," he insists. "Most are simply body shops, supplying computer-makers with supplementary programers but we're user-oriented. Sooner or later there will be a shakeout and only those firms with high technology will survive. The new kind of diversified computer service company will have three legs — software, hardware and communications — and we fully intend to be among the survivors."

CHAPTER 20

Database Services: Another New Growth Industry

A COMBINATION OF FORCES has conspired to transform database services from a slow starter to a high flyer with what industry leaders believe to have the greatest growth potential of all computer services in Japan. Among prime stimulants are advances in telecommunications coupled with rapid decreases in the cost of computer power and storage. The latent efficiency instincts of consumers of information, who want ever more data at less retrieval cost, have been the high octane fuel for launching this new information-age industry into orbit. Government, business, science and education are now joined by individual computer users in what appears to be an almost insatiable appetite for more, better information with greater economies of use and storage.

After a timid beginning in 1969 when New Japan Securities inaugurated its investment information service, database services gradually gained momentum as on-line facilities were introduced in the 1970s and Kokusai Denshin Denwa Co. (KDD), the public overseas communications monopoly, introduced a new data communications service called International Computer Access Service (ICAS) in September 1980. From 4.6 percent of the total turnover of computer services in 1973, database services rose to 7.5 percent in 1981. In the past two years, revenues of database service companies have doubled, growing at annual rates ranging from 36 to 40 percent. A new industry is off the information-age launch pad and into orbit, with the universe and human organization as the only limits to growth.

But the industry is still in its shakedown stage. The number of purely Japanese databases is very small, consisting mainly of JICST On-Line Information System (JOIS), a scientific literature and research database operated by the Japan Information Center of Science and Technology; NEEDS, an economic analysis database operated by Nihon Keizai Shimbun Ltd.; PATOLIS, a patent information database service of the Japan

Reprinted from "Database Services: From Slow Starter to High Flyer," *Far Eastern Economic Review*, 3 December 1982.

Patent Information Center; and QUICK, a securities and foreign exchange database service offered by the Quotation Information Center.

These pioneer services had all been established by 1974. It was in 1974, too, that a financial reporting service covering all companies listed on the Tokyo Stock Exchange was started by the Industrial Bank of Japan; and a market price information service covering all stock and bond issues on that exchange was begun by Jiji Press Ltd., which distributes its services through the Dentsu-GE Mark III network of Dentsu International Information Service. During the same year the Mitsubishi Research Institute and Micro Comshare also inaugurated their TSS, marking 1974 as the beginning of the industry's growth period.

Since the mid-1970s there has been a steady increase in the number of on-line database services produced in Japan. Database building through international exchange became an important feature of the industry. The Japan External Trade Organization initiated the exchange of trade statistics on magnetic tapes with the United States, France and Canada (other European countries having declined the invitation to participate) and the Japan Patent Information Center exchanges its data files with the International Patent Document Center in Geneva, as well as with the United States.

The Chemical Information Association has plugged into the global Chemical Abstracts Service database, providing inputs of literature produced in Japan. In cooperation with the American Chemistry Association, 27 Japanese academic societies and associations — along with public and private organizations in the field of chemistry — established this association for the specific purpose of building and distributing an international database containing information on the structures of millions of chemical substances. Similarly, the National Museum of Ethnology serves as the depository of the Human Relations Area Files, a database produced by Yale University for the purposes of developing a computer-readable retrieval system providing inputs from Japan.

Furthermore, some Japanese universities and research institutes have become international database centers. Tokyo University serves as a subcenter for the Protein Structure Database of the Brookhaven National Institute and Kanazawa Technical College as a subcenter for the Astronomical Information Database of the Strasbourg Astronomical Information Center. Similar arrangements exist for participation in and distribution of worldwide databases on active volcanoes and other bodies of scientific information.

Generally, though not always, the Japanese producer of a database also acts as its distributor in Japan. However, in the last half of the 1970s there emerged a new breed of service companies specializing in the distribution of database services, both domestic and foreign.

Three types of distributors emerged, in both the private and public sectors: producers which double as distributors, vendors which offer on-line services featuring databases produced elsewhere and those which act as database import or export agents.

Broadly speaking, on-line facilities of both producers and distributors are divided between center facilities and communication circuits. According to Japan's Public Electric Telecommunications Law, private vendors are not allowed to own communications equipment or facilities used for communications business purposes; database vendors must therefore rent circuits from Nippon Telegraph and Telephone (NTT) or, in the case of international circuits, from KDD. Producers, distributors and users, along with the Ministry of International Trade and Industry and other government agencies, have been clamoring for the liberalization of data communications circuits and an advisory body to the Ministry of Posts and Telecommunications has called for such action, but as yet the necessary measures have not been taken.

Among producers which also act as distributors of other databases, Quotation Information Center offers Reuters' Videomaster, JICST serves as a distributor of the Chemical Abstract Condensates and the MEDLARS database of the National Library of Medicine, and IBM Japan (which has been offering its own CALL 370 service since 1974) serves as a distributor of NEEDS, produced by Nihon Keizai Shimbun which, in turn, serves as the distributor for DRI's American and European economic statistics and industrial information.

The largest firm specializing in database distribution is the Dentsu International Information Service, with its GE-Mark III Service enabling customers, for a fee, to access many databases produced by other companies. In addition, Dentsu also provides a number of other databases produced in Japan, the United States and Europe. The Lockheed DIALOG Information Service is now distributed on-line or by batch service by Maruzen and Kinokuniya, major Japanese booksellers and publishers.

ORBIT, which like the Lockheed service offers a wide selection of databases on science and technology, economics and business as well as social studies and the humanities, is distributed by System Development Corp. of Japan, a joint venture of SDC of the United States. SDC-J also offers SEARCH-J, a global patent, pharmaceutical and chemical database that had virtually immediate success in the Japanese market.

In 1980, SEARCH-J was among the first services in Japan which

Table 1. Database Services Market in Japan

	1973	1974	1975	1976	1977	1978	1979	1980	1981
Total data-processing services	167,162	245,263	275,090	306,966	412,581	460,241	596,613	669,844	805,692
No. of establishments	1,105	1,322	1,276	1,276	1,640	1,672	1,761	1,731	1,801
Database services	7,620	13,046	14,376	11,972	23,930	27,154	31,620	44,210	60,428
No. of establishments	220	265	256	221	278	302	296	165	—
Share of database services (3:1) %	4.6	5.3	5.2	3.9	5.8	5.9	5.3	6.6	7.5

Source: Ministry of International Trade and Industry.

Table 2. Industrial Database Utilization (1982)

Type of Industry / Current Utilization	Mode of Utilization	Information Utilized	Number of Respondents	Documentary Data			Factual Data		Still Picture Data	Motion Picture Data	Audio Data	Cumulative Responses
				Scientific and Technical Fields	Patents	Industrial, Economic, Social and other Fields	Scientific and Technical Fields	Industrial, Economic, Social and other Fields				
	Batch*	Number of Companies	82	24	22	14	14	37	3	0	0	114
		%	100	293	26.8	17.1	17.1	45.1	3.7	0.0	0.0	139.0
	On-line*	Number of Companies	116	50	33	15	21	48	11	0	1	179
		%	100	43.1	28.4	12.9	18.1	41.4	9.5	0.0	0.9	154.3
	Actual Number of Users	Number of Companies	171	64	53	26	34	73	13	0	1	264
		%	100	37.4	310	15.2	19.9	42.7	7.6	0.0	0.6	154.4

* Figures for batch and on-line untilization include a certain number of users who use both.

Source: JIPDEC

allowed Japanese users direct access to information retrieval from their own terminals on an on-line basis at any time. Previously, a major bottleneck had been the huge time and cost inherent in database access via telex, a factor which had contributed to the delayed take-off of the database service business.

Significantly, SEARCH-J, DIALOG and ORBIT were particularly instrumental in raising the number of database users in Japan. There are currently a total of 2,200 subscribers to the DIALOG database service — 1,200 of whom have access to that service via the Maruzen system — and the remaining 1,000 are serviced by Kinokuniya. Japanese users of ORBIT number 900 in all.

If almost all leading databases are operated either by agents of U.S. database service companies or by joint ventures in which one of the parties is a U.S. firm, the number of Japanese-produced databases and their users is also increasing. Currently, JOIS has about 800 users and PATOLIS about 520. Both of these systems offer their services in kanji and kana (Chinese and Japanese characters), making for a more accurate and easy-to-understand service.

During 1982 a number of new Japanese-language databases were produced, including several science- and technology-related documents (as opposed to statistical or bibliographical) databases such as JOIS-II, business information databases like TSR-BIGS and newspaper databases such as NEEDS-IR and Technopac. The popularity of portable kanji terminals which can be leased for less than ¥20,000 (US$73.53) per month has contributed considerably to this phenomenon.

A further boon to the database industry is expected with the imminent expansion of ICAS services to gain access to the rich databases of Euronet. France's Telesystems Co. has already decided to offer its wide range of database services in the fields of science and technology, patent and legal information on the Japanese market. Once Euronet becomes accessible from Japan, the majority of databases currently in existence in the world will become available to the Japanese user.

Yet another technical innovation is destined to spur the growth of the industry in coming years. Just as the introduction in 1981 of direct ordering of hard copy via terminal-boosted database usage in 1982, the integration of independent databases in single systems will be an added stimulus. Distributed database systems that make use of a number of distributed host computers linked in a single network will make possible integrated utilization.

Construction of these kinds of systems began in 1981 under the aegis of a national project called Research into Advanced Utilization of Chemical Databases via Shared Networks, supported by the Japanese

Science and Technology Agency. This particular system will make it possible for governmental and related agencies to gather a wide range of information on chemical substances quickly and easily from a number of relevant databases linked to the same computer network.

Estimates of future growth of the Japanese database service industry vary widely but, at present rates of expansion, revenues will increase 10-fold during the 1980s. Most certainly, the economics of supply and demand augur well for such a rapid expansion. Database usage literally decimates information retrieval costs while increasing effectiveness.

On-line retrieval operations by in-house end users at Sony palpably illustrate these economies. A recent survey of the Technical Information Center found that time saved by utilizing on-line retrieval methods, when put in financial terms, would amount to some hundreds of million yen per year. Similarly, in an experiment conducted by the library section of Mitsui Metal Mining Industries, abstracts covering a number of specially selected themes were retrieved manually from Chemical Abstracts and then via on-line database: the manual search required 104 hours to search four year's abstracts at an average of seven hours a day for 15 days. Searching for the same materials using on-line retrieval methods required only tens of minutes, costing about ¥4,000.

Using the manual and a magnetic tape version of the same chemical abstracts, Sumitomo Chemical Co. assigned a dozen or so researchers the task of retrieving a specified set of documents pertinent to their work. As a result of their manual research they concluded that 42.8 percent of the documents they had searched for were not available. Then a group of retrieval experts were asked to search for the same materials using the magnetic tape and a computer. The results were convincing: the experts concluded that only 2.3 percent of the sought-after documents were not available. Based on these findings, management set about organizing the Sumitomo Chemical Information Center and is staffing it with qualified information specialists.

Matters of information retrieval are not always so clear, however. Retrieval costs, other than those incurred in transmission, are easy to compare since they are all listed on the price lists of database services. But in Japan, long-distance transmission fees are at present relatively expensive and no alternatives to use of NTT or KDD circuits are available. Both conditions are due to change, however, making database services increasingly attractive. The Information Network System introduced in 1981 by NTT will revolutionize data communications through extensive application of optical fibers and very-large-scale-integration semiconductor technology. Transmission costs will likely be one-hundredth of current costs and prices of terminal equipment will decline radically. Much of this

ON-LINE DATABASE SERVICES AVAILABLE IN JAPAN, 1982 *

Distributor (**)	Name of Service (System)	Principal Databases offered	Payment System
Japan Information Center of Science and Technology	JOIS-II	Science and Technology Document Database	Meter rate system
Japan Patent Information Center	PATOLIS	Domestic Patent and Utility Models and U.S. Patent Information (Document Database)	Meter rate system
Nihon Keizai Shimbun Ltd. Nippon Tel.	NEEDS-TS	Japanese Macro- and Micro-economic Information and U.S. Macro-economic and Stock Information (Fact Database)	Membership fee
Nippon Telegraph and Telephone Public Corp.	NEEDS-IR	Newspaper and Magazine Articles (Document Database)	Meter rate system
NTT	TSR-BIGS	Domestic Business Finance Overviews	Meter rate system
Quotation Information Center	QUICK	Stock Prices (Domestic and Foreign) and Foreign Exchange Information (Fact Database)	Membership fee
System Development Corp. of Japan Ltd.	SEARCH-J	Patent, Pharmaceutical and Chemical Document Databases from around the World	Membership fee
G.E. Information Service Co. (Dentsu International Information Service)	MARK-III	Fact Databases Containing Foreign and Domestic Macro-economic, Industrial Statistics and Stock Price Information	Meter rate system
DIALOG Information Services, Inc. (Maruzen, Kinokuniya)	DIALOG	Approx. 100 Different Scholarly Document Databases on Science & Technology, Economics, Business, Social and Humane Service	Meter rate system

		Spectrums and Toxicity	
SDC	ORBIT	Approx. 60 different Scholarly Document Databases on Science & Technology, Economics and Business, Social and Humane Studies	Meter rate system
The New York Times Information Service Co. (Nihon Keizai Shimbun, Ltd.)	*The New York Times* Information Bank (The Information Bank)	Approx. 90 different Newspaper and Magazine Document Databases including that for *The New York Times*	Meter rate system
CDC (Nihon CDC)	SBC Database Service (CALL)	U.S. Business Finances and Economic Statistics Fact Databases	Membership fee
Dr Dvorkovitz & Associates (Dvorkovitz & Associates, Japan)	DDA Service	Technical Information on approx. 40 thousand licences (Document Database)	Membership fee
DRI (Nihon Keizai Shimbun, Ltd.)	DRI	American and European Economic Statistics and Industrial Information	Membership fee
CAS (Chemical Abstracts Service) (Chemical Information Association)	CAS ONLINE	A CAS fact Database Containing Data on the Structures of Millions of Chemical Substances	Meter rate system
Japan Information Processing Service Co., Ltd.	JIP/BRS	Document Databases Containing World Medical and Electrical Engineering Information	Meter rate system
Interactive Data Corp. (Nomura Research Institute)	Interactive Service	World Economic Statistics and Industrial Information	Membership fee

* In the period from compilation of this data in 1982 to 1985, 74 companies have entered the database services industry.

** Organizations in parentheses are sales representatives.

Source: JIPDEC.

new system should be in place by 1984 by which time the proposed liberalization of data communications should also be a reality.

The combined effect of these changes in computer and telecommunications technology on the organization and conduct of business in Japan is enormous. Already database utilization is finding its way into the average enterprise and the spread of jobs related to that service is affecting all aspects of company performance, much more so than originally predicted.

The office automation boom is but one reflection of this phenomenon. As the utilization of personal and micro-computers as on-line terminals becomes commonplace, the use of database services is bound to increase. So, the availability of such services gives added impetus to office automation, especially where the cost and effectiveness of information retrieval are the critical variables of business performance.

As it has become relatively easy to develop in-house information-processing systems using time-sharing networks, the number of people in a company capable of operating computer terminals is increasing. This, in turn, is augmenting demand for convenient, easy-to-use company information retrieval and time-sharing systems using advanced database management software. Information specialists, who up to now have been primarily concerned with the task of documentation, are now being called on to construct new information systems combining knowledge of secondary sources and the skills of on-line retrieval.

Companies are beginning to view their libraries and information services as possible profit centers, providing trend studies, computational functions and the analysis of assorted information. Quite a few Japanese companies have been striving to apply database management-systems technology to their own collections of data to produce in-house databases.

Asahi Chemical, one of Japan's leading chemical companies, has taken the plunge into the database business starting with the supply of data on analyses of chemical substances via Asahi's subsidiary, Asahi Research Center and Dentsu's MARK III services. Asahi's database service is offered in such a way that Japanese clients can have dual access to:

• more than 43,000 standard spectrums edited at the U.S. Materials Test Association (located in the premises of the General Electric Computer Center in Cleveland) and

• the files of spectrum programs developed by Asahi using public communications circuits serving their terminals. The new database producer's next step is to expand the service to overseas clients.

Asahi's experience is especially instructive. Science and technology, as well as the industrial database services being developed for their effective management, are inherently universal, bearing witness to the stark reality that nationally produced databases are hopelessly inadequate. Information is universal and can be best managed only in a global context.

One of the most striking and significant features of the database industry is that Japan is destined to be an import market for the foreseeable future as it has been for information generally throughout its modern period. In the Information Age, Japan will most certainly be one of the most avid consumers of its representative product.

Already and not coincidentally, Japan has become a principal market for foreign vendors of huge databases, a fact which — JIPDEC (Japan Information Processing Development Center) observed in a recent study — is already having a considerable influence on the Japanese information industry: there is concern in some quarters that the large market share of strong foreign information supply companies may thwart the development of a healthy Japanese database industry.

Asahi's entry into the market with a global system, designed to utilize worldwide resources, melding its own with those of others to make the synergistic combination available to Japanese and foreign users alike, is not likely to be a unique experience but one of a kind. That said, even if Japan were to continue to run a 100 percent deficit balance of trade in database services, the Japanese and their economy would be the beneficiaries and, even if Japan were to try to play the reciprocity game in this instance, it could not because, given the reality of languages, there is little that Japan has to offer for export at this point. Although it would be harmful to consider international data flows in terms of international trade, in this case at least Japan's deficit balance of information flow holds prospects of healthy effects upon its international economic relations.

PART VII.

Factory Automation

CHAPTER 21

Japanese Factory 1990

THE JAPANESE FACTORY in 1990 will be quite unlike that of today. Unmanned production systems will work tirelessly 24 hours a day without interruption. Any mass production systems manned by assembly line workers that may still exist will be the last vestiges of a virtually extinct mode of manufacturing. New manufacturing systems, entirely computerized from product design to distribution, will be at once more flexible, more productive and more reliable than present automated mass production.

Unlike mass production transfer assembly lines, moreover, the new production systems will be suited to low- and medium-, as well as large-scale, production. The change is a big one. In Japan, over 70 percent of machinery output, for example, is medium volume, too small for mass production, yet too large for previous numerically-controlled production. Now, by combining robots with computer-controlled machine tools, automated warehousing and computer-aided-design, all the benefits of total automation are available to medium-size and smaller enterprises.

Robot versatility is the central and determinant feature of the new flexible manufacturing systems (FMS). By varying the order in which goods in process are transferred between different manufacturing operations, robots will give production systems of 1990 the flexibility to handle a variety of products. Economies of scale in machinery production will be obtained at much lower production runs, bringing benefits of automation heretofore limited to large-volume automobile and electric appliance production within reach of the vast majority of enterprises.

In 1990, factories consisting of a single entrepreneur and several robots will coexist with large factories employing hundreds of robots controlled by mammoth computers. Already, in 1980, cottage-level factories having only family workers began to use robots; by the end of the decade this will have become the prevailing pattern of production in small enterprises.

Originally published as "Robotics: They are Smart and They Never Need a Tea-Break," *Far Eastern Economic* Review, 4 December 1981.

The result will be an explosive growth of demand for robots and other elements of flexible production systems. According to industry estimates, which experience suggests will have to be revised steadily upwards, by 1990 the output of robots in Japan will exceed ¥1,000 billion, or 12.75 times 1980 production. At the same time, demand for numerical-controlled machine tools and computerized design equipment will continue to increase by 40 percent per year or more. The pace with which Japanese industry is moving in this radical transformation of production systems is remarkable even by Japanese standards.

In 1967, the first industrial robot was imported into Japan from the United States. A year later, Japan's first industrial robot was developed. Since then, the progress of Japanese robot technology has been spectacular. Within just 12 years, by 1980, Japanese industry employed an astonishing 70 percent of the total world robot population in its factories. Demand has outstripped the ability of industry to produce. More than 140 robot-makers have been scrambling to develop new products and strategies which respond to the booming market.

As a result, in 1980, robot production shot up 85 percent to ¥78 billion and in 1981 will soar to ¥120 billion, an increase of over 50 percent. Robot watchers now predict that sales of the industry will continue

Table 1. Industrial Robot Production Value

Year	Billion Yen	Million US$
1968	1.4	7.0
1969	1.5	7.5
1970	4.9	24.5
1971	4.3	21.5
1972	6.1	30.5
1973	9.3	46.5
1974	11.4	57.0
1975	11.1	55.5
1976	14.1	70.5
1977	21.6	108.0
1978	24.7	123.5
1979	42.4	212.0
1980	78.4	392.0
1981E	100.0+	500.0
1985E	500.0	2,500.0
1990E	1,000.0	5,000.0

Source: Daiwa Securities America Inc.
**Exchange Rate: Yen 200 = US$1.00

to grow by 40 percent a year well into the next decade as new types of robots are developed to perform an increasingly wide range of tasks.

At first, relatively stupid robots were introduced by Japanese manufacturers to do the dirty work of industry, relieving men and women of difficult or hazardous jobs. Over the past decade the main application of industrial robots has been for spot welding, arc welding and spray painting, especially in the automotive and electric machinery industries. But robots are getting smarter all the time, and some already have a sense of "touch" as well as "sight." With the development of more intelligent robots, assembly promises to be the most important task for future application.

The extensive use of robots for automated assembly will mark an important step forward towards the development of flexible manufacturing systems. Many companies are awakening to these possibilities since Fujitsu Fanuc opened its remarkable unmanned factory employing robots to produce robots. At present several dozen unmanned manufacturing systems are producing a diverse range of products from instant noodles to machine tools, with large computers controlling integrated production cells of microprocessor-directed machines.

A completely automated manufacturing system, like the one Fuji Electric has designed for instant noodle production or Niigata Engineering's diesel engine plant, may be composed of various combinations of seven major sub-systems, each made possible by recent rapid advances in microelectronics:

- Numerical-controlled (NC) machine tools and integrated, computer-controlled systems such as machining centers.
- Industrial robots for handling, pressing, welding, painting and other intermediate tasks.
- Totally automated processing or assembly lines employing intelligent robots.
- Conveyor systems and unmanned carrier robots.
- Automated warehouses.
- Sensors by which robots see, hear, feel and smell, to conduct the information content of production systems to microprocessors and central computers.
- Computer-aided design and computer-aided manufacturing (CAD/CAM) systems which control entire manufacturing processes.

The net result of this combination of technologies and systems amounts to nothing less than a revolution in industrial processes.

Most important, it promises to change the entire man-machine interface. In contrast to the traditional mass-production assembly line system in which the machine determines the activity of the operators, in the new computer-controlled factory systems the operator programs the system and its various sub-systems, including robots, NC tools and computers. Man is master of the machine. Psychological resistance to conveyor systems is replaced by a sense of confidence and satisfaction derived from the control over robots and entire production systems. Hazardous and noxious work is undertaken by machines, reducing the incidence of accidents and occupational diseases. And totally new horizons are opened for women and the handicapped, who are no longer disadvantaged by the physical requirements of the new technology.

Furthermore, the new production system, by its inherent characteristics as a man-dominated computer-robot-machine system, will lead to the creation and rapid development of completely new technologies and their application in biotechnology, nuclear and fusion energy, new materials production, microelectronics, ocean resource development and space factories.

The rapid diffusion of flexible manufacturing systems in Japanese industry will definitively undo the delusion, still cherished by some pundits and decision-makers in high places, that the organizational skills Japanese industry used so effectively in catching-up are probably not appropriate for creative innovation at the cutting-edge of advanced technology. Equipped with integrated and totally automated production systems, Japanese firms which are now putting their best resources and talents to work at creative pursuits can safely be counted among the more prolific innovators of the coming decades.

Ultimately, however, the economic effects of flexible manufacturing systems will determine the rate and extent of their diffusion. If humanization of working life by releasing blue collar employees from dangerous, tiring work and simple, repetitive tasks has increased the attractiveness of robots for men and women on the factory floor, the prospects of retraining and more challenging work assured under the Japanese lifetime employment system explains the enthusiasm with which Japanese workers themselves contribute their proposals for successive steps towards the unmanned factory. As skilled labor is becoming increasingly scarce and the needs for upgrading technical and managerial capabilities of the existing workforce are prevailing features of the current Japanese labor scene, both workers and managers have a compelling interest in expediting this process of change.

The manufacturing sector accounted for around 60 percent of the 840,000 skilled labor insufficiency in 1980, which meant an insufficiency ratio of 8.2 percent (just the reverse of the unemployment ratio in most other advanced industrial countries). Without extensive adoption of in-

Table 2.   Trend of Robot-Wage Costs

Year	Price of playback robot (A)	Annual wage per capita of companies with		(A)/ (B)	(A)/ (C)
		5-29 employees (B)	500 or more employees (C)	— (times) —	
1976	10,920	1,528	2,605	7.1	4.2
1977	8,640	1,827	3,017	4.7	2.9
1980	7,790	2,060	3,553	3.8	2.2

Source: Japan   Industrial   Robot   Association   and   *White   Paper   on   Labor.*

dustrial robots and other labor-saving equipment, the labor shortage in the manufacturing sector would become more serious, due to the decline of newly employable high school graduates, the exhaustion of rural sources of labor, the aging of the working population, and the increasing proportion of students who go on to higher education.

The economic advantages of flexible, unmanned manufacturing systems are clearly reflected in the changing relative costs of labor and robots. From 1968, when the first Japanese robot was developed, until the oil crisis of 1973, robots remained relatively expensive; demand was correspondingly low. But during the 1970s labor costs rose sharply, and the rapid strides in microelectronics kept the prices of robots more-or-less stable.

Indeed, the manufacturing cost of industrial robots of all types declined from 1970 to 1975. After 1975, however, the price of simpler manipulator and fixed sequence robots tended to rise, while that of more sophisticated robots continued to decline. As a result, the most versatile and efficient robots have become much more economical. Given the costs of robots and labor in 1971, it would have taken 22.5 years on average for a company to recover its investment in a robot if it were operated on a single shift basis and 9.4 years if operated two shifts a day. By 1980, industrial robots could pay for themselves in less than four years on a single shift basis and less than two if operated two shifts daily.

Although the actual expenses of robot installation and maintenance resulted in a somewhat slower rate of amortization, actual return on investment calculations provide increasingly convincing reasons for the purchase of industrial robots. By 1980, investment in robots was estimated by Nomura Research Institute to yield more than 43 percent for double shift operation and 10 percent for single shift.

Still, in the first year of robot introduction, total costs in higher depreciation, interest costs, installations and production slowdowns dur-

307

ing the period when the robot is being integrated into the total manufacturing system can be very high. Given the high interest rates and the increasingly shorter payouts sought by American managers, such high start-up costs have mitigated against the introduction of robots even to a greater extent than labor union opposition. In Japan, both the life-long employment system and the cost of financing capital investment in robots tend to make the longer payoff much more acceptable, and the absence of stock options to managers reinforces this tendency.

But if labor displacement is the central benefit obtained from the introduction of robots, it is far from being the sole factor in the quotient of economic advantage. Inventory reduction and fewer in-process products as a result of automation and the use of robotics achieve substantial economies. While model changes or modifications usually require changing or rebuilding entire special-purpose automated assembly lines, in flexible manufacturing systems where industrial robots are used, a mere change in the computer program is often all that is required. As product life cycles shorten, and markets are increasingly uncertain, the flexibility and versatility of industrial robots become increasingly attractive.

Moreover, as industrial robots, unlike humans, are free of fatigue from performing simple duties for many hours, the number of defective products is reduced and quality is improved by their use. The resultant savings are, of course, considerable, and the competitivity of products is reinforced in the marketplace.

But other contributions of industrial robots to resource economy, an increasingly high priority since the oil crisis of 1973, are also significant. Spray painting robots, for example, obtain materials savings of from 20 to 30 percent. Since industrial robots working in an unmanned factory do not require airconditioning, heating, ventilation, lighting, music to work by or tea breaks, substantial energy and other incidental overhead cost savings are possible through their use. And, since robots permit higher operating ratios through two and three shift production schedules, they result in lower investments in capital goods.

Despite these compelling advantages of robot usage, financial considerations and the fear of technical obsolescence combine to deter introduction of more sophisticated models, especially by small and medium sized manufacturers. To overcome these obstacles and hasten their introduction, industrial robots were officially designated as "experimental research promotion products" and "rationalization promotion products" with the promulgation of the special Machine Information Industry Promotion Extraordinary Measures Act of 1978. Robot production was identified as a major strategic industry for Japan's future, hence measures to promote their utilization were deemed necessary.

Accordingly, beginning in fiscal 1980 a series of incentives have been introduced to encourage investment in robots.

• Under MITI's guidance and with the support of the Japan Industrial Robot Association (JIRA), a robot leasing company, Japan Robot Lease (JAROL), was established in April 1980. Jointly owned by 24 robot manufacturers and 10 non-life insurance companies, JAROL's operating funds are financed by low-cost loans from the Japan Development Bank, the Long-Term Credit Bank, Industrial Bank of Japan and the city banks, enabling it to lease industrial robots under conditions more advantageous than ordinary leasing companies. Results have been much better than anticipated. During its first year of operation, JAROL's leasing contracts amounted to 1,150 million yen, which was considerably higher than the planned 700 million yen in robot leases.

• Between April 1, 1980 and March 31, 1983, purchasers of computer-controlled robots with more than five degrees of freedom and less than plus-or-minus 1 mm positioning repeatability (that is, sophisticated robots) were allowed to add to the original depreciation a special depreciation charge at 13 percent of purchase price during the first year. This means, in effect, that by installing a sophisticated industrial robot a firm can depreciate 53 percent in the first year, 13 percent plus 40 percent (5 year depreciation double declining).

• Under a special financing program, the Small Enterprise Loan Corp. and People's Finance Corp. make low-interest loans to small and medium-scale manufacturers to encourage robot usage to release workers from dangerous and unfavorable working conditions and to improve productivity.

• Local governments also have introduced loan and leasing programs for the modernization of smaller enterprises.

• MITI will undertake a ¥30 billion national robot research program beginning April 1, 1982 to develop intelligent robots for assembly work and robots for nuclear, space, oceanic and earth-moving industries. Top priority is given under this program to development of sensory perception, language systems and motional capacity.

The main thrust of R&D in robotics will continue to come from private industry, however. Although the number of robot research laboratories in universities and governmental research institutes rose from 43 in 1974 to 85 in 1980, in fiscal 1979 the total expenditures on robot research by these institutions was a very modest ¥320 million (about US$1.5 million). Expenditures of a single major robot manufacturer substantially surpassed this amount. Of the 107 robot makers surveyed by JIRA in 1979, twenty had a specialized robot research division in their ap-

Table 3. Industrial Robot Maker Distribution by Size of Capital

Less than ¥10 million	19 companies	14.3%
¥10 million — 100 million	36 companies	27.1%
¥100 million — 1 billion	23 companies	17.3%
¥1 billion — 3 billion	8 companies	6.0%
More than ¥3 billion	47 companies	35.3%
Total	133 companies	100.0%

Table 4. Industrial Robot Maker Distribution by Number of Employees

Less than 50	33 companies	24.8%
50 — 500	29 companies	21.8%
500 — 1000	15 companies	11.3%
1000 — 5000	25 companies	18.8%
More than 5000	31 companies	23.3%
Total	133 companies	100.0%

propriate laboratories, while another 52 had more than one member of their research staff specializing in robotics.

As demand for robots increases and quantum leaps forward in microelectronics make possible increasing refinements of function as well as reductions in cost, more manufacturers are entering the market and existing robot makers are expanding their product mix. By 1980, according to a Long-Term Credit Bank survey, as many as 149 companies, both large and small, were vying for a share of growing demand. Machinery makers, electric appliance manufacturers, shipbuilders and steel makers have moved into the market for differing reasons.

The giant electrical appliance and heavy machinery manufacturers have clearly been attracted by the high growth potential of industrial robots and entered the field to diversify their business. At the same time, however, production for outside demand enabled these firms to capitalize on heavy investments made in robotics for their own internal needs. A number of the leading robot manufacturers developed robots for their specific requirements, tested them, improved on the original models and then achieved important manufacturing experience to meet their own requirements — all before offering their robots in the marketplace. Some of the steel makers such as Kobe Steel and Daido Steel were also attracted to robots as part of their diversification progams.

This is not an industry which is the special reserve of large-scale manufacturers, however. Since robots often must be custom-made to the particular requirements of the user, and special knowledge of the industrial activity for which the robot is designed can be essential, there is considerable room for small robot makers. In 1980, as many as 63 industrial robot makers, or 42 percent of the total number, are currently capitalized at less than ¥100 million and 28 of these firms have equity capital of less than ¥10 million. If ranked by the number of employees, 77 of those firms building robots employed less than 500 and 33 had fewer than 50 employees.

In stark contrast to the U.S. industry, where two robot makers account for more than half of the market, the Japanese robot industry is widely dispersed, producing a broader range of automated machines designated as robots. And industry observers here see little likelihood of a significant change in the structure of manufacturing in the foreseeable future. Although robot production has expanded rapidly, and market entry has been relatively easy, as it was in the case of transistor radios, television sets, digital watches, calculators and video taperecorders; quite unlike consumer electronic products, robots must meet function- and systems-specific needs of the industrial end-user. And this requires special design and engineering capabilities of the manufacturer or vendor. In this sense, at least, the industrial robot is not just one more piece of electronic merchandise which is bought off the shelf.

Other structural features of the Japanese robot industry mitigate against the kind of concentration which has characterized the American industry until now. A large number of Japanese robot makers originally began producing robots to increase the productivity of their own manufacturing operations in an existing main line of products. (Hitachi, Matsushita, Toshiba, Sailor Pen, Pentel, Pilot, Okamura and Tokico among others). Some companies developed robots to sell their established mainline products, as has been the case of Aida, Japan's leading press manufacturer. Still others are producing robots for members of their own industrial group. Thus, Toyoda Machine Works provides welding and handling robots for Toyota, and Toshiba Machine Co. for Toshiba Corporation. Likewise, Mitsubishi Heavy Industries supplies robots to Mitsubishi Motors, its subsidiary automobile maker. These makers are likely to remain in the industry as independent producers.

Increasingly, too, each major industrial group is likely to have its own robot supplier, not only for the production needs of group companies but to develop and market total manufacturing systems. As an example, Fuji Electric built the robots for a completely unmanned computer-run dry noodle factory, including an automated warehouse,

Table 5.   U.S.-Japan Comparison: Industrial Robots
(U.S. Definition)

	Units		Value (million US$)	
	U.S.	Japan	U.S.	Japan
1980	1,269	3,200	100.5	180
1985	5,195	31,900	441.2	2,150
1990	21,575	57,450	1,884.0	4,450

Source: Daiwa Securities America Inc.

battery-operated cars, loading and unloading robots, automatic manufacturing and inspection systems and packing equipment — much of which was supplied by other member companies in the Dai-ichi Kangyo Bank group.

By the very nature of flexible manufacturing systems, they will be engineered and equipped as integrated units either by the user, a supplier or a plant engineering company. And in many cases the robot maker will be the main supplier around whose equipment the entire system will be designed.

In some cases, however, robot manufacturers are also makers of numerical controls and CAD/CAM systems, the other two critical elements of flexible manufacturing systems. These makers — including Hitachi, the Toshiba group, the Mitsubishi group, NEC and Fujitsu — will have a special advantage in the market. Moreover, since all five of these groups are also leaders in microelectronics, they can be expected to be numbered among the principal manufacturers of intelligent robots.

The expected rapid diffusion of flexible manufacturing systems will, therefore, not only spur robot applications in Japanese industry, but it will also give increased impetus to the already high-growth market for numerical-controlled machine tools. Moreover, the demand for new manufacturing systems is expected to reinforce a significant feature of the NC-tool market manifest during the 1970s.

During the last half of the seventies, Japan's machine tool production grew steadily, increasing up to 40.9 percent a year, despite two recessions. NC machine tools, however, showed an even sharper increase in output. From the trough year of 1975 up to 1980, production of NC machine tools grew by 54 percent a year, which means that the five-year growth of the machine tool industry was mainly attributable to the rapid growth of NC machine tools.

It would seem, from this performance, that the market for NC machine tools is far less cyclical than the demand for machinery in general. And since the production of industrial robots has tended to move in correlation to that of NC tools, in response to the demand for productivity improvement through labor-saving and production cost reductions, the spread of flexible manufacturing systems should have a synergistic effect upon the development of production, in terms of both its growth rates and its anti-cyclical pattern of robots and NC tools. And to some extent, at least, production and sales of CAD/CAM systems, automated warehouses and other elements of unmanned factories can be expected to follow a similar high growth pattern.

CHAPTER 22

Robotics and Flexible Manufacturing Systems

THE FACTORY OF the future now seems to be largely a Japanese affair: but it didn't have to be. In 1952, the American John Diebold described in detail the characteristics of a fully-automated flexible manufacturing system in his classic treatise on *Automation*. All of the critical elements of an automated factory were then known and technically feasible. In fact, the Rockford Ordinance Plan, designed during World War II and completed after the Korean conflict began, automatically manufactured 155 mm steel shell casings from raw steel stock. Only the absence of a central computer control mechanism — or "brain" — for the individual automatic machines distinguished it from science fiction accounts of the future factory.

Everyone assumed then that, just as U.S. industry had led the mass-production revolution earlier in the century, the fully automated factory of the future would be an all-American creation. But there was no particular hurry. Many people were not at all certain that factory automation was desirable or at all necessary to assure U.S. industrial leadership well into the 21st century. Things look quite different now. Although every major technology for the automated factory — machine tools, numerical controls, robots, flexible manufacturing systems — was indeed first developed in the United States, leadership in their use and production has passed, seemingly ineluctably, to Japan.

By 1982, Japanese industry had moved into the forefront of what now appears to be an industrial revolution of far more consequence than that triggered by Henry Ford's assembly line and the accompanying standardization of production. Mass production systems, in fact, apply to only about 25 to 30 percent of total industrial output in advanced countries. Most manufacturing, especially in metal-working and machine building, is conducted on a smaller scale, with production runs from 10 to 2,000 units. As societies become more affluent, consumers increasingly reject conformity in favor of greater variety and speciality in the products which

---

Reprinted from "The Factory of Japan's Future," *Management Today*, April 1984.

they use. As a result, fewer products are mass-produced, and shorter product life-cycles are the rule. Economies of scope are now needed, where scale economies once sufficed to assure competitive power.

That this fundamental change in the economics of production should have been felt first and most acutely in the rigorously competitive business environment of Japan is not at all surprising. Already in the early 1970s, much of Japanese industry had reached the upper limits of productivity with existing manufacturing systems. Labor shortages had become severe, while currency and energy crises threatened to erase the comparative advantage of Japanese manufacturers in world markets. Fierce competition at home put a new premium on efficiency and innovation.

Manufacturers found it increasingly necessary to build flexibility into their production systems so that they could supply more segmented markets with small lots of parts, components and finished products. Even more important, the growth of competition focused attention on the fact that many manufactured products spend up to 95 percent of their time in the factory just waiting to be processed. New production systems were needed to obtain radical economies between stages.

Flexible manufacturing systems (FMS) provided the answer to this need. Although the systems had been received with little enthusiasm by American managers or labor, the greater versatility and potential manpower economies of FMS provide irresistible attractions for Japanese managers. Skilled labor requirements in small batch production could be reduced by as much as 95 percent — and with greater diversification and higher quality of output.

What was seen as a threat in the United States offered greater benefits all round to Japanese workers and managers alike through the improved productivity and higher added value. FMS has added further value by reducing the necessity for sub-contracting production, and at the same time increasing the scope for diversification. Life-long employment and extensive in-company educational programs have made possible retraining of workers and engineers for the new more challenging type of work.

As a result, since the first FMS was developed by Ikegai Iron Works and Fujitsu Fanuc for the Japanese National Railways in the 1960s, there has been a steady move towards the ultimate stage of factory automation: the unmanned factory. By the beginning of the 1980s, Japan had 70-odd fully-automated factories turning out everything from instant noodles to robots. Some of these plants have become world-renowned showcases.

But even more important than these advanced systems (which include the world's first automated factory to be run by telephone from company headquarters) is the general aggregate level of automation in Japanese industry. The proliferation of numerically controlled (NC) machine tools and machining centers is especially significant, since they are the basic building blocks of automation. To make an FMS, they are linked together with robots, material handling systems and a central computer. By 1982, as many as 53.7 percent of all machine tools in Japan were numerically controlled, and machining centers accounted for 38.4 percent of the net annual output of new NC tools. The contrast with the United States is instructive. Although NC was developed first at the Massachusetts Institute of Technology as early as 1949, and U.S. industry led the refinement of the technology in the 1950s and 1960s, in 1982 only 5 percent of the U.S. machine tool park was NC-equipped.

Production of NC machine tools did not even begin in Japan until 1958. For almost a decade, NC prices remained extremely high, and usage was consequently confined to special purposes. But in the late 1960s, with the availability of cheap minicomputers for NC application, demand expanded rapidly. From 1971 to 1981, production of NC tools increased at an annual compound rate of 32.9 percent, and since 1977 the use of machining centers has been growing swiftly.

Unlike traditional machine tools, NC machining centers can perform a number of operations — such as milling, drilling, boring or attaching screws — as individual steps in a continous manufacturing process. Because one machining center can be programmed for various operations, a change in work sequence does not require an altered line-up of machines, as is often the case if the same work is done by a number of discrete, special-purpose NC machine tools. The result is a major saving of time and floor space, with large gains in productivity.

In 1980, Japanese output of machining centers was worth ¥111.4 billion (US$470 million), almost nine times the production value in 1975, and was rising at an annual rate of 69.4 percent — or more than twice the rate of NC machine tools all told. In response to this rising domestic demand, which has been accompanied by a 55 percent annual increase in exports in the two-year period from 1979 to 1981, the number of machining center manufacturers rose from 26 to 40, all of whom have steadily enlarged production capacity.

Since machining centers can constitute the main elements of flexible manufacturing systems, their increased use is a significant step in the direction of making unmanned factories a geneal reality. Fully equipped FMS, costing from ¥500-1,500 million (US$2.1 to US$6.4 million), involve large capital outlays which put them out of reach of all but the large-scale companies, with the result that cases of instant total conversion are relatively rare; alone, they hardly add up to a revolution in manufacturing. But since FMS are modular combinations of equipment, with NC tools such as machining centers as building blocks, they can be built in stages.

Total automation is the ultimate goal — and imperative for survival. So the step-by-step development process can be expected to speed up, with the demand for NC machine tools rising accordingly. Of the roughly 700,000 machine tools currently operating in Japan, 260,000 (37 percent) are more than 15 years old. Of these, only 500 are numerically controlled. Since this equipment will, perforce, be replaced in the remaining years of this decade, annual growth rates of NC machine tool production of at least 15-20 percent can be expected, assuming no further increase in the export ratio above the current 30 percent.

Given the systemic nature of factory automation, it is not at all coincidental that the growth pattern of robotics in Japan has followed the path of NC machine tools. Again the story begins in the United States, this time at a cocktail party in 1956, where Joe Engleberger and an inventor named George Devol asked themselves the heady question, "Why not make a robot?," or words to that effect. Two years later, the two men made history with the production of the world's first industrial robot; in 1962 Engleberger formed his company, Unimation.

But Engleberger had not anticipated the rest of the story. By the end of 1982, there were some 33,000 industrial robots, as defined in the United States, operating in Japan, as compared with only 6,301 in America itself. Annual production of robots, similarly defined, was 14,977 in Japan, compared to about one-tenth that number, or 1,601, in the United States. The stark reality is in even more vivid contrast, however. If the somewhat broader Japanese definition of a robot is accepted, including fixed sequence machines and manual manipulators, there were 103,500 robots installed in Japanese factories at the end of 1982 — growing at around 25,000 annually.

To say that U.S. manufacturers had greeted yet another major innovation in production technology with a yawn might be exaggerating. But the facts clearly witness that, despite the manifest enthusiasm for robotics in some quarters, major U.S. corporations have been reluctant to grasp the obvious long-run opportunity they represent. Opposition has come from management and labor alike. Managers with an eye on high interest rates and quicker returns on investment have opposed long-term capital commitments to robotics and some future stage of factory automation; while in the absence of measures to re-educate those displaced, workers have resisted investments which would cost their jobs.

In Japan, however, the common purpose of maximizing added-value in production, as an imperative of national survival and a ticket to a higher standard of living, has united workers, managers and bureaucrats in a concerted effort to introduce and integrate robotics in production processes. And the necessary funds were available. With prime interest rates of around 6.75 percent, the cost of financing robots has been reasonable. Moreover, special depreciation allowances, and the depreciation accounting methods in force, enabled firms to recover up to 53 percent of the cost of robots after the first year. And if outright purchase is unacceptable, flexible leasing arrangements are available through Japan Robot Lease (JAROL), which was established in 1980 by 24 robot makers with operating funds provided at low interest rates by the Japan Development Bank and other long-term credit banks. In addition, the Medium and Small-Scale Business Modernization Fund offers low-interest loans to encourage robot installation in manufacturing processes which are hazardous or critical to increased productivity.

Workers are involved in all decisions to introduce robots — in the labor-management consultation committees in the firm, in quality control circles, and in their daily work alongside production engineers on the factory floor. Thus, Nissan Motors, in signing a formal agreement, promised to advise the union in advance of the introduction of robots, to neither fire nor downgrade workers replaced by robots and to provide reeducation and training for those affected; it was publicly ratifying practices already standard in Japanese industry. What was left unsaid was the tacit understanding that workers and managers alike feel jointly responsible for developing new jobs for those displaced.

In sum, Japanese enterprises are organized for the robotics age, are able to finance the introduction of robots, and have resolved the problems of man-machine interface that have so bedeviled manufacturers in the United States and Europe. And the machinery manufacturers have been quick to respond to the resultant demand. Japanese industrial robot production grew at a 38 percent annually-compounded rate during the decade from 1971 to 1981, accelerating in value terms during the latter half of the period to 48.4 percent, while decelerating in volume as average unit prices rose.

Throughout the decade, remarkable technological advances in robotics have been made, both at the 85 universities and public institutions which now conduct basic research (a number that has doubled in the last six or seven years) and in the parallel applied research undertaken by the robot manufacturers themselves. A steady stream of innovations has produced higher speed, smaller size, lighter modular design, and robots with computer controls. Intelligent robots are being perfected and will probably enter widespread use by 1985. After these developments have been refined, it is expected that new uses for robots will be found in agriculture, forestry, mining, construction, medicine, nuclear power development, marine industries, and space. And, in the factory, the evolution of unmanned production systems will be accelerated.

In response to this vigorous demand, the robot industry has boomed. There were just 10 manufacturers in 1968; by 1981 there were 190, with no signs of shake-out on the horizon. Virtually every major electric machinery manufacturer now includes some kind of robot in its product mix. Machinery builders, precision instrument makers, material handling equipment manufacturers, steel-makers, and even textile firms have entered the market, along with a bevy of smaller firms, led by *wunderkinder* like Dainichi Kiko.

The industry may continue to expand as new firms enter to bid for the widening market. Growth up to 1985, according to Japan Industrial Robot Association (JIRA) estimates, will continue at the high rate of 35 percent a year, dropping to 14 percent for the rest of the decade. At this rate, total output will reach ¥450-600 billion (US$2.0 to US$2.6 billion) by 1990, or as much as 10 times the 1980 turnover. Projections by Daiwa American vice-chairman Paul Aron are even more bullish, however. Noting that JIRA forecasts in the past have proven to be overly conservative, Aron expects total robot production in Japan to top ¥1,000 billion (US$4.18 billion) by 1990. He believes that the 48.4 percent growth of the past five years will be maintained at least for the next two.

Linking Global Markets

But this is not the whole picture. Japanese robot makers are also moving quickly to rationalize their production internationally at an early stage of their products' life cycles. Fanuc's strategy is most far advanced in establishing cooperative relations with companies abroad having strong links with end-users and accumulated knowhow in robot applications. With the world's most advanced electro-servo technology (used in the production of servo motors for CNC and robots alike), Fanuc moved simultaneously in 1982 to establish a joint venture with GM (GM Fanuc Robotics) in the United States and to diversify its robot production to include spot welding, arc welding and spray painting robots for car manufacture.

Fanuc has established a strong overseas sales network linking global markets in the United States (where GM Fanuc will now market robots made by Fanuc in Japan), in Europe (Siemens, 600 Group, and Manurhin Automatic), and Asia (Tatung Co. of Taiwan and others). With these links, Fanuc is well on the way to becoming the world's most competitive and reliable source of robot hardware and services. But the goal is far from assured, for Dainichi Kiko is fast on its heels and fully determined to surpass Fanuc to become the world's largest robot producer. Since it has the backing of the Mitsubishi group, its intentions must be taken seriously.

If Japanese leadership in NC tools and robotics — two critical technological dimensions of FMS — seems assured for the remainder of this century, the contest for the third dimension is just beginning. This is the field of CAD/CAM (computer aided design/computer aided manufacturing), which unites all of the elements into a totally integrated manufacturing system. Once again, the key developments in the technology took place in the United States when researchers at MIT started drawing crude pictures of solid objects on television-like displays back in the early 1960s. Then, in 1965, Lockheed developed the first integrated CAD/CAM system, called CADAM (Computer Graphics Augmented Design and Manufacturing), whose improved version is accepted worldwide.

Computers capable of drawing straightforward two-dimensional layouts were snapped up by automotive, aerospace, semiconductor and computer manufacturers in the United States, and subsequently in Japan, for speeding the design of printed circuit boards, the horrendously complicated circuits of microchips, aircraft, spacecraft and new cars. But demand really exploded after the major break-through in the mid-1970s, when minicomputers combined with clever software enabled CAD/CAM systems to design both in three dimensions and in color.

The ability to design, draft and analyze with computer graphics displayed on a screen has radically changed the engineering process, reducing the human resources required to a small fraction of those formerly employed and telescoping the time taken to get a new product to market. Moreover, when combined with NC tools and robots, CAD/CAM can pay impressive dividends in increased production, better quality products and greater manufacturing flexibility. As a result, although much of the technology is still in its adolescence, by 1981 the U.S. market for CAD/CAM equipment had already surpassed the billion dollar level and was expected to reach $4.4 billion in 1986. Industry leadership remained very largely American, led by Computervision (34 percent), IBM (16 percent), Intergraph (12 percent), Applicon (11 percent) and GE's Calma (11 percent).

In Japan, leading car manufacturers and shipbuilders were first to launch CAD/CAM research and development in the latter half of the 1960s. In the 1970s, CAD systems were used in civil engineering, for designing buildings, ICs and their printed circuit boards. And from 1978 onward, small and medium-sized electronic component and metal parts makers, as well as clothes manufacturers, began to employ CAD/CAM systems. By 1982, the Japanese market had reached ¥40 billion (US$169 million), growing at a compound annual rate of 51 percent. At this growth rate, Japanese demand for CAD/CAM equipment will reach ¥217.6 billion (US$913 million) in 1986 — about a quarter of the expected U.S. market.

CAD/CAM Gets Hot

But this is just a beginning. If growth patterns of NC machine tools and robots can be taken as a guide, by 1990 the Japanese market will exceed ¥1,000 billion (US$4.18 billion). And this is enough to spur the development of CAD/CAM systems by Japanese computer-makers and other firms in the automation field. Since Hitachi's market entry in 1979, the other three mainframe computer builders — Fujitsu, NEC and Mitsubishi Electric — have all introduced systems. CAD/CAM suddenly became the hottest area of interest on Japanese stock exchanges, fueled by sharp upgrading of CAD/CAM systems following advances in large-scale integrated circuit technology, coupled with declining prices. The explosive chain reaction set alight any company that has any connection with CAD/CAM.

Further impetus to the boom has come in the form of major software improvements. Graphic software is advancing quickly towards common databases that can be used on any CAD/CAM system. Packaged CAD software is also attracting users, since the introduction of the system no longer entails heavy expenditures on software development. Even so, Japanese CAD/CAM systems for the moment still lag behind the U.S. leaders. The latter's advantage lies chiefly in software, especially in shape-modeling ability, which is the key factor in the evolution of CAD/CAM systems toward computer-aided engineering (CAE) and computer-integrated manufacturing (CIM).

The software lead may be far from permanent, however. Although at present CAD/CAM development tends to be judged mainly by the system's shape-modeling capability, the size of the common database and the aggregate level of automation in industry will eventually become critical determinants of systems develoment. Already, computer-aided manufacturing in Japan is running ahead of the United States, partly because NC machinery in Japan is 10 times more pervasive than in America. Also, Japanese companies are better able to finance the installation of total new manufacturing systems using CAD/CAM. To speed the process, systems designers are developing complicated Japanese language software for controlling every stage in the manufacturing.

At least 18 percent of all Japanese manufacturing enterprises are already using CAM for some aspect of production. If those experimenting with it or planning for its introduction are included, the proportion reaches 85 percent. Most important to the rate of CAM diffusion, the basic NC building blocks (which account for no less than 77 percent of the total cost of a system) are already in place. It only remains to add the computer and the system's software.

But the approach to building fully automated factories will be significantly changed by the seven-year Flexible Manufacturing Complex (FMC) Research Project, terminating with the construction of a prototype system which began operation in Tsukuba in April 1984. Unlike FMS, which are simply mixtures of conventional NC machine tools and automatic measuring mechanisms with automatic transfer and warehousing as auxiliary functions, FMC combines both total integration and modularity. Each FMC cell consists of five fully automatic, interactive operations: (1) parts fabrication using special flexible dies, (2) parts machining, (3) high-powered laser cutting, welding and treating, (4) assembly of the machined parts by robots, and (5) automated inspection. The entire operation is controlled by a hierarchy of computers, including a CAD/CAM system, which does without workers other than supervisors for standby systems.

Complex manufacturing makes possible diversified production of small quantities in a common space, with the same high productivity as special-purpose machines or transfer lines. Unlike mass production systems, FMCs are functional systems which need not be applied factory-wide. An FMC, for practical purposes, functions as an independent machine or group of machines automatically processing blanks into various finished products. Moreover, FMCs incorporate all the basic prerequisites for unmanned operations, using a battery of quite new manufacturing techniques such as laser-aided manufacturing, hot isostatic pressing, information processing, failure diagnosis, and accuracy compensation.

The complex assembly mechanism developed specifically for FMC systems processes sequentially those workpieces manufactured in the system along with bought-in parts or materials, adjusting itself flexibly and quickly to use of various assembly modes for different types of products manufactured in the system. Sub-systems inspect final assemblies, automatically diagnose failures and maintain product quality by making compensating adjustments throughout the system when and where necessary. These control techniques assure stable and continuous

FACTORY AUTOMATION

automatic operation and repair of any failed parts at an early stage. And since system stoppage is most likely to originate from tool failure, the diagnostic mechanism monitors continuously, suspending operations within a fraction of a second after a failure is detected.

Manufacture of such totally integrated automatic production systems is envisaged by 20 electrical machinery and machine tool makers who have cooperated with MITI in the R&D project; but the present suppliers of FMS are likely to be the first to offer the new integrated complex systems commercially. At present, Japan's three largest FMS suppliers are machine tool makers: Toyoda Machine Works, Hitachi Seiki and Toshiba Machine — all producers of uni-purpose machines automatically producing complex individual metal parts from raw materials. Although many multi-purpose machine-tool companies have constructed FMS to improve their own productivity, their lack of experience in producing continuous production systems has been a major deterrent to their supplying FMS.

Toyoda's Key Patents

A typical FMS currently comprises between two and five machining centers. The machining centers account for about 40-50 percent of the total cost, the remaining 50-60 percent going on automated conveyor systems and software. Because machine and computer software play important roles in FMS, suppliers should be familiar with both technologies. In the future, therefore, the Japanese machine tool industry may be divided into a group of total FMS and FMC suppliers and another group of tool makers or controls producers supplying components or modules for those systems. Historically, Toyoda Machine Works is the largest FMS manufacturer; it also leads in terms of technological prowess. With 30 basic FMS patents held by Toyoda, it is virtually impossible for other manufacturers to enter the market without its license and payment of the appropriate royalties. The company, which produces most of the basic elements of FMS itself, anticipates an order flow of five to 10 systems annually.

Hitachi Seiki, Japan's second largest FMS manufacturer, markets a variety of systems, which are currently being used by machine-tool companies, valve and compressor makers, as well as shipbuilders and agriculture machinery manufacturers. The company installed its own FMS in 1982 and plans to build its first FMC for a subsidiary company, to be used for demonstration purposes. Another leading machine tool maker, Okuma Machinery Works, constructed a large FMS, combining eight machining centers, automated conveyor systems and industrial

324

robots, at its Oguchi factory in 1982. In addition to a wide variety of machine tools, industrial robots and NC equipment, the company also produces and markets its own minicomputer, an advanced model of which controls FMS comprising up to 15 machine tools. In 1982, the company began preparing for a sharp increase in FMS orders.

Since FMS are the embodiment of mega-technologies, they require large-scale and diversified technological capabilities that are usually found only in the largest firms. An alternative industrial arrangement that is destined to be an important force in the FMS/FMC market is the consortium of companies in the same *keiretsu*. In the Furukawa Group companies, for example, Fuji Electric will supply monitor systems, automatic conveyors and automatic warehouses; Fanuc, the NC machine tools and industrial robots; while Fujitsu provides the computers to control complete systems.

With the completion of the FMC project, these systems builders are well-positioned to lead the race towards the factory of the future. Japanese manufacturers will set the pace in that race, not only because of their leadership in NC machine tools, robotics and flexible manufacturing systems, but for more fundamental underlying reasons, deeply rooted in Japanese economic instincts, social structure, cultural values and geopolitical realities. The logic of the Japanese enterprise, an educational system that develops appropriate human talents and a financial system that assures the supply of adequate and low-cost capital, are likely to be even more important determinants of industrial advance in factory automation using FMS/FMC than they were in the age of mass production.

PART VIII.

Communications

CHAPTER 23

Networks for
The Information Society

FOR THE PAST CENTURY, communications equipment and networks have been built for three basic systems: telephone, telegraph and broadcasting. Each system had its specific purpose, its separate network and frequently its special organizational structure.

Now technology is changing in ways which will extend and diversify telecommunications systems to embrace virtually every aspect of intellectual expression, social intercourse, economic activity and political behavior. The entire fabric of society in both developed and developing nations will be transformed as radically as it was by the advent of the steam engine, the internal combustion engine and mechanical power.

In Asia, this communications revolution coincides with rapid economic development which is catapulting nations, enterprises and their constituents through successive stages of industrialization which took advanced countries several generations to traverse.

The communications revolution also coincides with the end of an era in which cheap oil and cheap minerals, the basis of industrial society, were available in abundance. Now new forms of abundance are emerging. Microelectronic circuitry can provide limitless quantums of logic and computing power. Satellites and optical fiber links, which require relatively small inputs of energy and natural resources, assure communications capacity in profusion.

New telecommunications and computer systems now emerging qualify as epochal technological advances requiring a host of changes in business and governmental institutional arrangements to obtain their full benefits.

The challenge is not the scarcity of technology. Research laboratories of the world have never been more vibrant with new technology than they are now, nor have new technologies ever been so accessible. The main problem is rather that each stage of technological development has its

Reprinted from "Japan's Information Society: U.S. Manufacturers Are Now Clearly Becoming Alarmed," *Far Eastern Economic Review,* 7 December 1979.

unique institutional structure; the structures of one era are often ill-suited for the optimal development of new and qualitatively different technologies. Vested interests with heavy commitment to systems of the past systematically attempt to suppress technologies which threaten the existing order of things.

So it is with communications technology today. Not only does the proliferation of new communications and computer technologies defy the monopoly of state and private telephone and telegraph monopolies; these technologies require the fusion of point-to-point and mass communications systems now operated independently if the full advantages of these new technologies are to be realized.

It is now technically possible to replace both telegraphy and most first-class mail by fascimile services. Vast amounts of information long communicated by newspapers can now be transmitted more cheaply and instantaneously over telephone or cable TV circuits, Cable TV also can become a wholly interactive two-way communications system, providing far more diversified use of video technologies than the conventional TV broadcasting and motion picture industries combined.

Statellite transmission systems not only destroy much of the rationale for ensconced communications monopolies: they also render much of the infrastructure maintained by these monopolies obsolete. At the same time, optical fibers hold the promise for cheaper transmission of a wide variety of communications services which cannot be left exclusively to public or private monopolies to develop.

As a result, even after telecommunications technologies have proved economically advantageous, they take a long time to be developed into widespread use. As early as the mid-1960s, electronic switching, satellite communications, digital communications, the videophone and cable TV were all relatively well-developed technologically. But their introduction and commercial development have been more determined by relationships with existing communications systems operations and networks, by the availability of investment funds, by depreciation policies and by governmental constraints than by their technical maturity, efficiency or operational advantages. The question of which technologies will prevail is frequently confused with another critical issue: who will develop the technologies and commercialize the services?

Even in the most advanced industrial countries, at least 20 years is required for the introduction and full development of new technologies in a systems-oriented and highly-regulated industry such as telecommunications. In many developing countries, the time span has been much longer, while the most rapidly developing countries have succeeded in shortening the process appreciably.

Indeed, in an information-oriented industrial system, the pattern of economic development of all nations will be determined in large measure by the pace at which new communications technologies will be adopted and the efficiency with which they are rendered as useful services to business and individual subscribers. The relationship between the number of telephones in service and the level of economic development of a country has long been evident. In the future, more sophisticated indices of the use of a wide variety of telecommunications equipment and systems will measure the standards of living and quality of life throughout the world.

It is significant that technological advances have been introduced more rapidly in the less regulated industries and countries; and those countries which have adopted a rational information industry policy have been most sucessful in quickening the tempo of change.

It is in this context that two Japanese experiments in communications services based upon new technologies are especially important. In 1972, the Ministry of International Trade and Industry (MITI) and Nippon Telephone and Telegraph Public Corporation (NTT) took two decisive steps forward towards the information age of the 1980s,. launching parallel experimental projects in the development of "wired cities" to test all aspects of new communications services. In January 1976, at Tama New Town, a suburban community north of metropolitan Tokyo, 10 new communications services covering a wide range from VHF cable TV to wideband futuristic facsimile information services were introduced under the auspices of NTT using a whole new generation of equipment especially developed for the project over the previous four years. Some what more than two years later, on July 18, 1978, at Higashi Ikoma near Osaka, the MITI-sponsored HI-OVIS (Highly Interactive Optical Visual Information System) was inaugurated incorporating two remarkable features not to be found in conventional communications systems.

At Higashi Ikoma, for the first time in the world, optical technologies were employed throughout a communications system. Optical fibers, optical conversion equipment, optical connectors, optical switches, optical sensors, and holographs were installed in a total optical communications system. This system, in turn, made possible a fully two-way audio-visual communication function. Homes were equipped with keyboards, cameras and microphones in addition to regular TV receivers, enabling subscribers to select from a wide variety of entertainment, educational and informational services transmitted by a central station.

By means of the keyboard terminal linked to central and regional computers, subscribers can request TV programs, rebroadcast arrangements or any one of a variety of entertainment or educational video tapes.

A wide range of data is also available for transmission in still picture form, including current weather information, train and air schedules. stockmarket reports, time checks, local news and various announcements of general interest. Hard copies of video information can be obtained, through a facsimile terminal, which also transmits a home newspaper.

Subscribers do not have to leave their homes to shop or pay bills. Goods and prices are shown on the home TV screen, can be ordered through the keyboard terminal and paid for automatically from bank accounts. Gas and water meter charges are settled as cashless transactions.

Similarly, subscribers can make restaurant, theater and travel reservations through the system and pay for them from home: when they are away, they can regulate their home appliances, heating and cooling systems from outside their houses. Detectors are installed in each home so that the central computer can sense any abnormality, such as burglary, excessive heat or smoke, and respond as necessary. Alarms can be automatically transmitted to police and fire stations, instructions relayed to the household on emergency measures to be taken.

Computer-assisted instruction is available for both children and adults, using still pictures and video materials. Questions and assignments appear on the screen of the multi-channel home TV receiver; students respond on the keyboard and the computer immediately evaluates their answers and regulates the progression of teaching.

Information retrieval facilities and entertainment provide access not only to national and international events, but also to those originating locally. Professional producers provide a continuing flow of local programs through the central studio and mobile units which cover sports, school and other community activities. The same communications channels can provide medical facilities, some computerized and some via the TV screen. Health care and preventive medicine instruction are readily programmed into the system.

In the wired city, the cabling that provides these facilities is as important as its water pipes and electric network. When the HI-OVIS project was begun in 1972, plans were made using coaxial cables. Developers soon found, however, that the sheer volume, cost and relatively narrow transmission area per cross-section of cable made coaxial cable uneconomical for the geographical scope, multichannel requirements, two-way services and future flexibility of the communications net. By 1976 the project had come to a virtual standstill.

But Dr. Masahiro Kawahata, then deputy general manager of corporate planning at Fujitsu, convinced MITI that the adoption of optical fiber technology as the wider-band, lower cost and readily expandable alternative to coaxial cable, was the answer.

Subsequently invited to join the project as its managing director and chief engineer, he went on to mobilize the resources of Fujitsu, Matsushita Electric Industrial, Sumitomo Electric Industries and Toshiba to redesign and supply the major elements of the system.

The Higashi Ikoma pilot project is experimental not only in its evaluation of the performance of the system in operation and determination of the modifications necessary to transform it into a full-fledged profitable commercial communications operation. Its findings will also provide critical infrastructural data for the planning of small wired cities of no more than 100,000 residents each to be established on the periphery of the crowded population centers of the Kanto and Kansai plains.

Although the experiments at Tama New Town differ substantially from those conducted at Higashi Ikoma, the results provide important comparative data for the development of these future systems. While the HI-OVIS project, with the participation of the TV industry, is based essentially upon cable TV transmission technology, the Tama experiment was conceived and conducted by NTT as an exercise in diversified utilization of existing telecommunications networks, with necessary modifications of switching and transmission systems.

The basic assumption from which the Tama experiment began was that a telephone network employed only for voice communications is utilized for a small percentage of its total capacity. Since the Tama experiment did not include two-way visual communications, coaxial cable was quite adequate for its purposes. Data retrieval is obtained by using a touch-tone telephone: fascimile services are transmitted over telephone networks.

NTT's objective is to build a nationwide, high-speed information network to transmit national TV, facsimile and other forms of programed information. The Ministry of Posts and Telecommunications (MPT) wants to put a communications satellite into orbit for just this purpose in the early 1980s.

Obstacles to this scheme are formidable, however. Its success would require the full support and cooperation of the TV broadcasting industry, composed of powerful public and private corporations with heavy investments in existing transmission facilities which might more readily be supplemented by and integrated with a network of more or less independent local cable television systems. But there are also important financial constraints: before such a satellite system could be operated economically, a substantial base of facsimile receivers would have already to be operational and this remains problematical.

In the meantime, MPT has decided to move ahead with the development of an interactive videotex system in cooperation with NTT.

Christened the CAPTAIN (Character and Pattern Telephone Access Information Network System), the new system provides information stored in a central computer data base on request from subscribers via the existing public telephone and displayed on the TV screen. This system avoids the cost of laying expensive coaxial or optical fiber cables and putting a satellite in orbit, which would be prohibitive for NTT to undertake. Similar in many respects to Prestel, Teletel and Bildschirmtext systems now in the development stage in Europe, the CAPTAIN system differs essentially from its fore-runners in one important aspect: pattern transmission methods, rather than code transmission is used, with the installation of a character and graphic generator at the system's center.

At the end of April 1979, 130 companies and public agencies had joined to cooperate as information providers to the system which becomes operational on an experimental basis in Tokyo at the end of 1979.

This system is not incompatible with other systems such as that now on trial at Higashi Ikoma. The HI-OVIS project, unlike the Tama New Town system which required a national network, in its scope and organization arrangements is essentially a locally-based community-oriented information system not unlike locally-franchised cable TV systems which have developed in the U.S.

Since development of the complex mix of services would require massive investment in optical fiber transmission networks, funding might more easily be raised by a multiplicity of private entrepreneurs rather than by a single national corporation. Also, there is the lingering apprehension that NTT may be overly committed to its existing network and technology mitigating against the development of new technologies as aggressively as multiple private operators of numerous local systems.

The revolution in communications technology has much wider implications for Japan in the 1980s than that of communications itself. The fullest development of communications and computer technologies is vitally important for all Japanese industries and for the economy at large. Rapid development of communications technology is an imperative of the future economic development of Japan not only in its service aspects, but also because of the role hardware production must eventually play in Japan's knowledge-based industrial structural change.

Most important, the combination of optical fiber computers, audio-video and facsimile technology could ultimately prove to be the most important source of innovation in high added-value production for Japanese industry during the remainder of this century.

In the broader context of Japan's increasingly knowledge-based economy, communications and computers are key components of the technical infrastructure. The most dynamic arena of economic activity now and for the foreseeable future is predominantly occupied by producers of information goods and services. At the same time, the informational activities of Japanese private business and public bureaucracies — research and development, planning, control, marketing, record-keeping, coordination, monitoring, regulating and evaluating activities, and all the informational functions of government — constitute the increasingly important information overhead costs to the economy.

The primary information sector in Japan includes a growing multitude of industries that produce, process, disseminate, or transmit knowledge or messages: the secondary sector, particularly firms and government agencies which produce and consume a myriad of informational services internally, consumes a tremendous amount of both capital and human resources in the process. Taken together, they already account for a substantial share of Japan's GNP.

Consequently, major policy issues confronting Japanese business and government leaders are those which relate directly to the information sector. Decisions concerning arrangements of the information infrastructure itself — common carrier policy, competition in various communications markets and investments in new technologies — are no longer peripheral; they will determine the future course and rate of growth of the Japanese economy as well as the quality of life of the Japanese people.

Japan's role in the world economy is, at the same time, very much at stake in the emerging information age. As a prolific producer and exporter of electronic goods, Japanese communications and computer industries are clearly destined to be export-oriented. Although the foreign trade intensity of the Japanese economy can be expected to decline as it shifts from a heavy and chemical industry base to an information industry base, the importance of information industries in the overall mix of goods and services production will make it increasingly imperative that these industries supplement consumer electronics products with more sophisticated information equipment and services on the roster of Japanese exports.

Communications and computer markets have been characteristically less open than those for consumer products, and in the telecommunications and computer markets, U.S. industry still holds a clear worldwide advantage. U.S. manufacturers are noticeably alarmed at the prospects of losing market shares to Japanese competitors which have now attained parity in the forefront of communications and computer technologies. Their response has been to seek governmental measures to impose protective tariffs or nontariff barriers, thus undoing much that has been achieved in recent years by more liberal communications policies.

CHAPTER 24

The Changing Role of NTT

SINCE THE ESTABLISHMENT of the Nippon Telephone and Telegraph Public Corporation (NTT) in 1952, the government-owned service company has provided the main thrust for Japan's rapidly-developing communications equipment industry. With a continuing backlog of unfilled orders for telephone service, for more than two decades the public monopoly was a growing captive market for NTT family firms, its designated suppliers of electronic communications equipment, cables and a wide-range of other products. But, more than that, NTT has also provided enormous impetus to technological progress of the industry through its own extensive research and development activities and projects undertaken jointly with its various family companies.

Nurtured by steadily increasing orders and assisted by timely increments of new technology from NTT laboratories, Japanese communications equipment manufacturers have grown rapidly since the difficult days of the early 1950s. Major firms have attained strong financial positions and are now among the leading contenders for the world telecommunications equipment and engineering markets.

Still, the four largest suppliers — Nippon Electric, Fujitsu, Oki and Hitachi — together account for as much as 50 percent of NTT's equipment orders and individually rely on NTT for up to 40 percent of their turnover. The remaining 50 percent of NTT's annual equipment is supplied by some 250 medium- and small-sized family firms, many of which are dependent on the public monopoly for as much as 60 percent of their total sales.

When NTT's fifth five-year development program ended in March 1978, the number of telephone subscribers had reached 35 million, more than 22 times the number at the end of March 1953. During the intervening 25 years, NTT made investments totaling some ¥13,600 billion (US$57.1 billion), nearly 30 percent of which was used for procurement

Originally published as "All in the (NTT) Family, But All Is Not That Well," *Far Eastern Economic Review,* 7 December 1979.

of telephone or telephone exchange equipment supplied by Japanese communications manufacturers. In the fiscal year ending March 1979, NTT communications equipment purchases amounted to a formidable US$329.6 billion.

But despite this impressive performance, all has not been well with NTT. Prior to the end of the fiscal year in April 1979, for which the monopoly reported a healthy surplus, for four years running it had been operating at a loss. In 1974, demand for telephones, which had continuously surpassed supply since 1952, finally reached its peak and began to drop sharply in the following years. The backlog of telephones to be installed was quickly erased, leaving NTT with more than 400,000 unsold telephones on its hands at the end of fiscal 1975. Installations during that

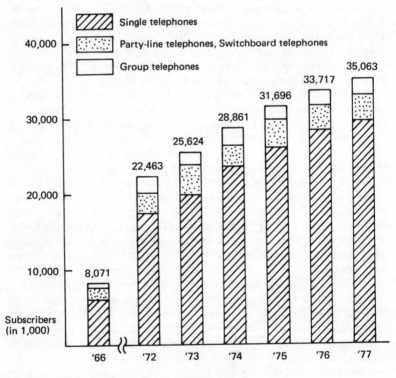

Trends in Telephone Subscription

Source: Ministry of Post and Telecommunications.

year had fallen far short of the projected 2.6 million telephones; and by 1978, the number of new telephone installations had dropped to about half that of the peak year of just four years earlier.

Also, in 1978, for the first time since the inauguration of NTT's telex service, its subscribers not only failed to increase but actually declined by almost 3 percent. To some extent, of course, the decline in demand for telephone services after 1974 was a reflection of the prolonged recession and the slow recovery of the economy. But the failure of demand to rebound in 1978 and the decline in the number of telex subscribers in that year suggests that a fundamental, structural change in the market has occurred.

It is not that the total demand for communications service has suddenly been satisfied, but rather that the demand has shifted to new forms of communications. Telephone, telegraph and telex communications are gradually being supplemented and replaced, especially in business, by data and fascimile communications.

Dramatic advances in telecommunications techology are bringing about radical changes in electronic equipment, the nature and organization of the market for services and equipment, the structure of the manufacturing industry and the role of NTT in the supply of services.

The most significant changes have been wrought by the digital computer, of course. The computer has generated demand for communications services, auxiliary equipment and finance quite different from those of the conventional telephone or telex. With scattered remote terminals connected to central computers and time-sharing of computers among multitudes of users, the requirements for data transmission cannot be met by a network, or by an organization, designed for voice communications. With recent advances in semiconductor technology, the proliferation of applications for remote data processing and other specialized communications services requires a wide variety of terminals. Those needed for handling airline or railway reservations systems are quite different from those required for verifying credit cards, for text editing, for engineering design or for high-speed fascimile transmission.

Markets for these products and services not only differ vastly from that of the telephone, but they vary substantially from product to product. While 66 percent of telephone subscribers in Japan are residential, the market for data communications, as well as for fascimile, is almost entirely institutional — business, government, research institutes and universities. And the market, in the case of data communications, is not essentially an equipment market, but a market for systems and service which often entails a substantial custom-designed software increment.

Although some designated suppliers of traditional telephone and telex equipment have emerged as major suppliers of data communications and facsimile equipment, their relative technological strength and market shares in the various product lines are quite different. Unlike traditional equipment, which is still supplied entirely by Japanese manufacturers, foreign suppliers have a strong position in the computer market. Also, while traditional equipment was usually supplied by specialized communications companies, data processing equipment is produced by a wide variety of manufacturers spanning the full range of the consumer and industrial electronics industry. Similarly, consumer electronics, business equipment and optical instruments manufacturers are competing with old-line communications companies for the rapidly growing fascimile market.

NTT has designated six traditional communications equipment makers as suppliers of home fascimile equipment to be introduced in 1980, and by so doing is certain to foster mounting resentment within industrial circles as well as in other agencies of the government. Pressures can be expected to increase, both from those Japanese facsimile manufacturers who have been excluded from the NTT family and from foreign manufacturers as well, to be allowed to make unrestricted sales of equipment directly to private users who would, in turn, interconnect with NTT facsimile networks.

Moreover, these disgruntled outsiders are not likely to find themselves alone. Other users find the arrangements originally designed for telephone and telex services ill-suited to needs of new communications media and equipment. Although NTT intends to purchase home facsimile for lease to subscribers just as it does telephone and telex equipment, users of data and specialized fascimile communications have been allowed to acquire their own equipment and lease communications circuits from NTT, as needed. But this experience has been far from satisfactory.

Industry and business circles contend that the present monopoly of the data communications lines by NTT obstructs efficient utilization of computers. Not only have corporate users of leased circuits not benefited from economies of scale, which is the main argument in favor of a single network, but NTT rates are generally estimated to be six times as high as comparable U.S. charges for equivalent data services.

Quite apart from this issue, however, advances in telecommunications technology raise questions about whether new and essentially different networks are required in addition to the conventional networks and whether these need to be supplied by a single public monopoly. For the main concern here is not just economies of scale for basic services; the focus is rather on economies involved in delivering specialized communications services, which are inherently dissimilar to the relatively homogeneous services of conventional telephone systems. With the increasing heterogeneity of data communications systems and terminal applications, large Japanese computer users as well as a growing number of nascent terminal manufacturers doubt whether the needs of the economy, and of society at large, are best met by having end-to-end communications service in the hands of one organization.

Major Japanese computer users complain that strict legal restrictions on the use of data communications cricuits make it difficult for them to expand their on-line networks as needed. Banks, for example, are not allowed to link on-line networks between more than two banks unless they use NTT computers and other telecommunications facilities, which they do not require. Other users have voiced grievances about the ban on message switching and NTT's rejection of applications by users wishing to install digital equipment from American suppliers, on the grounds that similar products would be forthcoming from Japanese manufacturers.

Some very important users have requested the government's Administrative Management Agency to investigate the data communications service of NTT, making it clear that they want the government to open use of the data communications circuits to private companies. In its paper published in early 1979, the Telecommunications Users Association, representing approximately 60 large corporate communications users, urged further that a more liberal interconnect standard be established, allowing Japanese companies to compete freely in the terminal market.

Thus an impressive alliance has emerged between large communications users and makers not included in the NTT family of suppliers, demanding that users be allowed to provide their own terminals and other equipment which could be plugged into the telecommunications transmission network, just as users provide their own home appliances and plug them into the electric power industry's grid.

This would not necessarily exclude NTT from the sale or lease of terminal equipment, of course. But it would force NTT's top management to transform the corporation from a dedicated monopolist, intent on preserving its privileges, into a vigorous marketing organization, responsive to customers and capable of thriving on unfettered competition.

An immediate effect of such a competitive environment would be the restructuring of that sector of the equipment industry referred to as the NTT family. NTT executives freely acknowledge that some of the smaller suppliers could not survive in the face of such open competition, and that some of these would undoubtedly have to be taken under the wing of the giants or allowed to fade away.

As things now stand, however, changes in this sector of the industry are to be expected in any case. Medium-scale and smaller makers of communications equipment have been especially hard hit by the sharp decline in NTT's orders for conventional devices; they also stand to benefit least from the emergence of new capital-intensive technologies with high economies of scale. With the rate of dependence on NTT orders as high as 60 percent for some firms in this category, marketing organizations and skills tend to be underdeveloped. Not only have many of these firms neglected to establish adequate sales networks abroad, but they are often not organizationally or managerially equipped to compete aggressively for private demand at home.

A number of firms in the family are managed by ex-NTT executives whose major function has been to strengthen ties with their former employer and now principal customer. Few have the experience, much less the inclination, necessary for success in aggressive marketing in an open competitive market. For some of the smaller firms the moment of truth is not far off. With increasing emphasis on electronic exchanges, data communications, facsimile and other equipment employing more advanced technology, all signs point towards greater NTT reliance on the Big Four equipment manufacturers.

This trend is clearly manifest in recent annual reports of NEC, Fujitsu and Hitachi, all of which report increased sales of telecommunications equipment to NTT. In fiscal 1977, Fujitsu reported that NTT orders, mainly for D-10 electronic switching systems and other digital equipment, were the highest ever recorded. All three leading communications equipment makers report rapid growth of facsimile equipment, while NEC has also announced the installation of a major experimental optical fiber cable system for NTT and another for a major electric power company.

Significantly, however, even as NTT's reliance on the Big Four seemed to be increasing, they stepped up their efforts to reduce their dependence on NTT procurement. Their strategy, in a word, is diversification — diversification of both product lines and markets.

To be sure, all four major communications equipment makers began their respective drives towards product diversification in the 1950s, and each of them took a major step in that direction by entering the computer field in the 1960s. But since 1972, when NTT orders began to level off, this movement towards greater diversification has assumed much greater amplitude.

For Hitachi, this came naturally. Already a highly diversified electrical machinery manufacturer, communications and data processing have never constituted a major share of Hitachi's turnover and in 1978 was only 17 percent of total revenues. By 1978, Fujitsu's sales of communications equipment had dropped to 25 percent of total turnover, having been replaced by computers as the company's major product line. Meanwhile at NEC, telecommunications equipment was steadily declining in importance, falling below 40 percent of total revenues in 1978, as corporate resources were channeled increasingly into data processing, semiconductors and consumer electronic products.

Since 1972 also, all four firms have expanded their overseas operations. In the manufacture of advanced communications equipment, as in computers and semiconductors, the economics of production and competitive power renders global startegies imperative. Once reliance on NTT was no longer a viable option for sustained growth and competition became the *sine qua non* of survival, the four major communications equipment makers began to develop export markets aggressively.

Exports have risen steadily through-out the 1970s, with sharp increases registered in 1975, 1976 and 1978. In 1978 overseas shipments of communications equipment reached ¥202 billion, equal to almost two thirds the total equipment procurement of NTT for the year. Since most advanced country markets remained closed to communications equipment, the major share of these exports was to developing countries. Exports of electronic exchanges to the United States have registered notable success, however, nudging total sales of Japanese communications equipment there to US$38 million in 1978, or about 3.8 percent of the industry's worldwide exports.

More important, by 1978 Japanese manufacturers had long since decided to take the investment route to the U.S. communications market rather than attempt to develop it mainly through exports of equipment from Japan or offshore Asian production facilities. All four major communications equipment makers are now committed to manufacturing electronic private branch exchanges (EPBX) in the United States, after the success of Oki Electric's assembly operations, begun at the time of its initial entry into the market in 1971.

Oki Electronics of America Inc., a wholly-owned subsidiary of Oki Electric with manufacturing facilities in Fort Lauderdale, Florida, intends to double its sales of EPBX equipment to US$45 million during fiscal 1979. Likewise, Fujitsu will double the output of Focus II EPBX switchboards by its joint venture, American Telecom Inc. in California, and has revealed plans to expand production to include electronic switchboards for telephone stations in the United States. This announcement followed shortly after a report that Nippon Electric is going to manufacture station-use telephone exchange equipment in its Dallas, Texas plant, where it also intends to double production of EPBXs in the current year. NEC is also preparing to install facilities in the state of Washington for production of ground station equipment for satellite communications.

Following this success of the three leading communications equipment manufacturers, the fourth, Hitachi, is reportedly now planning to replace its exports of telephone switchboards to the United States with local production. At present, Hitachi is marketing crossbar telephone exchanges and space-division electronic exchanges through RCA after assembly at the Atlanta office of Hitachi America. Hitachi's plan for expanding those operations into manufacturing of EPBX is based on a market survey which indicates that more than 70 percent of potential users of telephone switchboards in the United States want time-division electronic exchanges.

Before the June 2, 1979, reciprocal open door policy agreement between Japan and the United States, the threat of protectionist measures figured among the factors determining the decision to produce in the United States to supply the American market. It was not the main reason for the initial decision, however, and will have even less influence on future decisions if the general U.S.-Japan agreement to the principle of reciprocity is translated into meaningful arrangements.

Japanese communications equipment makers have clearly found that production of switchboards and other terminals in the United States has sound economic reasons. With the revaluation of the yen and soaring production costs in Japan, exports of completed switchboards to the United States have become increasingly less profitable. Hitachi has determined that costs can be cut by local production using semiconductors made in the United States. Apparently those ICs (integrated circuits) and LSIs (large-scale integrated circuits) used in electronic switchboards are available in the United States at prices lower than their equivalents in Japan.

It is in this general context of far-reaching changes in the structure of the Japanese communications equipment industry that the sudden emergence of NTT's procurement policies as a major issue in U.S.-Japanese economic relations must be seen.

Available evidence does not sustain common wisdom that trade, present and future, was the main purport of the NTT affair. Although it is true that trade between the two countries in communications equipment is weighted more than three to one in favor of Japan, the total amount of that trade in both directions is negligible in terms of overall trade flows between the two countries, and most certainly does not deserve all the attention it has been given.

Moreover, if we are to take seriously U.S. Department of Commerce prognoses in its recent study of U.S. trade opportunities in Japan, the picture is not expected to change much in coming years. The authors of that study state unequivocally that "although the government market may be somewhat more open in coming years, sales [by foreign suppliers] will continue to be limited essentially to specialized equipment and domestic suppliers will maintain a strong market position." The share of imports, the Department of Commerce predicts, will actually drop from 2.7 percent of the market in 1977 to 1.9 percent in 1982.

It has been observed that if NTT were to adopt open bidding procedures, in fact the main beneficiaries would very likely be those Japanese communications equipment makers that remain outside the NTT fold. American communications equipment manufacturers, on the other hand, usually prefer to locate production in their major markets. ITT, for example, has as many as 11 reed relay plants in Europe alone, because the realities of the communications business require a close working relationship between equipment suppliers and service organizations, and the latter are still structured on strict national lines.

The main problem for American communications equipment manufacturers, then, is not that they cannot sell to Japan but that they have not been able to invest in Japan except as minority partners of Japanese makers.

Remarkably, as in the case of color TV, American complaints against Japanese competition and alleged exclusion of U.S. products from the Japanese market began *after* the decision had already been made by major Japanese firms to invest in U.S. production facilities.

If we read between the lines of diplomatic rhetoric in these two cases, a most useful object lesson is to be found. Japanese firms in those sectors of industry characterized by an absence of American investment in their home market are likely to encounter serious resistance when they begin to invest in U.S. production. It is noteworthy that the lobbying action in Washington against Japanese color TV imports was led by firms with no major stake in the Japanese market, while multinationals such as RCA and General Electric did not take part in that action.

Similary, Texas Instruments, which has successful manufacturing operations in Japan, has refused to join in defensive political moves to stem the tide of Japanese semiconductor imports in the United States.

In the communications equipment industry there are undoubtedly opportunities for American investment in Japan that would directly benefit all parties concerned, and would simultaneously mitigate some of the concerns that NTT has about international procurement outside Japan. At the same time, such investments would have the very important effect of demonstrating in a concrete way that the Japanese market is just as open as that of the United States. Internationalism is a two-way affair: and internationalization of Japanese production necessarily entails visible signs of internationalization of the Japanese domestic industry.

To achieve any kind of internationalization in the communications industry, a positive understanding of and initiative by NTT is essential. And here lies the real importance of the June 2, 1979 agreement in which the principle of reciprocity was endorsed by the Japanese and U.S. governments.

Reciprocity is in keeping with the economic and technological realities of the communications industry. Manufacturers in both the United States and Japan must operate within the framework of these same realities, which among other things dictate that manufacturers of high-technology communications equipment must make the best return on investment through rational development of world markets, which entails production in major markets.

This is especially true for communications equipment which is sold mainly to public or private monopolies. Communications equipment in these markets is not sold off the shelf. It is designed to user specifications and usually entails intensive involvement of the user in all stages of this manufacturing process. Here again, the provision in the U.S.-Japan accord of June 2, 1979, by specifically declaring the intention of the two governments to foster business-government cooperation between the two countries in research and development in the field of communications, constitutes a most significant step forward towards effective internationalization of the industry.

The day of cloistered national communications industries pursuing mercantile business strategies clearly belongs to the past. The future of the communications equipment industries in advanced and new industrial countries alike lies in cooperative, global solutions. This is clearly the way to progress in the communications industries of tomorrow as well as to harmony in a communications-oriented global economy.

Major New Technologies for the 1980s

Communications satellites
Each successive generation of satellites has greater communications capacity and higher radiated power. More satellite power means that smaller earth stations can be used, a trend that will result in direct rooftop-to-rooftop communications in the 1980s.

Low-cost satellite earth stations
Planar microwave circuits make it possible to mass-produce satellite receiving equipment at very low cost. Satellite receivers cheap enough for home use have been installed experimentally in Canada, Japan and India.

Time-assigned speech interpolation (TASI)
Low-cost LSI microprocessors and memory units make possible more efficient use of satellite or other high-capacity transmission channels allowing them to be shared by multiple geographically dispersed users in a highly flexible manner; portions of channel capcity can now be allocated to users according to their instantaneous time requirements.

Helical waveguides
A metal tube has been devised that can carry 250,000 or more simultaneous telephone conversations, 200 TV broadcasts, videophone pictures, electrocardiograms, data between thousands of computers all at once. The bandwidth available for transmission through the waveguide, 40 to 110 gigahertz (GHz), is a greater bandwidth than all present through-the-air radio bandwidths added together.

Optical fibers
Thin flexible fibers made of extremely pure glass can carry up to one billion times as much informatinon as conventional telephone cables. Japanese manufacturers have perfected optical fibers capable of transmitting more than one billion bits per second. Given their small size, a number of lower-capacity fibers can be run together in a cable to meet virtually any transmission need.

Lasers
Laser beams, which can be amplified and have a frequency 100,000 times higher than microwave signals currently in use, have potential information-carrying capacity thousands of times greater than microwave. Used with optical fibers, laser semiconductors can carry several thousand simultaneous telephone conversations or their equivalent.

347

COMMUNICATIONS

Millimeter-wave radio
The new frontier of radio is in millimeter wave transmission. Radio at frequencies in the band above the microwave band can relay a volume of information greater than all the other radio bands combined. The payoff for those discovering how to use new millimeter-wave radio technologies will be enormous.

Packet networks
Generalized packet switching networks will be used widely in the 1980s for transmitting digital data and messages terminal-to-terminal for most major businesses and some professionals in their homes. Like end-to-end satellite communications, packet services are now primarily designed for business data users who can afford the required processors at each terminal. However, by the late 1980s, the costs of word processing and message transmission should be sufficiently low to make them attractive to all business and professional entities, large and small, substituting for much first-class business mail.

Toll-call signaling
Common Channel Interoffice Signalling (CCIS) is an innovation that illustrates the growing use of digital transmission in the switched-telephone network. A system for transmitting the information necessary to route a call through the toll network over digital circuits separate from those used for the call itself, CCIS is vitally important for data transmission that sends short bursts of pulses between computer terminals. Besides using toll circuits more efficiently, CCIS encourages the use of the switched network for both voice and data.

Cellular mobile systems
Mobile communications systems of the 1980s will make extensive use of microprocessors and associated LSI circuits for signal processing and control functions in order to use the limited available frequency sepctrum more efficiently. One innovation for expanding the number of mobile radio users is to reassign or "re-use" frequencies within a mobile service area by dividing a large mobile service area (MSA) into smaller cells ranging from one to eight miles in radius. Frequencies used in one cell — equipped with a low-powered base station transmitter operating at different frequencies than those in adjacent cells — can then be reused in other non-adjacent cells within the MSA. A cellular system can accommodate roughly 3 to 50 times the number of mobile units as the conventional multichannel trunked system, depending on cell size and the frequency-assignment scheme.

Packet radio
An alternative new technology, in a packet radio system, a mobile unit transmits bursts of digital pulses using the entire mobile frequency band, rather than transmitting continuously. The advantages of a packet mobile radio system are the system's ability to integrate voice and data communications efficiently, its flexibility in handling varying demand patterns, and its extensive use of LSI digital circuitry, which promises large cost reductions over the next decade.

348

Pulse code modulation

All signals, including telephone, videophone, music, facsimile, and television are readily converted into digital bit streams and transmitted, along with computer data, over the same digital networks with the use of pulse code modulation (PCM) links. PCM has the advantage of lower costs per telephone channel over short-haul lines and can greatly multiply the utilization of lines within city areas.

Codecs

A major thrust in communications innovation is the encoding of speech in a smaller number of bits. Codecs are circuits which convert signals such as speech, music, and television into a bit stream, and convert such bit streams back into the original signals. Codec design is expected to improve greatly in the 1980s, so that 96 or more conversations of excellent quality (instead of the standard 24) can be sent over a single channel. Codecs will thus become steadily more inexpensive and efficient.

Data broadcasting

A wide variety of information can be broadcast in digital form at VHF and UHF frequencies for reception on home TV sets, special terminals, or portable receivers. Stock market quotations, for example, might conceivably be received by an LSI wristwatch which picks up Dow Jones or Reuters services from a broadcast radio signal.

Computerized switching

The major switching innovation in the past decade, computer-controlled switching in local central offices and PABXs, is now surpassed by digital or "time division" switching that more fully integrates computer technology and is more compatible with digital communications. Smaller and lower-cost computerized exchanges make possible switching close to user locations, rather than in large centralized offices with vast numbers of cables linking them with subscribers. Switching, like computing, can now be done in mini- and microdevices of high reliability, as well as in large centralized units.

Word processors

The evolution of the typewriter into an intelligent word processor is the key step in the development of the office of the future. Not only will word processors containing a microprocessor and associated memory, a keyboard, a printer or graphic display become univeral in usage during the 1980s, they will interface with the communications network to permit transmission to a central facility for storage or distribution or to recipients outside the office, bypassing not only the postal services but also internal mail distribution systems at both ends.

Intelligent data terminals

The same technical trends will make data terminals less expensive, more powerful, and largely indistinguishable from word processors. Decreasing costs of microprocessors will not only encourage more data processing at the terminal; at the same time, computer users will routinely have access to remote information bases and computing facilities.

Intelligent copiers
Copying devices will evolve into "intelligent copiers" with the addition of a microprocessor, memory unit and multiple-font character generator incorporating facsimile scanners that use bandwidth-compression techniques to provide transmission of graphic materials over the communications network. By 1990, intelligent copiers will likely replace, in large measure, offset presses and duplicators, photo-composition equipment, stand-alone office copiers and commercial facsimile equipment presently in service.

Computer-based PABXs
In the 1980s, businesses will increasingly intermix voice, data, text and message information on the same communications lines. An intelligent, or computer-based, PABX will control these information flows.

Teleconferencing
Intelligent terminals in the office will permit an expansion of teleconferencing from audio only, which has been available for some years, to include audio and graphics, messages, or video. Text and other alphanumeric materials can be exchanged using graphic terminals that will be widely available in offices in the 1980s; special equipment for transmitting handwriting and other graphics will also be commercially available, while technology will reduce the costs of switched video service over the next decade.

Point of sale terminals
POS terminals with a low-data-rate interface to the communications line and relatively simple control logic will enable stores to give immediate check and credit authorization, debit or credit funds in customers' bank accounts, and maintain accurate sales and inventory records long-distance.

Banking terminals
Automated teller machines (ATM) and on-line teller terminals will provide such banking services as cash withdrawals or advances, deposits to checking or savings accounts, or transfers of funds among accounts, without a human operator. POS and ATM can be linked in integrated electronic funds-transfer (EFT) systems that will reduce the flow of checks and credit card paper work.

Videophones
Telephones combined with a picture tube enabling subscribers to see each other as well as still images will become economically feasible in the 1980s. Inexpensive digital logic and memory units in the videophone terminals now make possible signal processing and storage to reduce transmission costs. In the coming decade, the declining cost of LSI and bubble memories should permit bandwidth-compression techniques to improve storage capacity at each terminal while

charge-coupled-device cameras will also reduce the costs of video input terminals, further reducing overall costs of Videophone services.

Intelligent telephones

The 12-key pushbutton telephone enhanced by a microprocessor and associated memory, a slot for a magnetic card or some other simple data-entry device, and a liquid crystal or light-emitting diode (LED) alphanumeric line readout will serve as a low-cost intelligent terminal. Such an intelligent telephone could be used for conference calls, repertory dialing, call-forwarding, and other telephone-related functions as well as providing access to information and transaction services that do not require extensive graphic or hard-copy output, including financial transactions, stock quotation services, clock and calcuator functions. It could also serve as a dedicated home computer system that would, among other things, monitor and control energy use.

Intelligent television receivers

The addition of an LSI chip with memory and a character generator to a television receiver now enables viewers in Japan and a number of European countries to receive an increasingly diversified mix of programming and information transmitted over spare lines of the broadcast television signal. This is the first step towards interactive two-way television services which will provide subscribers with a much richer array of information services via telephone lines or a two-way cable television network. Video cameras, alphanumeric keyboards, and facsimile terminals all connected with the modified television set can potentially transform the television receiver into the "ultimate" home terminal that could be used for a wide range of services — image, voice, message, data and text.

Home facsimile

Either as an independent terminal or in parallel with modified television receivers, facsimile devices will become widely used during the 1980s for home terminal-to-terminal messages that could potentially displace a significant volume of first-class mail services. Other possible uses for home facsimile terminals include delivery of newspapers, periodicals and other bulky documents, but at present costs for home delivery would have to rise by a factor of ten before electronic transmission of second-class mail matter using available technology would become competitive.

Home printing terminals

Impact or non-impact printing terminals for the home, with full alphanumeric keyboard, microprocessor, memory and communications interface are currently being developed. Further technological advances will be necessary before the electric typewriter can form the basis for lowcost home printing terminals, analogous to the transformation of the office typewriter into a word processor, however.

Cable television

In the 1980s, cable television systems are expected finally to develop their long-awaited capacity for two-way interactive television and other services. Cable television terminals equipped with the basic logic and communications elements needed to send data and short messages from the home to the cable system studio or headend could be used for other data, information and message services not readily available by, or supplemental to, services provided by telephone lines.

Large TV screens
Conventional television receivers and display terminals can be replaced by TV screens that occupy a wall, if necessary.

CHAPTER 25

New Technology and
Industrial Structures

IF, INDEED, TECHNOLOGY is the main engine of industrial and social change, the mounting wave of telecommunications innovations being introduced in Japan is destined to radically transform structures of economic activity and society as a whole during the remainder of the 20th Century. The long-mooted Information Society is finally emerging from the misty realm of dreams to tangible reality. The imminent privatization of the Nippon Telegraph and Telephone Public Corporation (NTT), the launching of Japan's first operational communications satellite, and the meteoric takeoff of optical fiber production are but signals indicating the shape of things to come.

Changes now underway will have effects on human institutions and behavior as far-reaching as the invention of printing, the telegraph and the telephone. Office, factory and home are already being transformed.

The effects of new communications technologies are already apparent in the restructuring of old industries and calling forth of new ones. Japanese copper refiners, who depend on telecommunications for at least 12 percent of their total turnover, are confronted with declining orders as traditional wire cable is replaced by optical fibers in a widening range of applications. Cable makers are perforce diversifying from copper wire into glass fiber production, and a new sector of optical fiber component makers is emerging. As communications equipment shifts from the analog to digital mode, and software is increasingly built into systems firmware, large integrated electronics companies are gaining increasing shares of the market.

A new division of labor is developing between the major communications equipment makers and the smaller manufacturers gravitating into their orbits. At the same time, a new breed of software entrepreneurs whose services are required for advanced communications network design is shaking up established patterns of industrial organiza-

Revised text, originally published as "Japan: Electronic Highways," *Far Eastern Economic Review*, 10 November 1983.

tion, much as Sony, Casio and Pioneer did by spurring the earlier rise of the consumer electronics industry.

Japanese communications and information industry policymakers have been specifically and wholly committed to the development of information technologies and their societal underpinnings for well over a decade. As a result, microelectronic, computer and consumer electronics firms — along with their labor unions and bankers — are geared for the incipient communications revolution. Leadership in VLSI technology, fifth generation computer development and optical fiber technology — with special emphasis on their mass production and application — place the combined Japanese electronics industry in an appropriate posture to take full advantage of the successive waves of opportunity in the new telecommunications age.

The nucleus of this information industry complex is the so-called "Den Den Family" of 30-odd major telecommunications equipment suppliers subjected to NTT's rigorous performance specifications and quality standards. These same firms, of course, have also been the most direct beneficiaries of selective technological advances emanating from the public monopoly's three (recently joined by a fourth) research laboratories.

No less important, these designated suppliers have a substantial ready market for the products developed jointly within the Den Den Family. Although NTT accounts for a steadily declining share of the market for telecommunications equipment — approximately 33 percent in 1982 — its ¥2.5 billion (US$10.82 million) annual procurement provides suppliers with substantial relief from risks inherent in technological change and assures them of important scale and experience economies that serve them well in other markets. In private and export markets, too, the high performance specifications imposed by NTT become important elements of competitive power.

Despite political pressures from the United States and the projected privatization of NTT, there are likely to be few major changes in these arrangements. The Den Den Family derives its strength, and reason for being, from the inherent nature of telecommunications systems and technology. Their development requires close cooperation at all stages. Uniformity and systems compatibility are imperative prerequisites of all technologies and equipment. Continuing communications made possible through a finely-tuned cooperative relationship cannot be replaced by arms-length ad hoc supply contracts. There is nothing in this logic which requires that all members of the family be Japanese companies, of course, but in the past communications age Japanese have been equipment makers, the logical partners in assuring operation of telecommunications systems.

And the mutual obligations of family membership are likely to be even more important in the fifth communications age than in previous eras. Based largely on these arrangements, NTT has developed plans for the total overhaul of the telecommunications system over the 20 years from 1981 to 2000. This massive effort to develop a nationwide integrated Information Network System (INS) entails replacement of all existing cables with fiber optics, digitalization of networks, expansion of data and facsimile services, introduction of new video services, replacement of all existing telephones and other terminals, and the addition of a variety of new terminals for home, office, and factory.

The INS will unify all networks through fully digitalized systems, replacing separate systems for different modes of communications — telephone, telegraph, telex, facsimile and data. At the same time, the new integrated system will be equipped with enhanced capabilities to assure projected changes in services:

- From mainly voice to video-intensive transmission.
- From principally man-to-man to more machine-to-machine communications in which computers and automatic remote control systems are active performers.
- To high speed and broadband transmission to accommodate increasing amounts of information.
- To add communications processing functions such as temporary storage, media or size conversion, translation, and retrieval of messages as well as computations.

Total cost of this systems revision, in terms of direct capital outlays for equipment, is estimated at ¥20-30 trillion, plus ¥1.3 trillion annual depreciation, over the 20-year plan period. In addition to these outlays by NTT, which surpass those since its founding in 1952, derived demand generated by this system — for local networks, exchanges and terminals — is estimated by Nomura Securities to be as high as ¥60-70 trillion.

Given the rapid pace of telecommunications technological change, however, these estimates could well prove overly conservative. One need only look back to 1978 to understand the difficulties inherent in such forecasts. Rand Corporation telecommunications experts then predicted that 64K microprocessors would be perfected in the early 1980s and reach integration densities in excess of 100K by the latter part of the decade. True. But the timing was off by several years, and by the end of the 1980s, devices will be available with many times the capacity of those forecast. This more rapid pace in LSI and VLSI technology in turn speeds the process of change in telecommunications technologies.

COMMUNICATIONS

Optical fiber usage has been especially sensitive to the availability of more advanced LSI devices. Sales of optical fiber communications systems in Japan are outstripping earlier forcasts and demand continues to exceed supply even at currently relatively high prices. From total sales valued at ¥70 billion in the 1981 fiscal year, Daiwa Securities estimates that they will rise to ¥700 billion in 1985, reaching at least ¥7 trillion in 1990. Although Yamaichi Research Institute projects a slower growth in the 1980s than this forecast, even this more conservative view of the future foresees optical fiber communications-related sales of ¥12 trillion by the year 2000.

By the end of 1984, NTT will have completed installation of the first trunk line using the large-scale optical fiber communications system that has been undergoing commercial tests since 1982. By 1990, according to plan, the entire nationwide grid, including individual subscriber lines will be converted to optical fiber cable and the Transpacific Optical Communications System will be operating between Japan and Hawaii.

Telecommunications carriers are by no means the only market for optical fiber makers, of course. Major Japanese electrical power companies have been perfecting their own optical communications systems for the monitoring and control of power grids. Since optical fiber is unaffected by the magnetic field of high-tension cables, Tokyo Electric Power Co. began research on optical communications in 1974 and perfected cable featuring tensile strength 25 times greater than the conventional nylon-coated counterpart, improved heat resistivity and other properties making it possible to combine optical fiber with its suspension wire into a unitized construction. This fiber-reinforced plastic (FRP) cable was developed jointly with Sumitomo Electric Industries, Furukawa Electric, Fujikura Cable and Hitachi Cable, all producing it for domestic and foreign markets.

Optical fiber communications systems are also being developed by railways, subway systems and expressways. The most elaborate of these new transportation control installations is the new Integrated Digital Communications System adopted by the Tozai Line of the Sapporo Municipal Bureau of Transportation. Combining facilities for telephone and announcement services, operation control, power control and sales data processing, this integrated system is expected to serve as a model for other transport systems throughout the country. Major segments of the system have been developed by cooperation between leading firms in the field: NEC and Sumitomo Electric Industries have perfected the data transmission systems, Matsushita Communications and Fujikura Cable the image transmission systems, and Fujitsu and Furukawa Electric the power control networks. Private and municipal railways have begun in-

356

troducing surveillance and control systems employing various combinations of the new optical fiber technologies.

Similarly, optical fiber communications have been adopted by the Hanshin Expressway Public Corporation for monitoring traffic flow and highway communications facilities.

In broadcasting, NHK (Nippon Hoso Kyokai) and private operators are using optical fibers for their CATV networks. NHK is also continuing research with Hitachi Cable to develop a multichannel optical-fiber system for VHF TV telecasting. And videotex systems undergoing tests since 1979 have been based upon integrated optical fiber communications.

Other expanding uses of optical fibers include local area networks being established in commercial buildings, companies, universities, and research institutes to meet the demand for multimodal transmission, to link computers to peripherals and satellite earth stations to computer nodes. Manufacturers are now using optical fiber multiplex communications systems in cars, ships and elevators to control electrical equipment, engines and automatic systems.

To meet this increasing demand, cable makers have been doubling production annually, with the lead taken by the three firms (Sumitomo Electric Industries, Furukawa Electric and Fujikura Cable Works) which joined NTT in the cooperative research on the Vapor Phase Axial Disposition (VAD) method of optical fiber-making. With market demand running ahead of supply, NTT decided to make its patents in this field available to the other three large cable makers (Hitachi Cable, Dainichi-Nippon Cables and Showa Electric Wire & Cable) as well.

Although the last three had developed their own optical-fiber technology independently, the VAD method has been proven most advantageous. Larger diameter cables and greater length are obtained at one-tenth the production time needed for the most widely-used process, the Modified Chemical Vapor Disposition (MCVD) method developed by Bell Laboratories and Corning Glass. Hence, in the interest of uniformity and economics of production, NTT has licensed all major Japanese cable makers with its technology.

As a result, at the outset of FY 1983, optical-fiber capacity ratings of the big six cable makers totaled 36,500 km monthly. (See Table 1.)

Both Nippon Sheet Glass and Mitsubishi, with their own production methods, are also producing smaller quantities of optical fibers.

Three distinguishing features mark the thrusting Japanese optical fiber industry:

• Compared with other optical fiber industries, the Japanese industry is much less concentrated. In fact, there are as many major makers

COMMUNICATIONS

Table 1. Optical Fiber Production

Manufacturer	km/month
Sumitomo Electric Industries	10,000
Furukawa Electric	8,000
Fujikura Cable Works	8,000
Hitachi Cable	3,000
Dainichi-Nippon Cables	2,000
Showa Electric Wire & Cable	1,500

in Japan as in the United States and Europe together.

• Each of the Japanese optical cable makers have joined with communications and optical equipment manufacturers of their respective *keiretsu* (industrial or banking groups) in the development of optical fiber production, applications and marketing:

Sumitomo Electric Industries is linked with NEC Corporation and Nippon Sheet Glass, all key Sumitomo Group companies.

Furukawa Electric has close ties with Fujitsu, both being members of the Furukawa and Dai-Ichi Kangyo Group.

Fujikura Cable Works works in tandem with Toshiba Corporation and Showa Electric Wire & Cable. Toshiba and Fujikura both have Mitsui Group connections.

Hitachi Cable is a subsidiary of Hitachi, Ltd., and both have common links to the Sanwa Group.

Dainichi-Nippon Cables, with strong Mitsubishi Group ties, works with Mitsubishi Electric.

These linkages not only assure the necessary financial resources for the most rapid development of optical fiber production, but also serve to speed the process of optical fiber technology diffusion throughout group companies which embrace many of the main fields of application.

• The technological commitment and strength of this basic industry of the information age is formidable. The top three optical fiber makers acquired MCVD licenses from Corning Glass and pioneered the development, along with NTT, of the VAD method. The other five producers all developed their own technology, and three of these have also been licensed to use the NTT's VAD process.

Hitachi Cable has, in fact, developed two optical fiber production methods of its own. One of them, the Soot Deposition Method, is similar to the VAD process and has almost all of the latter's advantages over the MCVD method. In addition, however, Hitachi Cable has jointly developed a single polarization monomode (SPM) fiber with Hitachi , Ltd. which permits light beams to travel straight down the axis of the

358

fiber, reaching their destination before those which bounce down the usual cable. While tests on its wider application are still being tested, this new fiber has found many non-communications applications in high precision optical fiber gyroscopes, blood flow meters, interferometers, magnetic or electric field measurement as well as connections between circuits. SPM-type communications using very high quality optical fiber measuring 10 microns or less in diameter are envisaged by NTT in a new public telephone service to be inaugurated in 1987 or 1988.

By then the impact of optical fiber communications will already be far-reaching. Per channel costs of glass fiber transmission will be only a fraction of existing tariffs. And since higher information-carrying capabilities make possible transmission of all communications modes on the same circuit, costs will further be reduced through their integration.

Lower service costs, broader transmission capabilities and the speed of communications will also be enhanced by improvements in microwave communications, satellite services and digital switching systems. Japan's first commercial communications satellite, Sakura 2, which went into service in June 1983 with 4,000 telephone circuits, will be joined by a second in 1987 carrying 6,000 circuits and more advanced craft in the 1990s with up to 200,000 circuits. Although limited to backup emergency services at the outset, satellite transmission will be used as relay transmission lines working in conjunction with optical fiber cable systems on the ground to form a highly reliable dual network.

In addition to these communications carrier satellites, broadcasting satellites will provide improved services for existing public and private networks, as well as new cable television systems. A group of three companies — Marubeni Corporation, the Chunichi Shimbun and the Uny Co. (a trading company, a leading newspaper and a major supermarket chain) — are preparing to launch a combined pay and free cable television service for the Nagoya region, using a communications satellite, in 1985. By 1988, the third Japanese broadcasting satellite, the BS3, will be launched, joining the BS2, which is scheduled to go into service February 1984.

More diversified communications services at lower costs are being matched by similar developments in terminals made possible through application of successive generations of large-scale integrated (LSI) circuits and more advanced very large-scale integrated (VLSI) circuits. The result is a boom in telecommunications terminals for office and home that will gain further momentum after 1985 when the INS becomes operative. New high definition television receivers with flat screens and multiple functions, fully digitalized, will replace existing models. Data communications services will give added impetus to personal computer usage, speeding the convergence of television with computer terminals. All existing telephones and switchboards will be converted to digital models. Developments in mobile communications are opening new markets for automobile radio telephones, cordless telephones and ultimately fully portable telephones. And that powerful sleeping giant, facsimile communications, is now finally awakening to an era of general office and home usage.

Although the takeoff of new television services and digital receivers is still a few years off, car telephone services, begun in 1979, have been extended to approximately 200 cities with over 20,000 subscribers by mid-1983. More economical base station equipment and a compact, low-cost mobile radio put into commercial use in 1982-83 are expected to give a new boost to domestic demand. Meanwhile, leading Japanese makers, especially NEC and Matsushita Communications Industrial, are preparing for the huge U.S. market, where 1-1.5 million mobile units are expected to be installed in automobiles by the mid-1990s. The Middle East and Western Europe are also focal points of Japanese export interest. Matsushita Communications has already established its position in the car telephone market of the United Arab Emirates and Kuwait, and Mitsubishi Electric has begun exports to Western Europe, with initial exports to the promising Swedish market.

Portable radio promises to be an even brighter star in tomorrow's communications firmament. As many as 20 Japanese makers are already in the market, which industry analysts estimate will be worth ¥2.2 trillion annually by the end of the 1980s. A rush for portable high-performance radio sets was triggered by the Ministry of Posts and Telecommunications' action in 1982 to rewrite the Telecommunications Law, giving increased access to radio waves by private citizens.

This growth in mobile communications has been overshadowed by burgeoning facsimile equipment sales. While new models were improved in both efficiency and quality, prices dropped precipitously during 1982 as a result of increasing miniaturization and the introduction of advanced microelectronic devices. The response was electric. Unit sales rose 77 percent in 1982, producing a 30 percent rise in sales revenues.

Output of facsimile equipment in the fiscal year 1983 is expected to reach 350,000 sets, which in value terms will mean a total turnover for the industry of approximately ¥180 billion. But estimates of future growth vary widely. Admittedly conservative forcasts of the Communications Industry Association of Japan (CIAJ), predicting output of ¥257 billion in 1985, have been scrapped. A more sanguine view, held by EDP/Japan Report, estimates 1985 output at ¥300 billion, rising to ¥350 billion in 1986.

NEW TECHNOLOGY AND INDUSTRIAL STRUCTURES

Evidence supporting the more optimistic outlook is convincing. The current boom in NTT mini-fax rentals, which has steadily gained momentum since introduction of the service in September 1981, will be further fueled by Den Den's launching in 1984 of the new economical and compact Mini-fax 2 capable of transmitting an A4 copy within three minutes. By 1986, all major cities in Japan will be included in the Public Facsimile Network, which will then be integrated with other communications networks in a single digitalized optical fiber system scheduled to be fully operative by 1990.

More than 25 makers are in the race to develop new high-speed units, dual purpose fax-copying machines, and combinations of fax with word processors, microfilm and optical character reading equipment — all calculated to increase the attractiveness of facsimile services to users. With these and other improvements in the wings, the number of facsimile installations is expected to rise from 300,000 at the end of 1982 to 10 million in 1990. And since this estimate is based on an 80 percent saturation of the office equipment market and only a modest 10 percent of the home market, the wide scope for further growth in the 1990s is expected to sustain the sizable number of producers in the industry.

Exports will spur the internationalization of major facsimile equipment producers. Although overseas shipments rose only 7.6 percent in the first seven months of 1983, total 1982 exports shot up 83 percent to approximately ¥40 billion, representing a 68 percent increase in volume to 63,300 units. The higher increase in value of exports, exceeding that of volume, reflects the greater advantage of Japanese higher-speed terminals.

Significantly, the relatively high rate of exports, which amounted to almost 27 percent of total sales in 1982, has been sustained less by direct marketing abroad by Japanese firms than by U.S. and European OEM (original equipment manufacturers) purchases in Japan. These supply arrangements are listed in Table 2.

By comparison, leading Japanese makers are estimated to have shipped only 15,000 units to the United States and 5,000-6,000 units to Europe under their own brand names.

Indications are that this pattern of supply will continue for the foreseeable future. With rapid improvement in production efficiency through more automated mass production, Japanese makers have reduced per-unit prices by more than 26 percent since 1980. Technological innovations, especially in LSI design and applications, have not only brought miniaturization and higher quality, but have also lowered the costs of parts and components. Moreover, Japanese facsimile units are now equipped with levels of functionality unmatched by those produced

361

Table 2. OEM Supply by Major Manufacturers

U.S.	Units	Europe	Units
Fujitsu to Burroughs	5000	Fujitsu to ITT Europe	3000
Murata to Burroughs	3000	Mitsubishi Electric	
Oki Electric to		to Siemens	2000
Pitney-Bowes	7000	Ricoh to Kalle Infotec	2000
Toshiba to Pitney-Bowes	8000		
Hitachi to Southern		Oki Electric to DeTeWe	2000
Pacific Communications	2000	Oki Electric to	
Hitachi to Telautograph	2000	Muirhead Corporation	2000

in the United States and Europe. As a result, in 1982 the British Post Office and the Swedish Telecommunications Agency (STA), for example, concluded contracts directly with Matsushita Graphic Communications for supply of "mini-fax" machines. Since the STA is reportedly planning to use Japanese mini-fax equipment in its nationwide communications network redevelopment program, Japanese facsimile exports to Sweden alone could amount ultimately to hundreds of thousands of units.

Spurred by this rise in exports of facsimile sets, overseas shipments of telecommunications equipment rose a spectacular 57.1 percent in 1981 and continued upward at a 21.7 percent rate in the first seven months of 1983. Prospects are that, just as exports have led the growth of other electronic industry sub-sectors in past years, the growth in the Japanese communications industry during the 1980s will be spurred by overseas sales. By 1986, exports are expected to account for approximately 30 percent of total telecommunications equipment shipments.

At the same time, Japanese equipment makers will move towards increasing manufacture of major products in key foreign markets. NEC, Fujitsu and Oki have begun manufacture of PABX (private automatic branch exchanges) in the United States, and Fujitsu has announced plans for U.S. production of optical fiber communications-related equipment (the fiber itself excepted) in early 1984, in conjunction with an order from MCI for installation of a long distance optical fiber communications system between Washington and New York.

Parallel to the rise in exports, private demand for communications equipment in Japan will continue to grow faster than purchases of NTT, KDD and other public sector users. In 1986, equipment sales to the private sector are expected to top ¥697 billion, estimated at 35.3 percent of total production, or about the same as the 35.6 percent taken by sales to public communications systems operators.

Further liberalization of the market for telephones and interconnect regulations for value-added networks will have a significant impact on the market for new generations of terminal equipment. Not only will an increasing share of home telecommunications equipment be sold by the manufacturer to the consumer through a variety of retail outlets, but customer-premises sales of digital exchanges, key telephone systems and local area networks will grow rapidly.

Radical changes in technology, combined with changing patterns of demand are transforming the structure of the telecommunications equipment industry. With the development of optical fiber communications, cooperation between cable makers and equipment manufacturers is becoming even closer than it was in the age of copper wiring. Satellite communications, a macrotechnology susceptible to management only by large-scale enterprises, will increase the share of major equipment makers in transmission equipment markets. Similarly, digital exchanges, like mainframe computers, have high levels of minimum efficient scale of production, and are therefore likely to add to the power of major manufacturers in the marketplace.

Small and medium-sized telecommunications equipment makers will lose market share for cable, transmission equipment and exchanges to the larger diversified communications companies, on the one hand. And on the other, they will be faced by rising competition for terminal markets from consumer electronics and office equipment manufacturers.

Virtually all major appliance manufacturers and office equipment makers have diversified into telecommunications terminal production in recent years and are likely to be leading suppliers of personal computer-cum-data communications terminals, facsimile equipment, as well as television and other video equipment connected to the communications network. Indeed, the advantage of appliance makers is so strong that major communications equipment makers such as NEC are investing heavily in consumer electronics production to acquire the mass production capabilities that will be required to compete effectively in the terminal markets of tomorrow.

Quite understandably, foreign communications equipment suppliers view this trend with mixed feelings. While some specialized communications equipment manufacturers are confident they will be able to compete more effectively in the Japanese market after privatization of NTT and with the growing importance of the private sector, others are less optimistic. Just as improved technology, higher domestic performance specifications and quality standards, and superior mass production capabilities are serving to make Japanese communications equipment increasingly attractive abroad, foreign suppliers will have to compete with this combination of advantages in the Japanese domestic market. Private buyers tend to be just as concerned with performance and quality as the public services, and they are even more impervious to political pressures from abroad.

But that is not all. "There is the classic problem of interface between foreign manufacturers and the Japanese market that must be solved," a Tokyo representative of one leading international equipment maker confesses. "The home office blames the Japanese and the branch office for failure to penetrate the market, but are not prepared to do their homework or modify their equipment to meet NTT standards. We have to face the facts. NTT standards are tighter, and in other ways special, but they're not going to relax them simply to please or accommodate foreign suppliers."

The plain truth of the matter is that, for many of these firms, meeting Japanese (not only NTT) specifications at competitive prices and normal business risks is not feasible. The volume of sales, under the best of all possible conditions, often does not warrant the investment.

But there are signs that the market is changing for some foreign suppliers. NTT has established technological ties with both IBM and AT&T which should make possible the rational development of systems and equipment that are compatible, thus enabling these leading U.S. companies to share eventually in the Japanese market for the big ticket items which is their forte. Motorola has been selected as a designated supplier, not only for pocket paging systems, but also for cellular telephones.Rolms has received NTT approval of its digital private branch exchanges and private firms have been buying Rolm computerized branch exchanges (CBXs), which first appeared on the Japanese market as recently as April 1982. Fortified with this success, the company is resolved to capture 3-4 percent of the market for this type of equipment.

As a harbinger of things to come, at the beginning of October 1983, Ichio Kato, NTT's director of international procurement, announced in Washington the placement of orders for an advanced traffic observation and management information collecting system from AT&T International, a US$21 million super-computer system from Cray Research, and a US$4.3 million transportable digital switching system from Northern Telecom. Although sales of U.S. firms to NTT have been doubling every year since 1981, when the agreement between the United States and Japan was signed giving American firms the right to compete on equal terms for NTT procurements, this is the largest set of U.S. equipment purchases to date.

By current industrial standards, the pace of NTT's liberalization signals a major change in policy which is further reflected in the new open tender system of procurement. To expect radical changes in patterns of supply, even if the quality and prices of products are competitive with those being used, flies in the face of reality of telecommunications systems imperatives. All products must be compatible with the total system and with its various components, which usually means that individual products must be designed for the needs of the system if optimal results are to be obtained. This limits the number of equipment items that can be bought "off the shelf," and requires a rather lengthy lead time between the design stage, the placement of orders, and final delivery.

The question suppliers are now asking is: will NTT continue on its present course of liberalization after it becomes a private corporation? Although NTT will not be legally bound to comply with the procurement agreement once the status of public corporation is withdrawn, there are good reasons to expect that purchases abroad will continue to increase. As an executive of one American equipment supplier put it: "The important thing is that recent changes in the climate induce foreign firms to try harder." But, equally important, NTT will no doubt remain committed to the path of cooperation it has taken with recent technical tie-ups abroad.

There are also some sound technical and economic reasons for an increasing specialization and international division of labor as telecommunications technologies and products proliferate. And as uniformity in global communications systems becomes increasingly important, there is likely to be a growing trend towards standardization and compatibility of systems equipment. Closer cooperation between communications utilities and suppliers throughout the world becomes, therefore, one of the positive features of the fifth communications age. Liberalization of communications systems the world over will also enhance the importance of private sector markets and increase world trade in communications equipment. Although Japanese telecommunications equipment makers will be major competitors in this new global environment, effective competition in this field often requires cross-border cooperation. Increased international competition and cooperation thus become the logical pattern for the communications industry in the information age.

CHAPTER 26

The Den Den Family Under Stress

FEW DEVELOPMENTS OF industrial policy in the postwar period have been more significant than the reshaping of Japan's telecommunications, outlined in broad relief in three telecommunications business bills expected to pass the Upper House of the Diet and become law in December 1984.

Bringing to an end the government monopoly of telecommunications services, these measures close more than a century-long episode in Japanese industrial development from the introduction of telegraphy in 1869 to the completion of preparations for a new nationwide information network system (INS) and the first operational Japanese communications satellite. The opening of telecommunications services to the full thrust of the fiercely competitive Japanese market-place marks the beginning of a new era in which the convergence of communications, computer and consumer-electronics technologies will drive a massive transformation of Japan's industrial structure that promises to be even more far-reaching than the one following the postwar reconstruction and ascendance of Japanese heavy industries.

In a very real sense, 1984 was year one of the long-heralded Information Age of which the communications industry will be the vital infrastructure linking men, computers and a myriad of new products made possible by successive generations of microelectronic devices, enabling the full development of information as the basic commodity of the new era.

In stark contrast to dire Orwellian predictions, with the passage of these measures we are witnessing the withdrawal of direct government intervention in this vital human and economic activity. The new information age, if these measures are effective, will be marked by greater decentralization of economic structures and the diversity of modes of information-processing, unleashing new energies which will make possible the transformation of information, as the basic raw material, into great endless wealth.

Reprinted from "This is Year One of the New Information Age," *Far Eastern Economic Review*, 8 September 1984.

More immediately, the privatization of the hitherto public telecommunications agency, Nippon Telegraph and Telephone (NTT), will at once bring a measure of welcome relief to the national Treasury through the sale of the company's stock and enable the rationalization of what has become a gargantuan muscle-bound bureaucracy impeding more than facilitating the full development of communications technologies.

While privatization holds neither the prospect — nor the threat — of NTT's dismemberment after the fashion of the Bell system in the United States, by submitting the new organization to the full fury of competitive forces, it forces NTT to streamline its structures and obtain maximum utility of its vast technological potential and investment in physical assets. This in itself holds the promise of vastly improved communications services at lower cost to the user, with all that entails as stimulus for economic growth and industrial structural change.

At the same time, the new legislation opens the Japanese economy to greater internationalization of the vital communications sector. Explicitly, these measures enable foreign participation in the rapidly expanding data-communications sector as operators of value-added networks (VAN). But implicitly, the privatization of NTT and liberalization of telecommunications services in general open all kinds of possibilities for international competition and cooperation in both terrestrial and satellite communications. Clearly, if recent experience means anything, the liberalization of telecommunications services is at least as much an unleashing of Japanese industry, including of course NTT itself, for participation in the global communications industry as it is an opening of the Japanese market to foreign participation.

Some of the new shapes of the industry are already becoming apparent. Even before the ink was dry on the new legislative measures, a spate of new entrants lined up at the starting gate awaiting the opening of the marathon scheduled for April 1985.

• Anticipating the end of NTT's monopoly, a number of companies and government agencies have indicated their interest in operating common-carrier services in direct competition with the new privatized NTT. Spearheading the moves to create new communications utilities is the Kyocera Group, originally established with ¥1.6 billion (US$6.58 million) capital (enough to cover the cost of a good feasibility study) subscribed by 25 companies led by Kyocera and including such well-known high-flyers as Sony, Secom and Ushio. Equally important, if less visible, the new group also includes leading general-trading companies such as Mitsui, Mitsubishi, Marubeni and Sumitomo along with major commercial banks.

Since its establishment in May 1984, 40 more companies including foreign financial institutions have joined the new company, appropriately christened Daini Den Den. In its initial feasibility study, the planning group is weighing various options — optical cable, microwave or satellite — for linking major cities in a new communications grid to become functional around 1988.

• Similar services are planned by at least three other groups. A consortium of 37 member companies of the federation of economic organizations (Keidanren), all of them in the nation's key industries, plans to establish a joint company to operate a communications satellite in tandem with optical fiber or microwave circuits linking Tokyo and Osaka via Nagoya. In addition, the Japanese National Railways and Japan Highway Public Corp. have announced plans for entering the communications business, establishing networks using fiber cables installed along railways and expressways.

• Other possible entrants as common carriers include electric-power companies, which are reportedly considering optical fiber networks along electric lines — something which is now technically feasible since electric-power transmission does not interfere with optical transmission through glass fibers.

• Responding to the opening of value-added networking to private business, either domestic or foreign, the major communications-equipment and computer-mainframe makers, leading electrical-machinery manufacturers and at least three foreign firms in the information-processing field announced their plans for establishing large-scale VAN services once the new laws become effective. Fujitsu, NEC, Hitachi, Toshiba, Matsushita and IBM Japan are making rapid preparations to launch their wholly owned VAN services as soon as possible after April 1985, while AT&T and McDonnell Douglas Corp. have announced their plans for VAN services in joint enterprises with Japanese partners.

But among the radical changes in the shape of Japanese telecommunications for which 1984 will be renowned will not be the eclipse of NTT as the dominant force in the industry: quite the contrary.

The privatized NTT will, after all, be the largest stock company in Japan, with an estimated capital of ¥1 trillion, about 330,000 employees and annual revenues exceeding ¥4.5 trillion. None of the projected telecommunications firms are likely to be a match for such a colossus, nor would NTT's position be seriously threatened if, as the Ministry of Posts and Telecommunications (MPT) proposed, all the plans were integrated into one large-scale project.

The purpose of privatization and the creation of a new competitive environment through liberalization of the market for telecommunications services is, after all, to revitalize NTT, not to destroy it.

It is remarkable, then, that the only vocal opposition to the reorganization of NTT has come from the All-Japan Telecommunications Workers' Union (Zendentsu), whose leaders claim that the bills as drafted by the MPT were aimed at enabling the NTT management to restrict workers' right to strike. More likely, as employees of a private stock company, NTT workers will tend to unite with management in the face of competition to improve the performance of the company, thus reducing the external labor-federation influence. Most certainly, with competition for telecommunications markets becoming increasingly fierce simultaneously on many fronts, NTT must realistically face and resolve a number of serious internal structural problems, among them that of labor relations:

• Labor-management relations have definitely not been good at NTT. Employees work shorter hours than the average in private business and labor agreements have restrained relocation, hampering the streamlining of business operations,

• As personnel costs continue to rise, operating costs have continued steadily upward.

• Telephone service revenues, which still constitute 90 percent of NTT's operating income, are rising more slowly as the annual growth in the number of telephone subscribers declines.

• As a public corporation, NTT's budget has been subject to Diet approval, its business operations have been controlled by the MPT and employee wages have been governed by levels and administrative procedures which apply to all government agencies. Little incentive has remained for improvement of efficiency.

• All the combined drawbacks of monopoly and public bureaucracy have prevailed: lack of the work ethic, of spirit of service, market-orientation and willingness to keep pace with technology.

At the same time, NTT has suffered a serious erosion of its control over the communications-equipment market. Well before U.S. diplomatic pressures forced open the doors for U.S. suppliers, the so-called Den Den family of 300 equipment suppliers had come under heavy strain making fundamental changes in its structure inevitable.

Most important, NTT was unable to develop new systems fast enough to assure a market for the continuing flow of new or improved communications equipment as it developed apace with rapidly changing technology. As a result, equipment makers have had, in some instances, to introduce their products in foreign markets before Japan and major NTT suppliers have shifted resources increasingly to products salable in the higher-growth private market.

In 1980, the interconnect market was liberalized, following the example of the United States and in response to the mounting pressures within the Den Den family itself, as well as from firms outside the family for whom telecommunications terminals represent an important potential field of diversification. Under new regulations, only the first telephone instrument used by a subscriber had to be leased from NTT. All subsequent units could be bought from manufacturers or dealers, or from department stores where foreign, as well as Japanese, brands were to be found in increasing variety and profusion.

The effects were swift and stupendous. By 1983, NTT had a warehouse full of used business telephones of its latest design. As many as 690,000 units were in stock and the number returned by subscribers each year had mounted to 350,000 from just 180,000 in 1979. Inventory value at market prices had reached ¥1.2 billion and prospects of reduction became incresingly remote as more attractive instruments produced by private manufacturers appeared on the market.

By late 1984, NTT's share of second-telephone set sales had dropped to a low 9.9 percent, of the PBX market to only 14.3 percent, and of total fascimile-machine sales to less than 20 percent.

Overall, NTT's share of the total market for communications equipment has been declining steadily. During 1978-82, while NTT equipment purchases remained stable at more than ¥600 billion annually, the total market for telecommunications equipment grew by 60 percent. Private purchases almost doubled during the period, rising at an annual rate of about 20 percent to exceed total purchases by NTT for the first time in 1982. In the same five-year period, exports outpaced domestic demand in terms of annual growth rate, reducing even further the reliance of equipment manufacturers on NTT procurement.

The significance, for foreign suppliers, of this structural shift seems to have been lost in the political fog that has hung over the NTT procurement issue. In telecommunications as in other industrial sectors, the most rapid growth is in the private commercial market, not in the public sector. Although this distinction will no longer exist to the same extent, it should be remembered that in Japan, as in other countries, public-communications enterprise is a tough customer for any supplier. As Yoshimichi Yamashita, president of Arthur D. Little (Japan), Inc., has pointed out: "Even Japanese high-technology companies such as Toshiba, Mitsubishi Electric and Matsushita find the NTT market quite hard to penetrate . . . For foreign suppliers to attempt to captalize on current political moves by targeting the NTT market would be a serious mistake. Clearly, the NTT market will continue to grow . . . Nonetheless the commercial market should be growing faster than NTT's procurement": and this will be even more true after the entry of new common carriers and VAN operators in coming years.

Just how volatile change can be in this new competitive environment is reflected in the market for plain old telephones. For years, the domestic market was largely in the hands of five Den Den family members — Iwasaki Electric, Hitachi, NEC, Fujitsu and Oki Electric — which, as late as 1982, accounted for 95.2 percent of all telephones sold in Japan. In 1983, however, all five suppliers lost market share to newcomers who collectively boosted their share of the action from 4.8 percent to 16.3 percent in a single year.

Among the new contenders for the markets is Toshiba. Never a Den Den insider, in 1984 Toshiba won large orders for telephone sets and was among the new partners selected by NTT for cooperative product development to meet the needs of INS when construction begins next year.

Already, with mounting NTT orders and the increasing demands of the commercial market for second phones, Toshiba's telephone factory at Hino in the western suburbs of Tokyo has been operating at full capacity. To meet soaring export demand, Toshiba planners are considering locating a second telephone-manufacturing facility in the United States, where sales doubled in 1983 to ¥15 billion and are expected to rise to ¥25 billion in 1984.

Key telephone production is even more indicative of the shape of things to come. Although the market for key telephones first developed in the United States, where they are widely used as digital-switching equipment in decentralized automated offices, Japanese communications-equipment manufacturers grasped an early decisive lead in the market using to their advantage mass-production capabilities and experience accumulated from electronic-appliance manufacture. Given the production strength of Japanese makers, the U.S., European and Australian communications-equipment suppliers have turned to Japan for key telephones rather than produce them at home. To meet the rapid expansion of export demand, output by Japanese makers has quadrupled since 1977 from less than ¥43 billion to an estimated ¥175 billion in 1984.

The market for key telephones is, in several significant ways, unlike that for the interconnect communications equipment produced in Japan. Most important, the market has not been made by the NTT but by overseas demand. NTT, which began employing key telephones only in 1983, has in fact had little influence on the dramatic growth of production of this equipment.

Nor have major communications-equipment makers been prominent among the producers. Two medium-sized makers, Nitsuko and

Iwatsu Electric, have, until now, accounted for more than 50 percent of total production, with the remainder of the market shared by Matsushita Communications, Taiko Electric Works, Meisei Electric, Nakayo Telecommunications and Kanda Tsushin — for the most part smaller-scale producers of specialized communications equipment. Only Oki Electric among major equipment makers has managed to capture as much as 1 percent of the total market, domestic and export.

With NTT's purchase of the equipment, prospects for rapid and substantial increases in local demand are improved; some makers are predicting production valued at between ¥280 and ¥350 billion by 1987. In response to rising demand, major manufacturers — including Hitachi, Toshiba, NEC and Fujitsu — have all launched new products using their own microcomputers in a bid to wrest market share from the present front-runners.

Almost certainly, the structure of the market will change over the medium term. The new entrants, all vertically integrated manufacturers with their own integrated-circuit production, massive output capacity and global organizational structures have the necessary strength to gain market share rather rapidly: and, since most of the sales will be direct to users in Japan, these larger firms are better positioned to market a product line that is becoming as important as switchboards and facsimile machines and can be sold through the same channels.

If the market for facsimile machines becomes the pattern for key telephones, it will be a tableau with major Japanese equipment makers prominent in the foreground. The major difference, of course, is that the Japanese market has long been the world's largest for facsimile, accounting for almost one-third of the total number of machines operating globally in 1983.

The main reason for this is obvious: facsimile is far more suitable than telex or other forms of textual communications for the transmission of ideographs used in Japanese writing. Thus in 1983 alone, as prices declined, elastic demand rose dramatically pushing sales up to 216,088 machines — which amounted to a 20 percent increase in the total worldwide facsimile park.

At this rate, during the next three years domestic demand is expected to triple to 710,000 machines. As transmission costs decline further with the introduction of optical fiber and satellite systems and as unit costs of machines are reduced to reflect scale economies and the smaller machine size made possible by contact scanning, facsimile machines will enter the home as a common communications terminal. By the mid-1990s, NEC's Takao Matsushita predicts, as many as 100 million facsimile machines could be operative in Japan alone.

373

The global implications of such an expansion are indelibly clear. Already, the level of domestic demand has been sufficient to assure Japanese manufacturers a commanding lead in world markets, supplied mainly under original-equipment-manufacturer agreements with major communications-equipment makers in the United States and Western Europe. At least seven U.S. firms — including Xerox, 3M, Burroughs and Exxon — have sub-contracted production with leading Japanese makers, as have 11 European communications-equipment suppliers: and, in addition to these private-label sales routes, massive facsimile orders are being placed directly with Japanese makers by post-and-telecommunications authorities of many European countries.

To assuage the misgivings of key telephone and fascimile machine makers that the restructured NTT might seek to enter these markets with its own manufacturing facilities, using the combined advantage of its research and development (R&D) capability and domestic-market power as leverage, NTT president Hisashi Shinto has pointedly ruled out such a possibility. With apparent reference to the superior marketing expertise of equipment makers, Shinto has dismissed the stratagem as uneconomical and unlikely to succeed.

Instead, the main thrust of NTT's competitive strategy for the future will come in the form of enhanced and diversified services. In 1984, the company will have completed its preparations for implementation of the projected 20-year INS plan to digitize the entire telecommunications system using a combined nationwide optical-fiber and satellite grid that no competitor can hope to duplicate. Requiring an investment conservatively estimated at ¥20-30 trillion, the new network will enable NTT to obtain cost reductions of at least 30 percent on transmissions and up to 50 percent with the introduction of digital-switching equipment. If all goes according to plan, tariff schedules will be revised to eliminate the distance factor, eventually making possible uniform national charges for voice-, visual- and data-communications services.

In order to assure the efficiency of these various services, decentralization will be the order of the day at NTT in the future. Indications are that autonomous subsidiaries will be formed to operate specific services or in particular regions, enabling the streamlining of services, gradual reduction of personnel by 100,000 employees and elimination of cumbersome bureaucracy.

Along with improved network facilities, the new parent company will support its operating-service subsidiaries with enhanced technological strength. NTT's present R&D facilities, covering a wide range of research on material, semiconductors, computers, transmission and switching technologies as well as end-use terminals for home and office, already constitutes one of the world's leading technical organizations, surpassed in its field only by Bell Laboratories of the United States. In 1983 the fourth NTT research facility, representing an investment of US$100 million, was inaugurated at Atsugi, Kanagawa prefecture, to conduct advanced semiconductor-device research: and in 1984, making it clear that the new company intends to make technology trump in the new competitive game, Moriji Kuwahara, director-general of NTT's engineering bureau, announced that R&D expenditures would be increased by 30 percent over fiscal 1983.

As work on the INS optical-fiber pipeline and communications-satellite system progressed, NTT began in 1984 to introduce a wide array of new services:

• Up-staging its satellite-communications business, NTT will introduce commercial-satellite services before the end of 1984 using the CA-1B communications satellite placed in stationary orbit in August 1984. Ground stations, already in place in five major cities — Tokyo, Osaka, Nagoya, Sapporo and Fukuoka — will enable banks in these cities to transmit massive volumes of daily transactions and allow firms to maintain regular TV conferencing services and to transmit documents simultaneouly to branch offices at uniform hourly or daily service rates.

• A new ultra-high-speed facsimile-transmission system will make its debut as a part of INS services to begin in Tokyo's Musashino-Mitaka district before the end of 1984. This new system, using both fiber-optic and satellite-communications channels, will transmit A4 manuscripts at speeds of 2 seconds a page, compared with 60 seconds needed by conventional systems, upgrading NTT's existing facsimile-communications network service introduced in September 1981. The new service, using Minifax machines developed by NTT, is expected to give added impetus to the current rapid growth in the market for fascimile machines.

• At the same time, NTT's version of videotex, the character and pattern telephone-access INS (CAPTAINS) will begin operations on a trial basis at the Musashino-Mitaka pilot model, bringing the wired city of the future out of the realm of science fiction into reality. As a result, a giant home-information industry is taking shape, initially involving more than 300 information suppliers.

CAPTAIN Service Co., formed in February 1984 with a capital of ¥300 million — subscribed by NTT, information suppliers and terminal makers — will at the outset provide teletext information services to users within a 30-km radius of the center of Tokyo. In 1985, service will be extended to greater Tokyo, Kyoto, Kobe and Nagoya. CAPTAINS Service, which will blanket all major cities in Japan by 1986, will cost subscribers, who must buy their own terminals, only about ¥300 for a lifelong contract and ¥30 for three-minute access.

• NTT and NEC will also market a newly developed private CAPTAIN system designed for use by large firms and local governments as an internal information facility, competing with a similar system to be offered by Fujitsu and the Canadian Telidon System now sold by Mitsui, Nichimen and Sumitomo.

• NTT has begun a nationwide credit-card authorization system, called CAFIS (credit and finance information switching system) to serve American Express International, six Japanese credit-card companies and department stores with their own independent-consumer credit-card systems. CAFIS will compete head-on with a similar service to be offered by IBM Japan called CATNE (credit authorization terminal network), aimed mainly at independent credit-sales companies.

• Anticipating the entry of competing services and possibly to dissuade some, NTT began reducing long-distance telephone charges in April 1984 and extended the hours for off-peak discount calls. As a result, the 40:1 ratio between long-distance and local calls was reduced to 10:1, a move in the direction of uniform nationwide telephone rates.

Finance for NTT's further advance into the information age is to be provided by a special fund-procurement stipulation plan to be established in April 1985 with the issuance of ¥4 trillion in debentures, four times the company's declared capitalization. This mammoth funding, specifically earmarked for replacement of existing communications circuits with digital networks, combined with NTT's enormous technological capability, will go far to assuring the reorganized company's leadership of the information revolution well into the next century.

To speed the process nationwide, MPT and the Ministry of International Trade and Industry (MITI) have announced plans for model information or new-media cities where homes, businesses, hospitals and administrative services will be linked in the new INS network. To implement this new-media-community concept, one of the key MITI policies for fiscal 1984, support will be provided to facilitate the installation and use of computer mainframes, development of software and the enactment of two computer-related laws to protect privacy in the use of both. MITI's program is aimed at establishing the least expensive information system in keeping with special regional needs, by combining two-way cable TV, satellite communications, the new optical-fiber network and microwave communications.

The results of this technology-push-user-pull approach to the propulsion of communications-systems development will be nothing less than total industrial transformation, a dynamic process that now is entering a new phase with the creation of a more competitive market for communications services. The global implications of this process are stupendous. The likelihood is that, by the end of the century, Japan will emerge equipped with the world's leading information industrial system. Just as history was radically altered by the shift of the epicenter of the industrial system to North America, its course will almost certainly reflect that ascendancy of advanced information systems and societal arrangements for their more effective management in Japan and East Asia.

CHAPTER 27

New Media: Facsimile

DURING THE 1960s, an astonished world witnessed Japan take the leadership in an epoch-making electronic calculator revolution. Again, in the 1970s, Japanese copier manufacturers overtook the front-running American industry in both technology and total output. At the beginning of the new decade, Japanese industry emerged as leaders in the facsimile equipment market, promising a monumental breakthrough into a new era of high-speed graphic communications. All signs point to Japan's being the first country in the world to have an operative electronic mail system using a home facsimile service.

By 1980, the diffusion of facsimile systems in Japan already far exceeds that of the United States and Europe. The number of installed units per capita is approximately twice that of the United States and more than four times that of Western Europe. In the United States, although the number of messages carried by facsimile systems has been growing at an average 23 percent yearly, the number of facsimile machines installed has been increasing at an annual rate below 20 percent, and sales have been rising at an even lower rate. (This disparity is due to the high proportion of high speed machines used in the United States and to the sharp decline in the price of those machines in recent years.) While in North America and Europe facsimile systems are still limited to intraorganizational communications in government, newspaper publishing and manufacturing, Japanese medium- and small-scale firms, golf clubs and other services are already using facsimile equipment for intercompany communications.

Production of facsimile equipment in Japan grew at an average 30 percent a year during the 1970s, while sales in 1978 were already more than ten times those of 1971, reflecting the increasing share of more costly medium- and high-speed machines in the market mix. Thus by 1977, total sales of Japanese facsimile machines were in excess of US$160

Reprinted from ''Japan's Third Wave: The Facsimile Era,'' *Communications International*, July 1980.

COMMUNICATIONS

Japanese Facsimile Equipment Manufacturers

Major Communications Equipment:
Fujitsu, Ltd.
Hitachi, Ltd.
Nippon Electric Co. Ltd.
Oki Electric Industry Ltd.
Medium-scale Communications Equipment:
Iwatsu Electric
Tamura Electric Works
Specialized Communications Equipment:
Matsushita Graphic Communications Systems, Inc.
Murata Data Equipment Corp.
General Electrical Machinery:
Mitsubishi Electric Co., Ltd.
Toshiba Corporation
Home Electric Appliances:
Sanyo Electric Co.
Sharp Corporation
Business Machines:
Canon, Inc.
Casio Computer
Fuji Xerox
Ricoh
Computer Terminals:
Yamura Shinko

million, compared to total facsimile market revenues in the United States of just over US$120 million, including sales of systems imported from Japan.

More important still for long run growth, the Japanese industry is structurally much more extensive than that of the United States. While five firms (one of which is Japanese) account for 85 percent of the American market, there are about twenty manufacturers of facsimile equipment in Japan.

Of these, only two — Matsushita Graphic Communications Systems (known in Japan as Matsushita Denso) and Murata Data Equipment — are specialized facsimile equipment manufacturers. The others are either diversified communications equipment manufacturers, electrical machinery or appliance makers, or business machine manufacturers.

Most of the industry is highly integrated. Seven manufacturers pro-

Fig. 1 Trends of Facsimile Apparatus for Document Transmission (and Facsimile Apparatus by Group*)

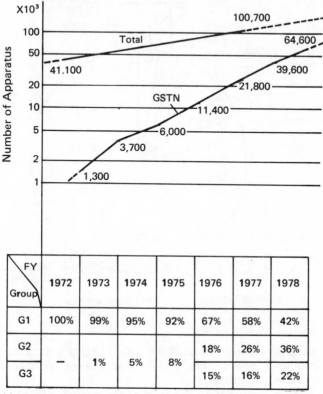

FY Group	1972	1973	1974	1975	1976	1977	1978
G1	100%	99%	95%	92%	67%	58%	42%
G2	—	1%	5%	8%	18%	26%	36%
G3					15%	16%	22%

*Apparatus are classified by group according to speed of transmission.

duce a rather wide range of related comunications and data processing equipment, and seven more are leaders of the burgeoning Japanese copier industry. All fourteen of these firms have extensive global marketing networks, and most of them are increasingly organizing their production multinationally, facilitating the rapid development of global sales, service and production strategies and structures.

The leader of the industry is clearly Matsushita Graphic Communications Systems, which has a market share of about 60-65 percent in Japan and approximately 40 percent worldwide. Matsushita Denso estimates that its overseas shipments have accounted for more than half of

Fig. 2 Facsimile Apparatus by Industry

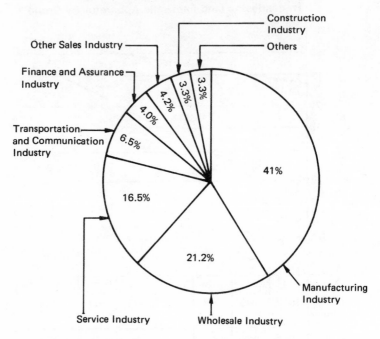

total unit production, or about 15,000 units a year as of 1978. In the six years prior to 1978, Matsushita Denso exported about 50,000 machines, mainly to the U.S. market.

Although most of Matsushita's sales at home and abroad have been of medium- and slow-speed models, the company has also been a world leader in newspaper facsimile systems. Matsushita Denso has installed about 200 newspaper fascimile systems in Japan and delivered similar systems to leading dailies of the United States, *Aftonbladet* in Sweden, *Pravda* of the Soviet Union and the *People's Daily* of China.

In 1977, Matsushita Denso established a joint venture, Panafax Corporation, in the United States with Visual Science Inc. mainly to market high-speed systems that transmit via microwave radio, voice-data lines, or ordinary telephone lines.

The Matsushita UB-2200 facsimile introduced in 1978 was not only fast, it also featured a number of auxiliary functions made possible by a Fairchild 6-bit F8 microprocessor. This was the first equipment developed to send the same series of documents sequentially to as many as 28 selected locations, with a timer starting operation after rates went down at night. Sequential transmission is clearly superior to simultaneous transmission since its speed can be optimized for the transmission line used, and error correction becomes possible when necessary.

The facsimile signal is stored in a digital memory from which it can be sent to multiple receivers. Then, on the receiving end, the memory makes it possible to print two extra copies of the message during the ten-second protocol interval before receiving the next message.

In 1979, the industry leader introduced a new two-color, two-minute analog facsimile system which prints the standard text in black, and corrections, revisions, or other information (such as the ubiquitous Japanese chop marks used in place of signatures) in red. The unit will also communicate with other terminals in black only, using CCITT three-minute standards.

At the same time, Matsushita Denso unveiled an on-line data entry facsimile system that links computers to facsimile machines, an important step in the direction of the intelligent facsimile system, a key element in the automated office of the future. In still another application, Matsushita Denso has contributed to the development of new Television Multiplex Facsimile to be produced and marketed by its parent, Matsushita Electric, once government and industry agree on new broadcasting systems and regulations.

But even without these needed basic changes in the regulatory environment, demand for facsimiles in Japan has been expanding at an explosive pace, attracting numerous new entrants and forcing existing manufacturers of facsimile equipment to expand and strengthen their production capability. Although facsimile production at Matsushita Denso's expanded Nagano and Shonan works continues to grow at 30 percent annually, Ricoh, NEC and Toshiba have set their targets at 40 percent or more, in a bid for a larger share of the market. Hitachi, Mitsubishi and Oki Electric are in hot pursuit. Fujitsu has also mounted an offensive with a new series manufactured in a recently constructed, specialized facsimile plant. Meanwhile, although Casio and Canon have fewer models in their facsimile lines, both diversified business machine makers are expanding their positions in the market.

Despite booming demand, prices have been falling at a pace which exceeds even that of the great calculator slide of the 1960s and 1970s. In 1975, the cheapest medium-speed (GII) device sold in Japan for about ¥1,300,000 (about US$4,333), but within a year the price was reduced to below a million yen. In 1977, products priced at ¥600,000 could be found on the market, a decline of 50 percent in about a year and a half.

Fig. 3 Public Facsimile Network Configuration

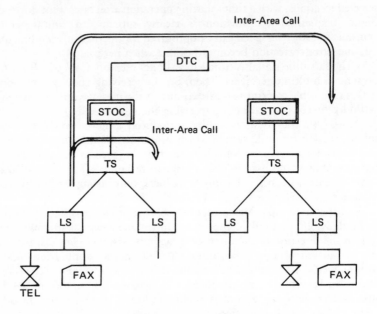

DTC   : Digital Transmission Channel
STOC : Facsimile Storage and Conversion System
TS   : (Electronic) Toll Switch
LS   : Local Switch
TEL   : Telephone
FAX   : Facsimile Terminal

And the contest for market share had only just begun. During the two succeeding years Mitsubishi Electric Corp., Oki Electric Industry Co., NEC and Matsushita Denso have all made heavy investments in extensive sales and service networks blanketing Japan and several have set up special marketing subsidiaries abroad, anticipating an escalation of competitive activity once Nippon Telegraph and Telephone Public Corp. (NTT) begins actively promoting its home facsimile service.

NTT set the standard specification for ultra-small home facsimile devices in 1978 and entrusted the production of the so-called Minifax machines to six designated suppliers: Matsushita Denso, NEC, Toshiba, Hitachi, Fujitsu and Tamura — all long-time members of the NTT family of communications equipment suppliers.

The simplified transceiver initially is expected to sell for less than US$1,000, with a target price of half that. To attain this lower price, however, each manufacturer would have to produce at least 10,000 units a month. This would mean sales at the rate of 720,000 units a year by the six producers, and not everyone in the industry is convinced that this is a realistic assessment of the Japanese market potential.

The specifications NTT has established for the new units include electronic scan by charge-coupled devices or an MOS linear array for pickup, and electronic scan by a thermal-head linear array for printing. Line density would be the standard 3.85 lines per mm, sent by amplitude-modulated and phase-modulated vestigial sideband, which is the modulation scheme used in CCITT 3-minute machines. Transmission time will be about 80 seconds for a page 182 mm long (a little over 7 inches). The home units will also have the line density and modulation scheme necessary to communicate with the 4-minute units used by the NTT at present.

Originally, NTT had planned to use telephone circuits for the carrier in this system. However, billings at the prevailing telephone rate would make transmission charges very expensive, especially over long distances, prejudicing the hoped-for rapid diffusion of facsimiles in Japanese households.

While telephone networks will be used for local systems, NTT will use newly developed digital circuits that make it possible to transmit information over long disances at faster speeds and lower costs, without intervening modem equipment. With this improved transmission system, NTT hopes to popularize facsimile services at costs comparable to present telephone services.

Research on an economical and intelligent public facsimile communications system reached its final stage during 1979 at NTT's Yokosuka Electrical Communication Laboratory (YECL). Newly developed storage and conversion communication (STOC) equipment receives, digitalizes and temporarily stores facsimile signals in a magnetic bubble memory after redundancy reduction of the data volume for storage and transmission to approximately one seventh. The coded signals are then transmitted at high speeds over a digital toll line to the receiving STOC equipment which stores the signals temporarily before reconverting the signal for transmittal to the receiving facsimile terminal. As the system transfers facsimile signals between STOCs at a 64 Kb/s rate, an A5 document can be transmitted in approximately 4 seconds, compared to about 90 seconds required to transmit the same size document over the public telephone network.

This new equipment significantly improves circuit utilization effi-

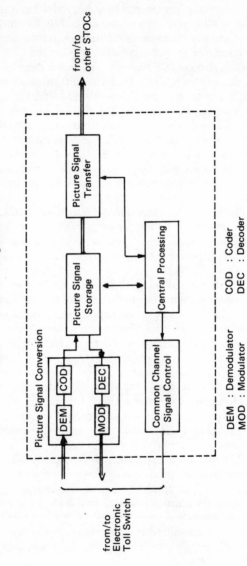

Fig. 4   STOC Configuration

ciency between toll sections, making possible diversified services at low costs. Facsimile signals can be received automatically, without the necessity of the sender ringing the receiver's bell. If the receiver's telephone happens to be busy at the time the message is sent, the network simply stores the signals until the line is free, and then automatically transmits the facsimile signal. Other features offered by this system include multiple address communication, transmission between otherwise incompatible facsimile equipment, confidentiality, dating, and insertion of the sender's number in the facsimile message.

More than any other factor, the development of this new communications system will set the stage for take-off of Japan's burgeoning facsimile industry. No matter what the capabilities of Japanese facsimile manufacturers, the market for their products will be largely determined by the available transmission facilities. Thus, the rapid growth of the Japanese facsimile equipment industry in the 1980s is predicated on the development and widespread commercialization of specialized transmission networks. Just as the freeing of NTT telephone cirucits for facsimile transmission in 1972 spurred the first boom in the industry during the remainder of the 1970s, the digitization of facsimile scanning, transmission and recording — and the resultant improvements in cost/performance ratios — will greatly enhance facsimile's competitiveness as a message option and will thus speed the spread of electronic mail systems.

To this extent, the Japanese facsimile equipment industry will be riding the crest of a monstrous wave sweeping over the communications industry. While communications facilities in other countries are hamstrung by regulatory hassles and other constraints imposed by general institutional sclerosis, the Japanese telecommunications system will be racing into the next century, ushering in a new era of facsimile and data communications.

The new economics of communications will generate new, powerful impulsion to effective demand for facsimile systems in Japan. But this revolutionary switch to digital systems is not the only economic force working in their favor.

Already, facsimile is more cost-effective in Japan than in other countries for compelling and inherent cultural reasons. Japan is the only modern society today in which the typewriter is not widely used. Virtually all correspondence, and all records, are written by hand. This means that there are fewer secretaries in Japanese offices than in their Western counterparts. It also means that teletype operators are more difficult to train, the more so since Japanese *kanji* must be romanized or rendered in *kana* before it can be transmitted by telex, or by telegram. As a result, facsimile has advantages in Japan that it does not have in other advanced industrial countries, thus making it more attractive for communications than elsewhere. This gives the Japanese industry a tremendous advantage and explains, to a great extent, the more rapid diffusion of facsimile technology and systems.

At the same time, of course, facsimile has become more cost competitive than ever before as a result of recent rate hikes for other communications services, especially mail and telex.

Other economic advantages derive from the structure of the Japanese industry. Since facsimile manufacturers are highly integrated, both vertically and horizontally, they will enjoy benefits of linkage in technology, production, logistics and marketing that are not always available to the same extent to foreign competitors. At the same time, competitive forces in the Japanese markets will spur the race towards higher value-added systems, higher quality and higher productivity.

These are indeed potent factors in the equation of competitive power, and, taken together, are probably sufficient in themselves to assure pre-eminence to Japanese facsimile production. But there are still other advantages which enhance and extend these factors.

Japanese industry not only is less inflicted with the institutional lag that impinges on technological change in the field of communications in other advanced countries. Japan also has the advantage of an industrial policy clearly committed to the allocation of resources to the rapid development of information technologies, among them facsimile systems. Likewise, Japan has a well-defined and consistent telecommunications policy, elaborated and executed through the close collaboration of the Ministry of Posts and Communications and the two major communications carriers.

As a result, Japan is better equipped than other advanced industrial countries to undertake the massive task of systems engineering that is necessary if full advantage of new information technologies is to be obtained. Telephone lines, electronic mail service, satellites, data links, mobile radio devices, and cable TV are interlinked in one gigantic system that can use the different facilities to their best advantage.

Tight systems engineering is possible, with optimum balance between the use of various message options. Old systems and plants are readily scrapped to be replaced with new equipment and technologies. The full advantages of computers and automation are obtained with the knowledge that with the resultant improvement in communications, the knowledge power of the entire nation is enhanced.

We are witnessing the execution of a massive communications systems design in Japan which parallels that of extraterrestrial space exploration. The rapid progress of facsimile systems development and installation, which is now unfolding, is but a part of a total Japanese information revolution. It is an important part to be sure, for it will not only hasten the era of instantaneous message delivery, but completely change the way which the information thereby transmitted is used.

CHAPTER 28

VAN — Birth of a New Industry

WHEN JAPAN'S Information Network System (INS) was announced at the third World Telecommunications Forum in September 1979, the futuristic plan looked like the ticket to a brave new world of communications. In contrast with earlier telecommunications systems, the value of the new network would be measured by the enhanced level of its information-processing capabilities, rather than by the sheer speed and fidelity of transmission and switching.

Advances in signal processing brought about by digital technology, accompanied by developments in voice, character and pattern recognition, would be made possible by very large integrated circuitry, computerized switching and optical-fiber transmission networks.

Now it is all happening. Not as it was originally foreseen by the INS designers, but in a more open and competitive business environment that promises to give more rapid and far-reaching effects to the revolutionary technologies the system was intended to exploit.

After months of wrangling between ministries and some unseasonably torrid criticism from Washington, the Japanese cabinet approved early in April 1984 two telecommunications bills that open the door to a new phase of the information age. The bills, when enacted, will end the 34-year-long domestic telecommunications monopoly enjoyed by Nippon Telegraph and Telephone Public Corp. (NTT) and liberalize value-added network (VAN) services to permit Japanese and foreign private firms to enter the market.

As a result, an important new information industry is emerging. This industry, which appropriately has been called network information services, will permit users to interact directly from remote terminals with one or more computers and associated data banks, or from one facsimile terminal to another, regardless of protocol compatiblity. Access to distributed information systems within and between enterprises, remote

---

First published as "Technology: VANguard of Progress," *Far Eastern Economic Review,* 3 May 1984.

transaction recording, data-base inquiry and computer conferencing will all become possible, bringing everyman a cornucopia of information.

Networks which make all this possible enhance the value of existing transmission links to the end user by packet switching of data through digitalized systems, altering format and encoding and decoding signals, facilitating computer communications. A value-added carrier does not construct new telecommunications links. instead, it leases links from other carriers and creates a VAN with sophisticated computer controls to provide new types of telecommunications services.

These new telecommunications services will not only transform the information infrastructure of Japanese society, they will radically alter society itself. Their widespread introduction will greatly enhance the availability of information, prompting changes in major information industries, distribution services, transport, education, law and medicine. Some observers have even likened the new era to that following the invention of printing, creating in its wake a wealth of entrepreneurial opportunities.

Clearly, a great deal is at stake. The information industry, which already has become the largest in Japan, will be given added momentum. Information production, processing and distribution — which currently involves an estimated 30-50 percent of the workforce — will offer more challenging job opportunities in both software and hardware engineering at a pace faster than education facilities can train people to fill them.

Many elements of the network information-service industry are, in fact, already in place, and many of the forces that will shape its future are in full motion. Major restructuring of communications will begin in 1985, with the general guidelines to be provided by the new legislation before the parliament.

In their present, and most likely definitive form, the telecommunications bills identify three categories of communications networks: common carriers that own and operate public-communications circuits; large-scale nationwide VANs (called "fundamental" services), and those VAN services limited to particular geographical areas, industries, companies, groups of companies or computer users. All three types of networks will be privately owned, with only the first confined to Japanese ownership.

At present, only NTT offers common-carrier network services. But this may change after liberalization in 1985. Three applications for approval of terrestrial common-carrier trunks already have been filed with the Ministry of Posts and Telecommunications and feasibility studies are reportedly well-advanced by other major operators.

What promises to be an explosive take-off for the new telecom-

Table 1. The Shape of Things to Come — Japanese VANs

COMMON CARRIERS	Intec Inc.
Nippon Telegraph and Telephone	Japan Information Processing Co.
The Kyocera group	Japan Information Service Co.
Mitsubishi Corp.	NEC Information Service Ltd.
Sanwa Bank group	Nippon Business Consultant Co.
	Tokyo Information Services Co.
LARGE-SCALE VAN	*Sogo Shoshas*
CARRIERS	C. Itoh
IBM	Marubeni
Network Business Association	Mitsubishi Corp. (with IBM and
(with AT&T)	Cosmo 80)
Fujitsu	Sumitomo Corp.
Hitachi	
NEC	*Wholesalers*
	Kirindo
SMALL-SCALE VAN	Pharma
SERVICES	
Computer service companies	*Transport companies*
Fujitsu FIP Co.	Seino Transportation Co.
Hibiya Computer Systems Co.	Nippon Express
	Yamato System Development Co.

munications era in Japan was sparked by a group of so-called upstarts, including Kyocera Corp., Sony Corp., Ushio Inc. and Secom Co., with their application in March 1984 for permission to install a large-capacity optical-fiber circuit between Tokyo and Osaka offering data communications, facsimile and telephone services. Radio and satellite communications also are being considered by the group, which Kyocera vice-president Singo Moriyama has called for other companies managed by "ambitious young executives" to join.

But young executives do not have a monopoly on entrepreneurial spirit. Less than a month after the Kyocera announcement, representative companies of two major Keiretsu groups made their bid for entry into the field, providing a foretaste of things yet to come. Mitsubishi Corp., backed by some 50 member firms of the Mitsubishi group, filed for permission to build and operate an optical-fiber communications system linking Tokyo, Nagoya and the Kansai area along lines of the Japanese National Railways (JNR), the expressways of the Japan Highway Public Corp. and the power lines of the utility companies.

The costs of laying a large-capacity optical-fiber cable linking these three business centers are estimated to range from ¥30-50 billion (US$133.3-222.2 million). Yet, despite this heavy capital outlay, Mitsubishi estimates indicate that handling telephone calls made by the Mitsubishi group enterprises — which account for 7 percent of the total peaktime telephone traffic along the Tokaido trunk line — would prove profitable with charges about 20 percent below NTT rates. One study claims that new entrants may be able to offer services at as low as 20 percent of current rates charged by NTT on Tokyo-Osaka circuits.

One unknown factor in the plans of all three groups is the strategy to be pursued by JNR, the Japan Highway Public Corp. and the utility companies, which are all said to be studying their own entry into the common-carrier field. JNR and the power companies have considerable network-management experience, which, when combined with their nationwide disribution grid, constitutes a major asset in the common-carrier communications business. And for the deficit-ridden JNR, the greener fields of telecommunications look like an inviting solution to nagging financial problems.

Whether other public corporations will be permitted to invade NTT territory after liberalization is still a moot question, however. Massive public funds have already been staked on NTT's INS, which is not scheduled to be fully operational until 1990.

Backed by this facility and a strong technological lead sustained by a 3,200-man research-and-development effort conducted at its three major laboratories, NTT is well-equipped to compete effectively with new entrants.

More problematic is its competitive strength in the market for fundamental VAN services. Although NTT is already in the VAN business, the challengers are formidable and are likely to overshadow NTT, even though they lease NTT lines.

Within hours of the government and the ruling Liberal Democratic Party's final compromise on the decontrol of VAN services, and before Cabinet endorsement of the two telecommunications bills, IBM Japan announced its decision to launch large-scale VAN services entirely on its own, instead of offering them through the company's joint venture with Mitsubishi and Cosmo 80, as had been widely expected. The IBM-developed VAN, called Information Network (IN), will initially use the computer giant's data-processing-center subsidiary to provide nationwide services at the earliest possible date after the proposed telecommunications business law comes into effect next year.

Under this plan, AST, the joint venture with Mitsubishi and Cosmo 80, in which IBM has a 42 percent stake, will be provided with IN control software to engage in small-scale VAN services. By supplying the same software to other owners of IBM computers, supplementing the IBM-designed Systems Network Architecture used for networking since 1975, IBM intends to build a system that will integrate small-scale VAN operators.

Still in the feasibility-study and negotiation stage is a second major large-scale VAN entry prospect, and one that keeps IBM executives awake at nights. A 12-company consortium, called the Network Business Association (NBA), has been formed under the joint leadership of Mitsui and Co. and the Industrial Bank of Japan to find a modus operandi for the introduction of AT&T's large-scale VAN network — AIS/Net 1000 — in Japan.

NBA, which includes major companies representing a broad spectrum of industries and several large industrial groups, intends to use AT&T's software advantage, which makes it possible to link computers of different makes into a universal system with extensive data-storage capacity and capability to develop applications software. The Net 1000 system would, if this project reaches fruition, be adapted for Japanese use by the combined software facilities of Mitsui, Mitsui Knowledge Industries and Nippon Univac (in which Mitsui has a significant stake).

Whether Nippon Univac will join the NBA consortium's large-scale VAN system is still, apparently, an open question. Nippon Univac Information Systems Kaisha, a wholly-owned subsidiary of Nippon Univac Kaisha, has been preparing for its own launching of VAN services and has announced plans to raise its capital for this purpose by selling shares to information-industry leaders, including Mitsubishi Electric Corp., Oki Electric Industry, Sanyo Electric and Toshiba Corp.

But whether this plan, which would have the effect of lowering the share of Sperry Corp. in the company, will now be carried out in view of the liberalization of foreign participation, is not yet clear. Although none of the companies named as prospective buyers of shares has yet shown signs of establishing its own VAN services, Mitsubishi Electric can be expected to participate in VAN projects of Mitsubishi group enterprises, while Toshiba seems likely to produce plans of its own for VAN operations. Still, the long-standing close ties of Nippon Univac with Mitsui and the interest which members have shown in AT&T's Net 1000, suggest a strong possibility of an eventual link with the NBA consortium.

Other announced contenders for the large-scale VAN services market include the three leading Japanese mainframe computer makers: Fujitsu, Hitachi and NEC. Fujitsu president Takuma Yamamoto has made clear his company's intention to enter this field in April 1985, through a crash three-stage development effort. In the first stage, Fujitsu will install an integrated intra-company information service linking the company's

COMMUNICATIONS

new Tokyo headquarters with its Osaka office, Numazu, Kawasaki and Oyama factories and systems laboratory in a 6.3-megabit-per-second digital network.

In a second stage, early in 1985, the Nagano works and all sales branches throughout the country will be added to the network. And finally, after liberalization becomes effective, Fujitsu subsidiaries, 24 affiliated software houses and Furukawa group companies will form the base, along with 108 computer-service companies belonging to the Facom customers' association, for a full-scale nationwide VAN service.

Fujitsu is not new to the VAN business. A subsidiary, Fujitsu FIP Co., was one of the first entrants into the small-scale VAN field when it was opened to private operators in 1982. One of the largest VAN services at present, the Fujitsu subsidiary links more than 400 supermarkets in its Varnet system.

So far, Hitachi has been considerably more modest in its announced plans. The Tokyo-based group entered the VAN service market on 1 April 1984 with the establishment by Hitachi Ltd. of a wholly owned subsidiary, Hitachi Information Network Co., to operate a nation-wide Hinet VAN service, mainly for Hitachi and its affiliated companies. The new enterprise, capitalized at ¥300 million, will establish packet-switching facilities in nine major cities throughout Japan, initially to serve four group companies.

Once the limitation on VAN services is lifted, the Hinet service will be extended to other Hitachi group companies.

A similar pattern of development already is taking shape at NEC, a major force in both the computer and communications-equipment markets. NEC Information Services (NEIS) already has launched its NEISNET, designed to serve mainly small and medium-scale businesses through its 24-member group of affiliated computer-service bureaus which provide a nationwide network extending from Hokkaido to Okinawa. Now limited by existing law to small- and medium-scale users of VAN services, when provided as a component of the company's data services, NEISNET can be offered to all users.

Developing new computer-network systems for expanded disributed data processing and on-line systems has been a key element of competitive strategies for all computer mainframe makers since the mid-1970s. By 1977, four leading Japanese makers — NEC, Fujitsu, Toshiba and Mitsubishi Electric — had all developed network architecture in response to IBM's multi-system network facility with advanced communications functions and the distribution communications architecture introduced by Nippon Univac.

By developing systems software and the solution to problems encountered by users owing to lack of continuity among communications networks, networking capability became the key to competitive strength in the sales, not only of standard computers, but also of communications processors, distributed processors, packet-switching processors, line connectors and an expanding range of terminals.

Now this networking capability must be made to serve as a competitive strategy not only at the intra-firm level, as in the past, but on an inter-firm or universal-utility level, Development of network systems software and the operation of large-scale networks themselves are seen as critical to survival and growth in the computer market, which explains the heavy pressure of the U.S. Government for complete liberalization of VAN services and IBM's immediate announcement of its own independent system.

With the leading Japanese makers positioned to follow IBM's lead, it remains to be seen whether other United States computer-hardware makers with networking systems — such as DEC with Decnet and Xerox with Xten — will enter the Japanese VAN market solo or in joint ventures with Japanese partners.

Their strategy will be determined, at least in part, by developments in the small-scale VAN market, where the pattern is much less clear. Estimates of the number of these network services, which link specific industries or service functions, runs into the 10,000s.

Among the early entries into this small-scale VAN market have been computer-service bureaus, transport companies, sogo shoshas and wholesalers. In addition to those operated by data-processing subsidiaries of Fujitsu, Hitachi and NEC, special VAN services have been introduced by other leading computer-service bureaus, such as Intec Inc., Japan Information Processing Co., Toyo Information Services and, recently, Hibiya Computer Systems.

VAN services in the distribution sector hold the promise of a revolution in wholesale and retail services. Sogo shoshas, with 50 percent of their turnover in domestic trade, have been quick to grasp the lead in this radical transformation. In April 1984, both Marubeni Corp. and C. Itoh announced the establishment of "textile VANs," linking textiles manufacturers, garment makers and wholesalers in on-line enhanced-communications networks.

Sumitomo Corp. has formed a food-industry distribution-net project team with the same objectives in view. And other industry-wide nets are in the planning stage, to be established by sogo shoshas in the near future to service the particular value-added communications needs of the automotive, steel and other basic commodity industries.

But the sogo shoshas are not the only entrepreneurs determined to

397

shape this revolution. Pharmaceuticals wholesalers, particularly Kirindo and Pharma, have established networks to provide market-and-supply information to chemists. A gift-marketing development study group has been formed by 30 leading companies, headed by Descente and Nippon Suisan, a garment maker and consumer-credit company respectively, to explore possibilities for combining VAN, CATV and CAPTAIN services into a multi-media version of the old mail-order business.

These early entrants represent but the beginning of the incipient network-information service industry which, in the coming decade, will expand to include a widely diversified range of services for business, industry, consumers and government.

Electronic mail networks, now in an embryonic stage, are likely to grow rapidly in tandem with private video, audio and computer-conferencing networks. Already, Marubeni Corp. has announced the installation of a multifunction communications system designed by the Dallas-based VMX Inc., featuring a system that combines store-and-forward benefits of electronic mail with the convenience, speed and efficiency of voice communications.

By the end of the 1980s, such systems are expected to deliver as much as 50 percent of all intra-corporate messages. This will then lead to the development of storage-and-retrieval data systems that also will provide inter-corporate electronic-message delivery.

If business offers the most immediate demand for network information services, consumer-network services to residences over networks via terminals already available in the home constitute the most potentially profitable network market. Some 36 resident-based consumer-network services have been identified and are under study. Once NTT's INS is completed, at the end of the decade, household-network services are expected to provide banking, shopping, security, general-information, education and entertainment services.

The major challenges then to be addressed will be the development of specific services consumers need. Providing these network services will require production facilities that will themselves be on networks. Large distributed networks of computers, software, terminals and databases will constitute the new "factories," manufacturing new services by adding value to existing communications services.

The technological inevitability of the development of the network-information service industry is now fully demonstrable. Accelerating progress in large-scale and very-large-scale integration makes possible increasingly intelligent terminals with enhanced memory, computational and communications capacity to deal with intermixed signals — data, image, video and voice — as required. When linked by optical fibers and laser beams, these terminals will reduce communications costs greatly while increasing the range of available services.

Demand for computers and communications equipment, in turn, will expand exponentially. For every yen spent on the transmission networks themselves, at least four more will be paid for processors, terminals and software. This means a vast new market not only for the makers of mainframes and complex communications equipment, but, because services and their users are so diverse, numerous small, innovative manufacturers of special equipment are also finding new opportunities.

Although the size of the Japanese market for information networks defies meaningful estimation at this stage in its development, the growth rate in coming years is expected to be in the neighborhood of 50 percent a year and could be as high as 100 percent in some years.

CHAPTER 29

LAN — The Net
Results of Automation

SINCE THEIR DEBUT in the United States in 1980, local area networks (LANs) have been widely proclaimed as the key to the electronic office of the future. Telephone circuits which transmit information at relatively slow rates increasingly were inadequate as communications links between computers with an information-processing capability many times greater. Thus, when networks were developed with transmission rates up to the tens of megabits-per-second range, the missing link which would make the promise of office automation (OA) a reality seemed to have been found.

Coaxial cable, used in Cable TV to carry video signals and optical fibers, were both adapted for office communications employing newly developed packet-matching techniques to connect minicomputers, data terminals, word processors, printers, copiers, facsimile devices and intelligent telephone sets. The logic of such local links was compelling. Apart from more efficient communications among them and with other systems, networks serving an office, factory, company, hospital, or university campus would enable cost-effective sharing of information within the organization, enhance information-systems control and hence assure more efficient information management.

Nowhere did this logic have greater attraction than in Japan, where the convergence of communications and computer technologies constitutes the organizational rationale for leading office-equipment manufacturers. Without exception, they rushed to acquire licenses for each new networking system, developed mainly in the United States. At least 27 makers of computers or communications equipment have entered the market with closed systems, using their own devices in bewildering variations of network configurations with different transmission media; systems architecture, interface, access methods and control mechanisms.

Partially as a result of the ensuing chaos, users went on buying stand-alone equipment, relying where necessary on telephone circuits for their

Reprinted from "Japan: Cost and Cacophony Delay LAN's Takeoff." 6 September 1984.

communications needs. Although the introduction of open network systems developed in the standard star, bus, or ring configurations allowed the connection of equipment of various makers, it did not assure greater utility or efficiency. A data network is not unlike a telephone system: you can call Beijing, but if you do not speak Chinese, the connection does not assure communication.

Nor does a network make it possible for machines with different protocols to take advantage of this new technology. The electronic translation of information flow between devices of different makes is necessary and this process has been costly. When Japanese versions of Ethernet — the Xerox LAN that has virtually become the world standard — first appeared, the cost of connecting a terminal to the net varied from the equivalent of US$3-4,000. Today, the price of connecting devices on a pair of chips has dropped to a tenth of that and is expected eventually to decline to as little as US$25. But this only assures connection with a network; it does not resolve translation problems, which require special black boxes that function between communicating devices and translate protocols. In their absence, the result is likely to be networked cacophony.

But cost and cacophony are only two of the reasons LANs have not yet taken off in Japan. Recognition of the value of networking is itself only an intellectual exercise: the principle is simple and sound, but the practice is fraught with complexities. Companies and institutions must have the equipment needed to communicate. They need to have had considerable organizational experience in the use of a variety of office machines in all conceivable applications. They must also be able to estimate with some accuracy their equipment and systems' needs for three to five years ahead; and they have to be able to decide which network system is best for those particular needs.

This is clearly an evolutionary process which may require three to five; even 10 years or more. The handling of specific tasks must first be upgraded through the introduction and utilization of single-function, single-unit machines. Then multifunction machines usually can be introduced, standardizing office tasks for greater efficiency and linking the various machines to widen their scope of application. Ultimately, as the number and diversity of terminals for personal use multiply, they may be interfaced to form an integrated system through intramural local networks — provided, of course, the additional investment is functionally cost-effective.

When the successive steps should be taken and which systems should be used are complex questions requiring clear organizational and managerial understanding about who will use the products and systems, what they will use them for and why they and the organization will be

402

better off for using them. Integration, though seemingly logical itself, does not bring synergy; it only makes synergy possible and assumes major organizational importance only when function is materially affected. The key word here is function, and this has had quite different operative implications for the adoption of LANs in Japan than elsewhere.

In the first place, the communications problem poses itself quite differently in the Japanese office. Since there were no Japanese typewriters to begin with — and *kanji* word processors are coming into wide usage only now — there has been neither the need for, nor experience in, information processing that was common to offices in the United States and Europe where these devices have for years been used extensively. Likewise, since the Japanese are not generally trained to use typewriter keyboards and executive keyboard allergy is more widespread than elsewhere, there are limitations on the diffusion of personal computers and terminals within many companies. Since it was precisely this combination of word processing and personal computing which made local networking necessary and cost effective in the United States and Western Europe, the absence of it has given networking a lower priority in investment decisions in Japan.

Then, too, the inability of Ethernet — the most widely-used baseband network — to transmit video signals constituted more of a decisive disadvantage in Japan than in other advanced countries. Optical readers, video and facsimile, tend to be used more in Japanese offices, where keyboard interface is more restricted than elsewhere. Optical transmissions, of course, can be carried by broadband cable which generally uses frequency-division multiplexing that allows several independent streams of information, including video, to be mixed or multiplexed on the coaxial cable, but connector costs for each station are sufficiently higher than for links with baseband cables to limit this option.

Other limitations of the Ethernet-type LANs have inhibited their use in manufacturing, a development which might have served as a further inducement to their introduction in the office. Ethernet was not designed for the continuous information flow needed in manufacturing, but for fast infrequent communications between machines. Moreover, a major drawback with Ethernet — which uses a bus typology with a single path connecting any pair of machines — is its information-delivery probability. Since no two stations can transmit simultaneously, each station on the net must listen before transmitting to assure that no one else is doing so. If so, on detecting the problem, a station would wait a short time before transmitting. While this is not a serious problem in many office applications, it is a fatal drawback in manufacturing where precision timing is critical to all automated production systems: and, given the importance of finely tuned, just-in-time production systems in Japan, 99 percent information-delivery probability is unacceptable. Again, this problem can be solved by using the so-called token-passing ring network with a broadband system which, though highly reliable, entails considerably higher set-up costs.

Function, of course, also depends on the Japanese management system. The main concern is in the productivity of the whole group — not just individual workers — and since secretaries, as such, are not numerous and do not play a significant role in the Japanese office, the purpose of OA has not been to improve the productivity of this particular group.

Whether LANs are cost-effective in a Japanese office will be determined only marginally by the number of secretaries that are replaced. Rather, the critical factor will be the extent to which information flow is enhanced for the entire managerial team from bottom to top. The problem, then, becomes how to improve individual efficiency without disrupting harmony within the group.

This means that LANs in Japan are usually large systems and large systems require large investments. The upshot is that few LANs systems have been established in Japan to date, despite the number of Japanese and foreign manufacturers offering various types of networks. According to one information-systems engineer, only 10 Ethernet systems are presently operating in Japan. Dedicated, closed LANs sold by computer makers and using their own equipment, generally in a star configuration with a mainframe as a host, are more numerous, but there are only about 40 of these. Japanese users, it appears, still have much to learn about LANs.

There have been exceptional firms, such as Okamura Corp., a medium-sized manufacturer of steel furniture in Tokyo, which have introduced total office systems. This enterprising company has linked its six factories, 58 sales offices and 2,200 employees in a nationwide on-line system, fully equipped with computer terminals, personal computers, microfilm filing systems and facsimile machines.

All in-house communications, many that would normally be by telephone, are relayed by computer unless the matter is unusually complicated. Similarly, all other announcements and communications are routed through computer terminals.

Supplies are ordered by computer. As a result, long lead times between the factory purchase order and headquarters action are avoided. The purchasing department head, upon receipt of a purchasing request, can give immediate approval after inventory levels, identity of suppliers, prices and delivery have been checked on his terminal.

Only five staff handle the voluminous head-office accounting tasks, including recording the movement of funds, preparing financial statements, making stock and tax payments. Before the installation of the integrated office system, at least 30 accountants were needed in this department. Similar labor savings have been achieved in the personnel department where only three people handle the affairs of 2,200 employees, recruit new members of the firm and assure monthly salary payments.

Okamura did not achieve this transformation overnight, however. Step by step, over a long 17-year period, this integrated system gradually took shape. The first on-line marketing and distribution system was established in 1966. In 1969, a computer center was established at headquarters and a production-information system was designed. Today, the company's network links the host computer with 330 automated on-line machines, making it possible for employees to obtain needed information concerning company operations instantaneously.

But Okamura's pioneering experience in integrated office automation was not widely emulated. On the contrary, it was a matter of derision for some cynics. As one management expert commented: "Okamura was like a farmer who had just purchased an executive jet."

Now, however, Okamura finds itself in the best of company. In April 1984, Toshiba inaugurated a new headquarters building — a veritable monument to integrated OA systems — destined to set the fashion for major and minor companies alike for the remainder of this century. Rising 40 stories on the site where the company's original predecessor, Shibaura Seisakusho, was founded in 1882, the ¥40 billion (US$164.41 million) edifice features more than 1,000 computers, word processors, facsimile machines, copiers and other OA equipment linked by a multi-dimensional LAN using both coaxial bus and optical fiber ring configurations.

Like the nervous system of the organization, information transmission channels with a speed of 10 Mb/s interlace the offices of each floor in a bus-type LAN network, integrated into the total grid by a multi-channel, ultra-high speed, 100 Mb/s, ring-type LAN. Organized as a hierarchical information system, equipment in each department is tailored to meet the needs of various levels of management. Each department is linked directly, not only with all other departments in headquarters but, through Toshiba's total on-line system, to 4,000 terminals in 120 locations, including factories and branch offices.

At the summit of this hierarchical system, all Toshiba executive offices are equipped with special terminals programmed to supply 300 separate data menus, including up-to-date sales, inventory, production and profit information, as well as current domestic and international economic and business trends. But executives are not passive information recipients of assorted information. They interact with the system and, using a market-simulation menu, can perform complex business analysis on their own terminals.

The entire system converges on a 38th-floor, futuristic computer-presentation room equipped with state-of-the-art audio; video and computer equipment in a simple and warm decor. Meetings are informed by full-color graphic presentations of the latest information which flows to the room through multi-channel, multi-level local area, wide area and large-scale value-added networks linking the Toshiba group of companies in a total cohesive on-line information system.

The entire structure exudes an aura of the information age. Although the information system itself accounted for only 10 percent of the building's costs, the spaciousness of the offices — each with its characteristic OA corner — bears witness to a pervasive harmony of man, information and technology.

Gone is the usual clutter and congestion of the traditional Japanese company office. No longer must each important conversation be recorded in a handwritten memorandum. Along with all other company records and communications, memoranda are drafted, communicated and filed automatically on word processors. Letters, which were also handwritten in the past (if in Japanese) are now drafted on a word processor in the OA corner or at the executive's desk and printed out at the 22nd-floor printing center.

Without exaggeration, the system has transformed the entire image of the company for those both inside and out. As one executive put it: "People working at Toshiba feel like they have been part of the rebirth of the company and, in fact, in the transformation of their entire working lives."

The system has something to make the work of every member of the headquarters staff not just more efficient, but more pleasant. From ID cards — serving as part of a new time-keeping system free of cumbersome and time-consuming paperwork — to a firm banking system and a cashless cafeteria, to automated delivery systems and electronic phone-directory and mail systems; all seem designed to assure a new interface which makes OA not just an instrument of higher productivity, but the servant of all those working in the organization.

That productivity has been enhanced there seems no doubt. "From the outset, it was a no-risk undertaking," Shigenori Matsushita, chief engineer of the information-system group reflects: "All we had to do to assure a return on our investment was to obtain a 3 percent increase in efficiency in the first year, dropping to a 1 percent improvement in the fifth.'' But, according to some calculations, productivity increases as high as 30 percent can be expected.

Add to that the boost in morale, the new image of the company and the demonstration effect of the integrated-information system. In sum, this investment — largely in the company's own equipment and software — must be about the best use of resources Toshiba has made in over a century of existence.

The effects on OA sales were immediate. Before the new machines had really been broken in, Alps Electric, a leading electronics-components maker, placed an order for an entire system with a reported price tag of ¥1.5 billion, Visitors continue to flow through the building, keeping a reinforced staff of guides and demonstrators busy full-time.

By 1985, in one short year, Matsushita estimates, 10 percent of Toshiba's OA business will be LAN-related: OA is itself the fastest growth sector in the company's vast range of products, networking is destined to become almost instantaneously an important source of revenue.

Sales of LAN systems will be mainly direct to the customers, rather than through dealers or agents. Detailed knowledge of the customer's needs is required by network designers and they in turn must have a combination of technical knowledge about both hardware and software. Software houses or computer-service bureaus, both of which have systems-design capability, must ultimately rely on the manufacturer for hardware technology.

Toshiba's LAN sales will be organized along user lines, with departments for each branch of business — manufacturing, financing and servicing. While, in the past, users relied mainly on their own computer divisions for systems development and applications software, this practice is fading. Instead, OA-equipment makers are expected to provide these services as part of a total systems package and this trend will increase as the use of LANs expands.

Prospects are that, following Toshiba's example, many OA firms will build showcase office buildings in the near future. Before the end of 1984. Fujitsu will move to a new 22-story facility in the financial district, where a LAN will link headquarters operations to five major factories and research centers in a wide area network that will itself be a part of the projected Fijitsu large-scale value-added network.

By 1933, LANs will be widely used by most major Japanese enterprises and they will be followed by a large number of medium-sized companies. As a measure of what this means in business volume, if only companies listed on the first Tokyo Stock Exchange were to spend as much as

Alps Electric during the next decade, there would aleady be a market of well over ¥1 trillion and this almost surprise-free scenario is only a beginning of things to come.

In keeping with all predictions about the growth of the OA market, this one is likely to prove a gross underestimation. Even without the impetus of LAN, demand for OA equipment is far outstripping earlier industry forecasts. According to the most recent long-term projection of the Japan Office Equipment Industry Association, total industry sales of ¥5.4 trillion estimated for 1990 will be surpassed in 1988: this amounts to a two-fold increase in five years during which production is expected to rise on average 14.7 percent annually. And, by 1993, gross production value of business machines and OA equipment will reach ¥9.5 trillion, if output grows at the projected average 10.6 percent annual rate for the intervening five years.

In the medium term, Japanese word-processor sales will lead the field, growing at an average annual rate of 47.7 percent in the 1983-88 period. Over these years, production of facsimile, personal computers, office computers and micrographic equipment will all register respectable growth of over 20 percent annually. This proliferation of machines, increasingly used as personal or decentralized work stations, will be then required linkage in LANs to achieve the full potential of their utility.

Ultimately, the latest OA white paper predicts, LAN and other forms of computer networks will come to make up 80 percent of the OA market. From 1988 to 1990, industry analysts expect that 55 percent of all office-equipment sales will be LAN-related, a trend that will steadily increase throughout the 1990s.

If the evolutionary model explains the slow progress during the past five years in Japan of OA's great phantom — LANs — it also provides an appropriate paradigm for analysing its prospective development in the coming decade. A LAN becomes necessary only at a certain point in the evolution of an organization's information systems: when the number of terminals, work stations and hosts requiring interconnection has reached a criticial mass. Prior to that the LAN is simply unnecessary cabling and connections: after that point is reached, local networking becomes imperative for enhanced efficiency.

PART IX.

Regional Integration

PART IX

Regional Integration

CHAPTER 30

East Asian Electronics:
System and Synergy

EAST ASIAN INDUSTRIALIZATION is reshaping the contours of the world economic system, providing a steadily increasing impetus to its continued growth. The perimeter stretching from Japan and Korea to Singapore has become the global epicenter for a growing array of traditional industries, demonstrating a remarkable capacity for producing the best for the least. But, even more important, East Asia is emerging as the heartland of the global electronics industry and a major center of innovation in advanced electronics technologies which provide the main source of future wealth creation, and hence sustained rapid economic advance.

The East Asian electronics industry, comprising some 40,000 firms employing 2.5 million people at the beginning of the 1980s, is not only the world's fastest growing one; taken as a whole, it is at once more highly diversified and integrated than those of North America and Europe. Soundly based on the manufacture of consumer electronics products, the industry embraces the world's widest and most rapidly changing range of electronic products. In no other region does such a large percentage of the firms in the industry pursue global stategies. By optimizing global markets, leading East Asian electronic firms in all sectors of the industry attain the necessary economies of scale and experience to sustain rapid technological innovation and diffusion, resulting in a steady shift in technological leadership from North America to East Asia.

As a result, East Asian pre-eminence in consumer electronic technology and production at the end of the 1970s was without challenge. Three out of every four radios produced in the world were East Asian; the largest number were made in Hong Kong. Three out of every five car radios were produced in the region, mainly in Japan. More than 60 percent of the world's monochrome television sets were made in East Asia, where Korea was the leading producer. More than 87 percent of all tape recorders were made in Japan or South-East Asia. Japanese manufac-

Reprinted from "Asia's Electronics Revolution," *Euro-Asia Business Review,* Vol. 1, No. 1, October 1982.

turers alone accounted for as much as 94.2 percent of global video taperecorder production. And when Hong Kong replaced Switzerland as the world's largest exporter of watches, East Asia's position as the center of the world's timepiece industry was assured.

If East Asian consumer electronics products are now standard household equipment the world over, East Asian leadership in microelectronics is at once less obvious and more recent, and therefore less well understood. It follows, however, that the region producing most of the world's solid-state consumer appliances will be one of the largest markets for semiconductors, capable of sustaining a large share of their production. And indeed, since the inception of the semiconductor industry, Japan has led the world in the production of semiconductors for consumer applications. Moreover, since 1963, when Fairchild began producing transistors in Hong Kong, virtually every major American semiconductor manufacturer has shifted transistor, or the final stages of integrated circuit, production to East Asia. As much as 95 percent of the integrated circuits produced by U.S. industry are wired, bonded and packaged abroad — mainly in the newly industrialized countries (NICs) and less-developed countries (LDCs) of East Asia.

A further development has added to the region's prowess in this critical sector of the industry since circa 1977, when Japanese industry began its spectacular ascendancy as a major supplier to world markets of integrated circuitry for computers and other industrial electronics equipment. In a short span of five years, comparative advantage and predominant market-share in state-of-the-art memories shifted from the U.S. to Japanese industry.

Japanese semiconductor manufacturers are now reputed to have captured as much as 70 percent of the world market for 64K random-access memory chips, currently the msot advanced microelectronics device, thus threatening a sector in which the American industry has held undisputed leadership since the early years of solid state electronics.

Equally important, other East Asian industries are now establishing their own semiconductor industries, independent of the Japanese and American front runners. In 1981 — at a time when Britain was still struggling to assure the future of its sole venture into microelectronics after massive injections of public funds and the Italian industry was floundering on the brink of disaster — Hong Kong entrepreneurs independently launched three vertically ingrated semiconductor ventures to produce advanced microelectronic devices and Taiwan's Electronics Research and Science Organization (ERSO) developed the technology of integrated circuit manufacture now being employed by United Microelectronics, located in the Hsinchu Science-based Industrial Park. Already, in 1979,

Table 1. Production and Consumption of Major Electronic Products by Region, 1978

(Unit: 1000 sets, %)

	Radio		B&W TV		Color TV		Radio Casette Taperecorder		Car Radio Car Stereo		VTR		IC*	
	P	C	P	C	P	C	P	C	P	C	P	C	P	C
Europe	10,620	24,675	4,539	6,115	10,262	10,040	5,961	14,146	5,930	10,673	90	297	433	1,100
	(11.5)	(27.7)	(18.5)	(27.2)	(32.0)	(33.8)	(11.5)	(25.2)	(11.9)	(21.5)	(5.8)	(21.8)	(6.5)	(21.6)
North America	2,062	38,130	1,024	6,394	8,423	11,135	204	20,800	10,237	23,500	0	426	4,582	2,304
	(2.2)	(42.7)	(4.2)	(28.5)	(26.3)	(37.5)	(0.4)	(37.0)	(20.6)	(47.3)		(31.3)	(68.5)	(45.3)
Latin America	5,223	6,423	3,078	2,728	1,100	1,350	395	995	1,740	1,776	0	15	n.a.	n.a.
	(5.6)	(7.2)	(12.6)	(12.1)	(3.4)	(4.5)	(0.8)	(1.8)	(3.5)	(3.6)		(1.1)		
Asia	62,754	4,758	10,788	2,846	3,205	681	23,712	4,882	5,617	543	0	53	n.a.	n.a.
	(67.8)	(5.3)	(44.1)	(12.7)	(10.0)	(2.3)	(45.6)	(8.7)	(11.3)	(1.1)		(3.8)		
Japan	7,927	3,720	4,567	870	8,549	5,630	21,730	6,920	25,691	9,920	1,470	400	1,195	1,413
	(8.6)	(4.2)	(18.7)	(3.9)	(26.7)	(18.9)	(41.8)	(12.3)	(51.8)	(20.0)	(94.2)	(29.3)	(17.9)	(27.8)
Others	4,000	11,500	475	3,500	500	892	nil	8,500	400	3,300	0	172	482	267
	(4.3)	(12.9)	(1.9)	(15.6)	(1.6)	(3.0)		(15.1)	(0.8)	(6.6)		(12.6)	(7.1)	(5.3)
Total	92,586	89,206	24,471	22,453	32,039	29,728	52,002	56,243	49,615	49,712	1,560	1,363	6,692	5,084
	(100.0)	(100.0)	(100.0)	(100.0)	(100.0)	(100.0)	(100.0)	(100.0)	(100.0)	(100.0)	(100.0)	(100.0)	(100.0)	(100.0)

Notes: (1) P: Production,   C: Consumption = Production + Import—Export
(2) Asia: Korea, Taiwan, Hong Kong, ASEAN countries.
(3) Unit: IC, Production Value. Million US$

Source: Nomura Research Institute

East Asian countries produced more than four times as many semiconductors as Western Europe. Now, with the steady advance of the state-of-the-art by Japanese industry and the emergence of local producers of integrated circuits in the NICs, the ascendancy of East Asia at the highest levels of microelectronic technology is a prospect which leaves no quarter to complacency in the boardrooms of European electronics firms or the once seemingly invincible innovative front-runners of Silicon Valley.

Indeed, given the experience of the past 25 years, the diffusion of successive generations of microelectronics technologies will surely be faster in East Asia than in either North America or Europe, honing the sharp cutting-edge of the industry in the region. As the Japanese industry shifts its accent from consumer to industrial electronics, leading makers are gaining market share in computers, copiers, and communications equipment. More than 135 robot manufacturers are building microcomputers into a growing array of automated production equipment and systems, which means that by the end of the 1980s, virtually every factory in Japan, whatever its size, will be quipped with flexible manufacturing systems using microcomputer-controlled machines. Already, in 1982, Fanuc, the world leader in numerical controls, is producing servo and spindle motors for automated systems in a new factory 'manned' by thirty people (including managers and janitors) and 101 robots for round-the-clock production.

Combining microelectronics with optical fibers, Japanese firms are taking the lead in telecommunications, automotive electronics, medical electronics and office automation. These applications combine with factory automation, data processing, and the massive East-Asian consumer electronics industry to provide a ready market for incremental improvements in semiconductor devices, sustaining a rapid pace of technological advance in microelectronics.

Demand for semiconductors is expanding exponentially as the East-Asian electronic industry broadens its base in Malaysia, Thailand and the Philippines, and the NICs follow Japan into more sophisticated products and automated production systems. By 1985, Taiwan manufacturers will be mass-producing robots of their own design to meet the growing demand at home and abroad. Domestic demand, estimated at 10,000 annually, will be supplemented by the increasing requirements for robots in neighboring countries such as Hong Kong, where factories producing a profusion of electronic watches are beginning to switch from labor-intensive to fully automated production. Currently, a whole new generation of high-technology electronic manufacturers is in the gestation or embryonic stage in the new science parks of the region.

At work here is a dynamic interaction between three tiers of a highly

Fig. 1   Production and Consumption of Radios by Region (1978)

(Unit: 1,000)

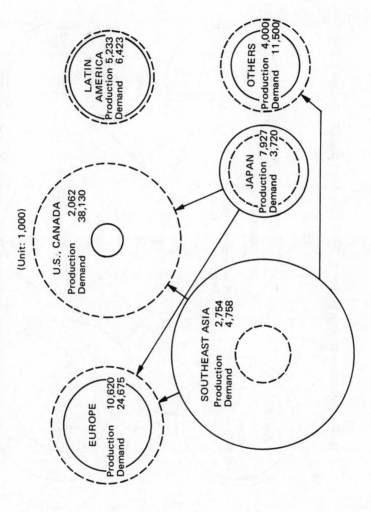

U.S., CANADA
Production   2,062
Demand   38,130

LATIN
AMERICA
Production 5,233
Demand   6,423

JAPAN
Production 7,927
Demand   3,720

OTHERS
Production 4,000
Demand   11,500

EUROPE
Production   10,620
Demand   24,675

SOUTHEAST ASIA
Production 2,754
Demand   4,758

Fig. 2 Demand and Production of B&W Television Sets by Region (1978)

(Unit: 1,000)

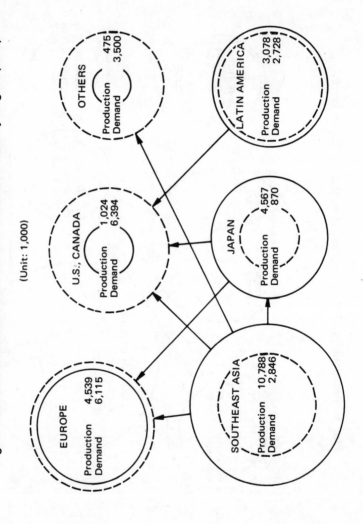

integrated regional electronics industrial system. The cadence of this systems advance in the past has been set by the rapid pace of innovation and the production efficiency of Japanese industry. But the emergence of the Asian NICs as important centers of consumer eelectronics and semiconductor manufacture in the 1970s has increased the competitive pressures on the Japanese industry itself, adding further momentum to the pace of innovation by leading manufacturers. The pace quickens as electronic manufacturers in the NICs are now confronted with the loss of comparative advantage due to rising labor and land costs at home, and competition from new export-oriented electronic industries in the developing countries of ASEAN. They are forced, therefore, to rapidly shift resources to high-technology production in direct competition with advanced Japanese industry.

But the forward thrust of this intricately interactive East Asian electronics complex cannot be understood in simple terms of inter-industry or inter-firm competition, fierce as that competition is.

Nor do ideas such as the product life-cycle theory take us toward an understanding of the dynamic processes of growth and geographical expansion of the East Asian electronics industry system. Transistor radio assembly and planar transistor production began in Hong Kong at the very early stages of the product life-cycles, shortly after their initial introduction in Japan and the United States, respectively, and for reasons which do not lend themselves to precise quantification.

Added to a robust business environment and comparatively low costs of production is a Darwinian struggle for survival among the countries of East Asia whose constituents equate the continuity of their respective societies and cultural integrity with the development of the electronics industry. Each of the four NICs, following the example of Japan, has elaborated industrial policies which link the survival of their industrial systems specifically to the mastery of electronics technology. Deprived of raw materials and energy, peoples of these countries understand the stark realities of their economic situation and are therefore more inclined to cooperative effort in the management of rapid technological change. Attitudes and institutional arrangements, therefore, conspire to sustain the rapid advance of the electronics industries as a vital part of a communal effort for survival.

Government policies have been designed specifically to assure the development of entrepreneurship and the other human resources required to transform an otherwise austere economic reality into an efficient, highly-geared, wealth-creating industrial system with rapidly advancing electronics technology as its main driving force.

The resultant growing might of the East Asian electronics industry, as

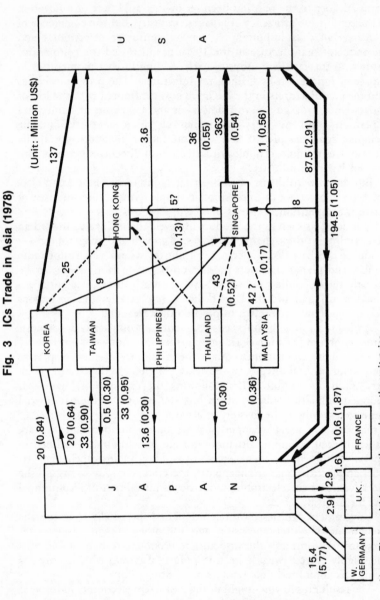

Fig. 3 ICs Trade in Asia (1978)

(Unit: Million US$)

Note: Figure within parentheses shows the unit price.
Source: Foreign trade statistics of each exporting country

is readily apparent, has already had its impact on its counterpart in North America and Europe. But how the loss of comparative advantage and institutional and cultural impediments to the rapid diffusion of advanced electronics technology in these Atlantic communities conspired, ironically, to further fuel the progress of the East Asian industry is less obvious.

First, prewar international cartel arrangements governing manufacturing which had effectively restricted trade in both receivers and components, were still intact in the postwar period. As a result, the U.S. industry was structured along national lines with essentially domestic manufacturing and market strategies. This left large open fields for development by a thrusting East Asian industry.

Second, the traditional policy of major U.S. manufacturers to refuse to supply to mass merchandisers whose pricing policies they could not control drove mass merchandisers to seek overseas suppliers. As a result, U.S. mass merchandisers became the natural allies of cost-efficient export-oriented East Asian manufacturers, providing them with 'instant access' to the vast American market without significant investment in marketing organizations and operations. Thus, East Asian electronics industries developed, at the outset, without international marketing strategies or structures, gaining large market shares at minimal cost.

Third, U.S. consumer electronics manufacturers moved their production to East Asian countries primarily in response to the rising market power of mass merchandisers, and not, as common wisdom has it, mainly in response to aggressive penetration of U.S. markets by Japanese manufacturers. This move provided a timely boost to the Taiwan and Korean electronics industries at a critical point in their shift to export-oriented production.

Fourth, faced with industrial and institutional lethargy at home, U.S. semiconductor makers have frequently been forced to seek application for their new devices in Japan and other East Asian countries. The dramatic ascendancy of East Asian calculator and electronic watch production was fueled at the take-off stage by just such an injection of new technology from American firms.

Finally, fierce competition among the leading U.S. semiconductor manufacturers for world markets, and the labor rigidities at home, spurred the shift of assembly operations to East Asia, providing timely infusions of capital and technology, especially for the Singapore and Malaysian electronics industries.

Industrial Dynamics: Stage One

The course of events and economic logic ordained from the outset that the East Asian electronics industry should be founded firmly on the bedrock of production for consumer markets. Cartelization of the radio receiver industry in the prewar period had produced a patchwork of national industries in Europe and the United States dominated by monopolies or oligopolies. In supreme command were Philips, Siemens-AEG-Telefunken and RCA, which had successfully conspired to divide world markets among themselves through an agreed patent pool, based on the old 'imperial' logic.

Philips, therefore, held sway in the Benelux countries and their colonies, and shared in the markets of the British and French empires through subsidiaries in the respective 'mother' countries. Only RCA, originally a joint venture and patent pool of General Electric, Westinghouse and AT&T, was subjected to the perennial gale of competition in its home market and U.S. territories, as a result of continuing pressures of the Federal Trade Commission and court action by RCA licensees. What was left of the world market was carved up among Philips, the German duopoly and RCA. Competition in the world marketplace was in the perennial doldrums.

Although Japanese manufacturers were tied into these cartel arrangements through patent arrangements and foreign investments in the Japanese radio and phonograph industries, as outsiders they were more victims than beneficiaries of the restrictive Atlantic combination. In the postwar period, therefore, the Japanese industry had every reason to launch a full-scale offensive against the remnants of a system not of their making. Two developments facilitated the Japanese challenge to the prewar structure of the world industry.

Most important was the decision by Bell Laboratories in 1952 to offer licenses for the newly-developed semiconductor technology on reasonable terms to anyone who wanted them. As a result, with the stroke of a few pens and the payment of modest advance royalties, the advantage held by Philips, RCA and the Germans, through their tightly held patent pool covering vacuum tubes and their applications, was eliminated. In the United States, the door was opened for a new breed of entrepreneur who founded the semiconductor industry in the south-west, quite apart from (and on organizational and behavioral principles quite different from) those of the prewar radio and vacuum-tube makers. Comfortable with their existing market power and underestimating a possible Japanese challenge to the system, none of the American or European consumer electronics manufacturers saw the full promise or the threat of the new technology.

Nor, indeed, did the established Japanese radio manufacturers, who were all locked into the existing patent system. Once Sony, then an unknown upstart called Tokyo Tsushin Kogyo K.K., had proved that there was a market for transistorized radios, however, Japanese competitors blossomed like cherry trees in the spring-time. The market Japanese makers found abroad, especially in the United States, was beyond all expectations. Although not a single Japanese radio manufacturer had an overseas marketing organization in the 1950s, that proved to be no major problem. Buyers beat a path to their doors.

Mass merchandisers, to whom leading manufacturers in the United States and Europe had traditionally refused to sell as part of their price maintenance system, emerged as a vital link with the United States market for the Japanese industry. Sears, Roebuck, J.C. Penny, Montgomery Ward and other leading chain stores in the United States placed massive contracts with Japanese makers supplying designs, specifications and quality control. Contary to common wisdom, the Japanese entry into the United States market was not achieved through aggressive marketing strategies, but rather as a result of aggresssive purchasing by highly efficient American mass merchandisers. Since retailers such as Sears, Roebuck controlled as much as 10 percent of the United States retail sales of consumer electronic products, large shares of the vast American market were obtained at little or no marketing cost to Japanese makers. Only Sony insisted on selling its products under is own trademark in overseas markets at the outset.

Thus, largely by default, American consumer electronics manufacturers lost large sectors of the market for transistor radios to Japanese makers, and by so doing they provided the cornerstone on which the Japanese electronics industry, and eventually the entire East Asian electronics industry, were built. Although all major U.S. consumer electronics manufacturers eventually undertook transistor radio production, none came to terms with the mass merchandisers. Nor was there a concerted strategy to develop markets outside the United States; indeed, both the European and Asian markets were left almost untouched. In the early years of import substitution assembly in the Philippines and Vietnam, RCA made a half-hearted attempt to meet Japanese competition but withdrew from Asian markets when Sony launched its first multiband transistor radio in 1959. Both the major U.S. retail chains and the Asian markets were left to Japanese makers by the American industry, and from the European quarter there was no contest.

With channels into the American market provided by an increasing number of mass merchandisers and an open field in the Asian region, Japanese electronics manufacturers found in consumer products the initial market pull for rapid growth. And further momentum was added by leading mail order houses, department store chains and supermarkets in Europe.

Further impetus was provided from a most unexpected quarter: American and European manufacturers themselves. Unable to compete with the rising tide of Japanese transistor radios introduced to the market by mass merchandisers, both U.S. and European manufacturers gradually abandoned production of radios. Resources were shifted to supply the rapidly growing market for monochrome television sets, and production of transistor radios was subcontracted to Japanese radio manufacturers. Rather than invest in the development of innovative radio designs, U.S. manufacturers opted for abandoning the field to East Asian producers.

In fact, mass merchandisers and manufacturers in the United States, and eventually Europe, did not confine their purchases to Japan alone for long. As soon as Hong Kong assemblers demonstrated that they could trim costs below those of Japanese manufacturers, even with imported Japanese components, large American buyers began shifting the purchase of more price-sensitive products to the colony. In time, Hong Kong became the world's largest producer of transistor radios, a distinction it still holds today.

Although the first transistor radios assembled in Hong Kong, by Champagne Engineering (later renamed Atlas Electronics) as subcontractor for Tokyo Tsushin Kogyo KK (later renamed Sony Corporation), were intended for export to Commonwealth countries which accorded Hong Kong products Imperial Preference treatment, the main interactive forces which propelled the development of the Hong Kong electronics industry were the pull of the U.S. market and the push of Japanese industry as supplier of components and technology. And Hong Kong entrepreneurs, given free rein, proved themselves remarkably ingenious in combining these two forces to develop a thrusting industry highly sensitized to both market and technological changes. As U.S. buyers shifted their orders from Japan to Hong Kong, the flow of orders for semiconductors and passive electronic components by the rapidly expanding Hong Kong radio assembly industry flowed back to Japan. Hong Kong entrepreneurs were quick to 'reverse engineer' the latest Japanese radio models and produce near-replicas which were sold at prices considerably below those of Japanese makers. Rather than attempt to inhibit this process, Japanese radio manufacturers, also producers of important com-

Table 2. Direct Overseas Investment by Japan's Electrical and Electronics Industries, 1960–1979

(Unit: case)

Industry	Establishment year	Korea	Taiwan	Hong Kong	Malaysia Singapore	Thailand Philippines Indonesia	S.E. Asia Total	North America	South America	Europe	Others	World Total
Consumer Electronic Products	–1965	0	2	1	1	2	6	2	2	0	1	11
	1966–1969	0	3	0	5	5	13	0	2	1	7	23
	1970–1974	6	6	0	11	6	29	3	7	6	9	54
	1975–1979	2	2	1	5	3	13	10	3	7	6	39
Electonic Parts	–1965	0	3	1	1	1	6	1	0	0	1	8
	1966–1969	1	13	0	0	1	15	0	3	0	4	22
	1970–1974	31	25	2	16	1	75	1	7	1	4	88
	1975–1979	8	9	2	17	4	40	5	3	3	1	52
Industrial Electronic Products	–1965	0	2	0	0	0	2	2	0	0	1	5
	1966–1969	0	1	0	0	0	1	1	2	1	1	6
	1970–1974	1	2	0	2	0	5	2	3	0	1	11
	1975–1979	2	0	0	1	0	3	4	1	1	1	10
Total		51	68	7	59	23	208	31	33	20	37	329

Note: Survey by Electronic Industries Association of Japan in 1979–80.

Source: Electronic Industries Association of Japan, *Report on the Directions and Effects of Internationalization in the Electronic Industry*. 1980 (in Japanese).

ponents, generally rode with what they came to accept as the natural course of events.

Lacking experience abroad and constrained by capital export resrictions at home, Japanese manufacturers did not rush to establish their own subsidiary assembly plants in Hong Kong. As this new intra-regional pattern of production and trade began to take shape, Japanese consumer electronic firms were content to reap the rewards of the steadily increasing flow of component orders and the demand for other Japanese electronic products which was primed by this flow. And, indeed, those rewards were considerable. Not only did the development of the Hong Kong industry mean expanded component sales, but, as a result of the larger market for semiconductors in particular, it was possible for Japanese firms to move more rapidly down the learning curve and obtain advantage of scale production which ultimately lowered the costs of their own finished product output.

Moreover, this pattern, once perfected in Hong Kong, was readily adaptable to Taiwan and later to Korea, Singapore and Malaysia in that chronological order. As each successive generation of radios was introduced by Japanese manufacturers and firmly established in the market-place, production developed in off-shore assembly sites which grew in rapid profusion, mainly under the initiative of East Asian entrepreneurs. But in the late 1960s a major change took place in this triangular relationship among the United States, the off-shore assemblers and the Japanese suppliers of components. Both U.S. and Japanese consumer electronics manufacturers began investing in their own production facilities in Taiwan, Korea and Singapore.

Industrial Dynamics: Stage Two

Five factors conspired to bring about this fundamental change.

(1) By the early 1960s, Taiwan and Korea had shifted the accent of their industrial policies from import-substitution to export-oriented production, identified the electronics industry as a key export industry for development, and adopted liberal foreign investment policies which included the establishment of export processing zones with attractive incentives for foreign firms. Then, in 1965, Singapore emerged as an independent city-state determined to assure its survival by following much the same route as that taken by Hong Kong, transforming its entire territory into a free trade zone with an economic environment conducive to the development of export industries in high-growth sectors and to the attraction of export-oriented foreign investments.

424

(2)   In the late 1960s, labor shortages began to affect both Japanese and Hong Kong electronics manufacturing, forcing wages steadily upward. As a result, comparative advantage in the labor-intensive assembly stage of production shifted first to Taiwan, where a substantial local electronics industry had already developed and the Kaohsiung Export Processing Zone was established in 1966, and then to Korea and Singapore.

Once again, U.S. and European mass merchandisers played the catalytic role in this shift. Competing fiercely for market shares in their respective countries, their response to changes in comparative advantage in the production of price-sensitive items at the lower end of the product spectrums for radios, taperecorders and an increasing range of audio equipment was rapid. And since they were accustomed to placing large orders on an annual basis, they enabled local entrepreneurs in Taiwan and Korea to move to the export stage of production without heavy investment in design facilities, production technology or overseas marketing organizations.

As in Japan and Hong Kong, mass merchandisers provided their own designs and engineering support for subcontractors to assure quality production, as well as the ready-made marketing channels and, in some cases, logistic services to expedite regular services. Once again, aggressive and highly efficient purchasing by the growing number of mass merchandisers in North America and Europe, where the service revolution was in full swing, assured the introduction of products from these new manufacturing centers. Aggressive marketing strategies of East Asian merchants or manufacturers played a secondary and subordinate role.

(3)   In fact, East Asian electronics manufacturers were surrogates in a continuing competitive struggle between mass merchandisers and manufacturers in the principal markets of North America and Europe. And in the late 1960s, the U.S. mass merchandisers were increasing their shares of the market for monochrome television receivers made under subcontract by leading Japanese manufacturers such as Toshiba, Sanyo and Sharp.

The strategic response of U.S. consumer electronics manufacturers to this competition was quite different from the case of radios and taperecorders. Radio production they were prepared to forfeit, and the taperecorder market, which seemed to offer little prospect for growth, was ceded almost without contest. But too much was at stake in television technology. To obtain the full benefits of the incipient market for color television receivers, U.S. consumer electronics firms had to retain their share of the monochrome market at all costs. And to do this they were prepared to move their production to offshore facilities, mainly in Taiwan. There they were able to combine skilled and strike-free low-cost

425

labor with Japanese and Taiwanese components to compete effectively with receivers which mass merchandisers were importing from Japan. While U.S. television receiver makers were in this way able to extend somewhat the return on their investment in monochrome television technology and sustain a lackluster rear-guard defense against competition from imported Japanese receivers, in so doing they provided the Japanese and Taiwanese industries with a broader market which resulted in lower costs for television components used in their receivers.

(4) After the liberalization of capital exports by the Japanese Ministry of Finance in 1968, Japanese makers responded to U.S. monochrome television production in East Asia by moving their own production off shore to Korea, Taiwan, Singapore and Malaysia. Although leading Japanese manufacturers had already established a few import-substitution joint ventures to serve local markets in South-East Asia, beginning in Thailand, then in Taiwan, in the early 1960s, the new Japanese export-oriented production in the NICs was either in wholly-owned subsidiaries or in subcontracts to local manufacturers. Local component content for this production usually had to be sufficient to assure the advantages of local certificates of origin for finished electronic appliances. And to assure the quality of locally supplied components, Japanese component manufacturers began transferring some of their production to these countries, especially after the combined effects of the revaluation of the yen and higher energy costs had eroded comparative advantage of Japan-based output.

(5) As U.S. and Japanese television production gained momentum in Taiwan and Korea, local producers such as Tatung, San Po, Gold Star, Samsung, and Taihan began manufacture for export as subcontractors for mass merchandisers or those Japanese and U.S. manufacturers which had not invested in their own offshore production facilities. As a result, by 1978 offshore monochrome television manufacture in East Asia accounted for 44.1 percent of total world production, surpassing by far the 22.7 percent of Europe and North America combined.

Throughout the 1970s this trend continued, with some important differences. As Japanese color television sets sold directly or supplied to mass merchandisers gained market share in the North American markets, U.S. manufacturers began supplementing or replacing their offshore production of monochrome television output with color television assembly. But in this case, the Japanese did not respond as they had in moving monochrome production offshore; rather they chose to automate manufacturing processes in Japan, moving very rapidly towards the production of all-solid-state receivers, which became feasible with the development of large-scale integrated circuitry.

Table 3. Growth of Electronics Industry in Asia, 1970–1979

Year	Japan		Korea		Taiwan		Hong Kong		Philippine		Thailand		Malaysia		Singapore		Indonesia	
	GVA (Bil. Yen)	EX	GVA (M. of US$)	EX	GVA (M. of NT $)	EX	GVA (M. of US$)	EX	GVA (M. of US$)	EX	GVA (M. of Baht)	EX	GVA (M. of US$)	EX	GVA (M. of SD)	EX	GVA (M. Rp)	EX (M. of US$)
1970	3,397	862						269	269	0	314		10.6		137	215		
1971	3,322	997	138	89	22,586			321			339		13.9		184	327		
1972	3,788	1,146	208	142	33,335	19,538		409			400		25.7		349	632		
1973	4,555	1,330	463	369	51,468	29,446	213	546			475		76.9		509	1,143		
1974	4,782	1,622	814	518	62,409	37,545		686			602	72	112.2	83	656	1,650		
1975	4,329	1,682	860	582	57,528	28,132		581	1,447	47.2	676	49.1		126	593	1,480		
1976	5,803	2,694	1,303	992	75,899	48,750	388	874	1,586	84.1	1,164	30.0		213	807	2,099	28,983	1.1
1977	6,012	2,682	1,758	1,051	87,190	56,445	497	940	1,801	113.3	1,516	1,160.9		337	930	2,418	44,691	0.5
1978	6,379	2,639	2,350	1,396	122,170	74,242		1,027	1,223	202.7	1,915	2,204.7		733	1,162	2,913	53,615	26.1
1979	7,050	3,154		1,845	147,379	94,958			1,545	313.9	2,671	3,114.6		782	1,600	3,429	58,177	75.9
Growth rate '70-'75 (%)	5.0	14.3	58.0	59.9	26.3	12.9	—	16.7	40.0	—	16.6	—	80.4	—	34.1	47.1	—	—
'75-'79 (%)	13.0	16.1	39.8	33.4	26.5	35.5	23.6	20.9	1.7	60.6	41.0	182.2	—	57.8	28.2	27.6	26.1	70.5

Source: Institute of Developing Economies

Using modular designs which substantially reduced the number of components and simplified assembly, these new automated systems enabled Japanese manufacturers to produce color television sets which were both of higher quality and price-competitive with those produced in Taiwan and Korea, where more labor-intensive technology is used. At this point, unable to compete with the Japanese through relocation to lower-cost labor locations, and ill-prepared to follow the Japanese in the new production systems, the U.S. television industry responded with legal and political action to impose restrictions on imports from East Asia.

Industrial Dynamics: Stage Three

The reaction of the Japanese industry was swift. Having read the protectionist handwriting on the wall correctly, leading Japanese manufacturers began shifting their advanced production systems to the United States. Sony led the way with their wholly-owned production facilities in San Diego; Matsushita followed with the acquisition of the faltering Motorola television division; and Sanyo responded to Sears, Roebuck's offer to take over Whirlpool's share in their joint venture at Warwick, Arkansas. Once again, U.S. mass merchandisers, concerned for continued supply from price- and quality-competitive East Asia manufacturers, played a key role in this most recent stage of the industry's internationalization of production. Other major Japanese manufacturers (Toshiba, Hitachi, Mitsubishi and Sharp) all moved production to North America, to be followed by both Taiwanese and Korean makers, a strategy which has been readily adopted by all three industries as a response to European protectionism as well.

Significantly, however, this massive move of East Asian color television production to principal markets in North America and Europe has not meant a reduction of output by East Asian manufacturers at home. In 1981, not only did Japan, Korea and Taiwan record their highest output ever, but Hong Kong emerged as a member of the export league, while output rose sharply in Thailand, Indonesia and Singapore.

Meanwhile, throughout the 1960s, 1970s and into the 1980s a dynamic interaction in semiconductor production has followed much the same pattern as color television. U.S. manufacturers moved labor-intensive assembly and packaging operations to East Asia, early in the product life-cycle to obtain market share advantages, while Japanese semiconductor makers responded by developing more automated production at home, with all the critical advantages of quality and reliability which these improved production technologies entailed.

428

As a result, beginning with 16K RAM devices in the late 1970s, Japanese semiconductor manufacturers penetrated deeply into the world markets, supplying as much as 30 percent of demand. Then, with the next generation of 64K RAM devices, production of which Japanese makers succeeded in perfecting in advance of the U.S. industry, that market-share was dramatically boosted to a reported 70 percent, and Japanese investments in U.S. and European production were boosted in anticipation of protectionist moves by the respective industries.

In Japan itself, production of integrated circuits climbed from 35.0 to 48.9 percent from 1978 to 1980, declining to 20.8 percent in 1981 due to sluggish exports of devices to recession-ridden Europe and the United States and stagnation of demand for calculators and watches. Also, following the pattern of color television production, local entrepreneurs in Taiwan, Hong Kong and Korea began to mount their own fully-integrated production facilities for integrated circuit manufacture; in the case of Taiwan, doing so with wholly indigenous technology.

The stage is now set and the curtain has risen for the next act in this remarkable drama. Mass merchandisers have begun shifting their purchases of labor-intensive consumer electronics products to the less developed countries of the region, with Japanese manufacturers of these products moving facilities to these areas in response to changing patterns of comparative advantage. Reacting to these same changes, manufacturers in the NICs are steadily shifting to higher value-added production using more sophisticated and automated systems. Output of telecommunications equipment, computers and peripherals, word-processors and copiers has been gaining momentum at varying speeds in each of the four countries. And, once again, both U.S. and Japanese manufacturers are responding to incentives offered by these countries to locate offshore production of high-technology products, and even some development stages, in these rapidly advancing countries.

Technology Flow

East Asian electronics industries have been successively vitalized by technological flow, first from the United States, but then even more substantially from Japan. Quite contrary to paradigms which dominate discussion of technology transfer in most international fora, the electronics technology that has been so rapidly diffused throughout East Asia has been massive, appropriate and remarkably cheap.

Moreover, East Asian experience destroys two other cherished assumptions of much of the rhetoric and literature on technology

429

transfer. Most important, it is clear from Japanese experience of technological transfer for over a century, experience which is confirmed many times over in other East Asian countries, that the flow of technology is ultimately governed by the recipients and not by the original source of technology. And, those countries which understand this best also have learned that technological dependence does not perpetuate economic underdevelopment, but rather vitalizes and speeds the processes of industrialization.

Clearly, the differential flows of technology in the East Asian region can be explained only by varying attitudes and policies in the various countries themselves, and not at all by the strategies of American, Japanese or European multinationals and mass merchandisers. Technology flow has been inhibited by government policies in those countries which, jealous of their 'economic independence', restrict foreign investments and pursue the indigenization of industrial enterprise. Yet, remarkably, it is representative voices in just these countries which protest the loudest that it is the foreign firms they exclude or otherwise restrict that are responsible for the unsatisfactory flow of technology, an ambivalence which can be afforded only by those countries possessing sufficiently abundant resources to assure their survival.

No such protestations are heard from the NICs, where governments share with private entrepreneurs the understanding that survival is intrinsically linked with the mastery of increasingly higher levels of technology. Governments in these countries also, like the Japanese before them, have understood that the flow of technology is ultimately determined by private entrepreneurs, foreign or domestic. And in their eagerness to overtake the Japanese, governments of the NICs have been even more liberal in their policies regulating foreign investments. Thus western and Japanese multinational corporations have become major vehicles for much of the technology flow to these countries, and the NICs have learned how to foster the most rapid and extensive technological assimilation and diffusion by their indigenous corporate systems. All have managed to find the technology they needed, at costs they could easily afford.

Indeed, much of this technology has been obtained royalty-free. The *quid pro quo* has been efficient and quality production at the lowest possible cost, with prompt and reliable delivery. Just as European basic technology has flowed to North America where production systems and factor costs combined to assure higher efficiency in application, so the technology has streamed in an ever-mounting torrent from North America and Japan to the NICs and those less developed countries which can effectively manage dynamic changes in comparative advantage.

Channels for these flows have been multitudinous. From Japan,

Table 4. Major Semiconductor Manufacturers in Southeast Asia (Semiconductors and ICs)

		Korea	Taiwan	Hong Kong	Philippines	Thailand	Malaysia	Singapore	Shipments in 1978 (Million US$)
U.S.	Texas Instruments		X		X		X	X	920
	Motorola	X		X	X		X	X	680
	National Semiconductor			X	X	X	X	X	420
	Fairchild	X		X				X	380
	Intel				X		X		300
	RCA	X	X	X			X		240
	Signetics	X	X						205
	General Instruments		X						n.a.
Holland	Philips		X	X		X			450
W. Germany	Siemens						X	X	200
Italy	SGS						X	X	n.a.
Japan	NEC	X					X	X	570
	Hitachi		X	X			X		440
	Toshiba	X					X		415
	Matsushita							X	210

Sources: *Denshi Shijo Yoran* (1980), *Denshi Kogyo Nenkan* (1979).
United Nations, "Transnational Corporations in the Consumer Industry of Developing ESCAP Countries," ESCAP, 1978.
UNCTAD, *International Subcontracting Arrangements in Electronics between Developed Market Economy Countries and Developing Countries*, 1975.

manufacturers eager to supply components to assemblers have been prepared to include production know-how as part of the sales/service package. Likewise, machinery makers and materials suppliers have been equally ready to supply manufacturing technology for electronic components as a condition of sale. Massive amounts of so-called intermediate technology, as well as more advanced technology, have flowed to indigenous entrepreneurs at very little cost.

Similarly, Japanese manufacturers as well as leading trading companies and mass-merchandiseres have provided design and production technology to local subcontractors to assure the most efficient and highest quality production. Combined with the technology which flowed freely through commercial channels, including that derived from reverse engineering of Japanese products by ingenious local engineers, this direct flow of technology in subcontracting arrangements spurred the rapid proliferation of indigenous electronic enterprises in each of the NICs, and more recently in the lesser developed countries as well.

In addition, Japanese manufacturers have provided more sophisticated technology under licenses assiduously sought by entrepreneurs and encouraged by governments in the NICs. Between 1962 and 1980, Japanese manufacturers were a major source of know-how agreements signed by Korean licensors. And a similar flow of know-how was primed by leading Taiwanese manufacturers seeking to obtain parity with Japanese manufacturers in advanced electronics technology.

Equally important, electronics industries throughout the region were vitalized by technology flows accompanying Japanese foreign direct investments in electronics manufacture. From 1960 to 1979, Japanese firms established as many as 208 wholly-owned or joint ventures in offshore East Asia, training vast numbers of technicians and skilled workers, many of whom now provide the technological underpinnings of indigenous enterprises. To the extent that these ventures rely on local subcontractors for components and materials, they have further contributed to the diffusion of technologies through the multiplier effect of their investment.

Parallel to the flow of technology from Japan to other countries of East Asia, there has been a substantial stream from the United States and Western Europe. As indicated above, U.S. mass-merchandisers, and to a lesser extent those of Europe, have provided valuable, timely, and appropriate increments of design and production technology for the manufacture of consumer electronics goods. Larger mass-merchandisers maintaining regional purchasing offices staffed with multinational engineering teams served to relate the needs of the market to the most efficient production facilities, identifying skills and organizational capabilities in the various countries of the region, thereby obtaining optimal prices, quality and delivery conditions. And, in addition to invaluable production technology, these merchant houses also provided important tutorial functions in logistic and other managerial technologies.

Unlike foreign manufacturers investing in these countries, merchandisers have no vested interest in any particular source of supply. So long as a supplier can deliver the required products on time at the most competitive prices, these merchant houses are prepared to assure market access and considerable technical assistance. But once comparative advantage of a given country declines, they are quick to switch their orders to new suppliers. As a result, they have served to transfer labor-intensive technology to successive tiers of developing countries, thereby speeding the process of diffusion throughout the East Asian region from Japan to the NICs and then to the LDCs. Moreover, at the same time they play an important role in spurring the move to successively higher stages of technology by up-grading their purchases in Japan and the NICs from radios and taperecorders to color television receivers, video-taperecorders and personal computers.

One effect of this pattern of procurement by mass merchandisers has been to force the pace of U.S. and European investments in consumer electronics production in East Asia, thus priming the further flow of technology to the region. Large scale production units introduced by RCA, Zenith, Philco, Admiral, Philips and Grundig have served as conduits of assembly and component manufacturing technology to East Asian countries over the past two decades.

Quite independently of this widening channel for consumer electronics technology, and beginning even earlier, U.S. semiconductor manufacturers, attracted by relatively low-cost, skilled labor and auspicious business environments, have provided an important training ground for technicians and managers at the frontier of advanced microelectronics.

Following Fairchild's lead in Hong Kong in 1963, eight major U.S. and three European semiconductor manufacturers established a total of 34 export-oriented production facilities in East Asia during the 1960s and 1970s to reach a total output in 1978 valued at more than US$3,795 million. By the end of the 1970s, governments of Malaysia, the Philippines and Thailand had succeeded in attracting a considerable portion of this production to their countries, broadening the flow of advanced microelectronics manufacturing technology throughout the region. Only Indonesia remained outside the mainstream.

Although some pundits have heavily discounted the value of the flow of technology accompanying these investments, the recent emergence of local manufacturers of integrated circuits most certainly

433

owes much to the experience gained by managers and skilled workers in the plants of foreign firms. Moreover, since this industry is marked by a relatively high labor turnover, especially in Hong Kong and Singapore, there have been times when foreign manufacturers had the impression that they were indeed in the business of training skilled workers for the local industry.

Undoubtedly, this broadening of skilled manpower in the NICs has served to attract further foreign investments by U.S., Japanese and European computer and telecommunications equipment manufacturers to the region. These investments in turn add to the flow of advanced technology, preparing the way for local manufacture of an increasing array of industrial electronics equipment.

The object lesson of this flood-tide of electronics technology flow to and within East Asia is clear. Technology transfer, like communications, is ultimately governed by the power of the receiver. Just as a powerful radio receiver can pull in the weakest broadcast signal, so imaginative and robust entrepreneurs tend to locate and attract the needed technology by mastering the complementary elements of efficient production. Indeed, as the ingenious Yankees of an earlier age demonstrated, the ability to apply technology tends to attract it as if by magnetic force. Those societies and public policies which assure the most salubrious conditions for the development of entrepreneurship, national and foreign, have, by induction, developed the necessary magnetic pull to sustain the massive flow of technology needed for continued rapid growth of the electronics industry.

System and Synergy

Since its inception in the mid-1950s, the East Asian elecronics complex has evolved into an intricately interactive system comprising three tiers of constituent countries: Japan, four NICs and four developing countries. The fundamental economic reality which has moulded the system is the dynamic complementarity of its constituent countries, a complementarity which is induced by competition for the same global markets. As comparative advantage shifts within the region, export-oriented manufacturers have no alternative but to rapidly abandon production which is no longer competitive abroad, developing new products and technology suitable to the changing costs of production. The result has been a thrusting, expanding highly-integrated system the whole of which is greater than the sum of its parts.

At the center of the system, providing its main technological drive, is Japanese industry. From Japan flow technology, components and processed materials critical to the electronics industries of the NICs and LDCs, all of which eventually emerge as competitors of Japanese firms producing finished consumer and industrial goods as well as the suppliers of components and materials themselves. Competition from other countries in the region in turn forces the pace of technological advance in Japan, where the industry must stay ahead to survive. Following Japan, the NICs also become sources of technology, components and materials to the LDCs as production know-how of electronics industries in the respective countries becomes increasingly sophisticated and economies of scale and learning are attained through an expanding presence in world markets for consumer and industrial electronic products.

As the system has no political boundaries, it is an expanding economic reality. New entrants are now poised for entry, with potentially major consequences for synergism of the system and its global impact. The Chinese electronics industry on the Mainland is poised on the launch-pad ready for take-off into global orbit. Export of consumer products and computers has already begun, with important assistance from both Japanese and Hong Kong industries. At the same time, on the periphery of the system, Sri Lanka is seeking entry which, if successful, would extend that system further westward into South Asia where the Indian industry has greater potential.

Seven significant features distinguish the East Asian electronics system.

Integration and diversification

The East Asian electronics industry, with its broad base in consumer electronics is the most highly integrated and diversified electronics complex in the world. The integration of the system provides its constituent firms with critically important external economies not readily available in either North America or Europe. Broad-based consumer electronics production sustains rapid advance in semiconductor output and technology as well as the essential underpinnings for an innovative passive component sector. Moreover, as the distinction between industrial and consumer electronics hardware diminishes with successive advances of semiconductor technology, the East Asian consumer electronics industry has been steadily shifting its thrust to the production of computers, copiers, communications equipment, and robots.

As a result, the East Asian electronics industry is transposing its lead in consumer electronics into pre-eminence in one industrial electronics product group after another.

435

National industrial policies

The pace of development of the various electronics industr es of the region, the rapidity with which they have moved into export production and the time-span of the successive phases of their 'internationalization' has depended in large part upon national industrial policies of the respective countries. Japanese industrial policy-makers led the way in 1957 by identifying electronics as a key sector of the future industrial structure, replacing heavy and chemical industries in the last quarter of this century. Hong Kong, with less interventionist mechanisms, followed suit almost immediately. At the same time other East Asian countries introduced industrial policies designed to substitute imports of consumer electronic products and their components with local production.

Since the early 1960s they have, one after another, shifted to export-oriented industrial policies with electronics targeted as a priority industry for promotion. Measures have differed, depending upon conditions in each country, but in general they have provided a package of benefits designed to channel scarce resources into the electronics industry. And to assure the necessary capital, technology and foreign market access, governments have progressively opened the door to foreign manufacturers, providing a variety of benefits, often on a par with domestic firms. In addition, most countries have adopted specific export promotion measures, including the establishment of export processing zones. These measures, coupled with incentives for investment of both domestic and foreign capital in key industries. were at first most successful in Taiwan, then South Korea and Singapore, and more recently Malaysia, where there has been a rapid development of the electronics industry during the past two decades.

International catalytic agents

East Asian countries, including Mainland China, have come to share the recognition that, properly managed, foreign investments in the electronics industry bring not only the advantages of additional capital, but also the benefits of new technology, managerial skills and marketing networks which assure accelerated growth and rapid technological advance. Japanese, American and European electronics manufacturers have responded to the resultant incentive in a competitive race to develop consumer electronics, semiconductors and industrial electronics production, serving as an important catalyst to the industry's development.

In South Korea, at the end of the 1970s, there were more than 200 joint ventures between foreign and domestic partners in electronics manufacture, almost one-third of the total number of enterprises in the

Korean industry. Singapore and Malaysia followed with more than 150 foreign firms operating wholly-owned electronics manufacturing operations in each country, while Taiwan and Hong Kong both had from 70 to 80 and other countries a smaller number. Japanese enterprises have led the field in the number of investments in the NICs and LDCs, with 290 as of 1979. Only in Hong Kong and the Philippines have American investments exceeded those of the Japanese electronics industry, while European investors are significant mainly in Singapore and Malaysia.

Recent surveys revealed that the ratio of national to foreign investment in the electronics industry stands at approximately 7:3 in Taiwan, 6:4 in Thailand, 4:6 in Indonesia and 3:7 in Malaysia. In Singapore, where the ratio of national to foreign enterprises is approximately 2:3, the capital ratio is estimated to approach that of Malaysia. These investments have not only spurred the pace of electronics production through the transfer of technology and value-added to the region, they have also provided important channels for exports from East Asia to the advanced countries and added an important integrative force for the regional electronics industry.

Global strategies

With the exception of those firms producing for domestic markets in less developed countries of the region, which account for a small and diminishing share of the region's total output, East Asian electronics manufacturers typically pursue global strategies. Since these manufacturers are subjected to the 'perennial gale of competition' in a marketplace beyond the potentially protective reach of their respective governments, they must compete by developing flexible responses to a wide range of different and changing market preferences. Ultimately, this requires a finely-tuned organizational effort to assimilate or develop new technologies. It also means that each producer must necessarily source its various material supplies, components and factors of production where they can be obtained at optimal advantage, transferring production internationally as comparative advantages shift and artificial impediments are introduced to change the external economies of manufacturing and logistics. It is precisely these global strategies which render the entire system so highly interactive and assure its powerful synergism.

Global structures

Each industry in the region has undergone the same three-phased development of global structures. In the first phase, characterized by almost total reliance on foreign distribution channels, resources are concentrated on production, the mastery of technology and the development

437

of managerial skills required for mass production. As managerial and organizational skills are developed, and financial resources are accumulated, leading firms of the first phase seek to increase their competitive strength and their share of the value-added by integrating forward into their main world markets, with distribution networks, selling products of their own brand in competition with continued supplies of buyers-branded products to mass merchandisers and manufacturers in those markets.

Parallel to the development of international marketing organizations, management of production technology has been developed to a high level of efficiency and sufficient capital resources have been accumulated to render feasible the internationalization of production. Moreover, at about the same time there have been parallel increases of factor costs at home and protectionist pressures abroad which force the relocation of production closer to the market. Depending on cultural differences, political considerations, market size, financial resources and managerial capabilities, these overseas production operations may be either wholly-owned subsidiaries or joint ventures with local capital.

Competition for world markets

Marcoeconomic trade flows indicating that Japanese and other East Asian industries are competing directly for world markets, while true, do not depict the total reality of that competition. Industries do not compete, manufacturers and merchants do. Since historically both Japanese and East Asian exports of electronic products have developed mainly through OEM channels, trade flows are determined initially and largely by competition between U.S. and European manufacturers and mass merchandisers.

This phenomenon is clearly shown by the ascendancy of South-East Asia over Japan as the world's leading supply center for both popular household electric appliances and semiconductors. In 1978, eight Asian industries taken together outstripped Japan in output and exports of radios, monochrome television sets, radio-cassette recorders and integrated circuits, even though those industries had much less well-developed international production and sales organizations.

Moreover, since many East Asian manufacturers are competing for relatively few markets, it is the buyers in those respective markets, through their enormous market power, which ultimately set the prices and other conditions of sales as well as the character and quality of products exported by those countries. This clearly suggests that much of the rhetoric about East Asian invasions of U.S. and European markets has little relation to reality, which no doubt explains why public policies designed to deal with the perceived problem have had little positive effect.

Rapid technological advance

Since East Asian electronics manufacturers are competing fiercely for the same world markets, those which are losing comparative advantage because of rising factor costs are forced to upgrade technology to survive. This vital linkage between survival of firms, and ultimately of entire industries, and technological advance has become a powerful force in the electronics industry throughout the region. A perpetual 'catching up' syndrome operates to exert increasing pressure on the Japanese industry which must accelerate the pace of innovation to survive the steady technological sophistication of electronics technology in the NICs.

The four industries of South Korea, Taiwan, Hong Kong and Singapore are themselves poised on the threshold of an era of rapid progress in high-technology in their struggle for survival with the emergent export prowess of manufacturers in the LDCs and on the Mainland of China. Both Hong Kong and Taiwan and already producing computers, and are expected to become major centers for computer production in the near future.

The implications of this rapid pace of technological change were cast in stark relief in July 1982, when Beijing Computer Industry Corp. and the Beijing branch of the China Electronics Import and Export Corp. established a joint venture with Hong Kong partners to export Chinese minicomputers and import technology. Since Hong Kong is already facing a shortage of 800 computer engineers and technicians, by 1983 competitively priced Chinese minicomputers and programs are expected to flow through commercial channels of which the Crown Colony constitutes an important junction.

As China enters the East Asian electronics system, the complementarity which has been one of its distinguishing features is further reinforced. The tempo of technological change will be quickened with the extension of the regional agglomeration to the Mainland, intensifying the frequency of interaction and the synergistic effects of the system on its various constituents. If the world has seldom seen such a remarkable performance of sustained industrial advance as that evidenced by the East Asian electronics industry in the past quarter century, all signs are that this is but a beginning of things yet to come. Since this performance is propelled by a rapid pace of technological change, with a new accent on creativity, prospects are great that the East Asian electronics industry will be a leading source of innovation in the future. And, if the past is to tell us anything, leading firms around the globe will find it increasingly necessary to participate in this process in order to survive the rising tide of competition.

INDEX

acquisitions, 12, 14, 15, 20, 144, 151, 163, 288
Administrative Management Agency, Japan, 341
Administrative Procedures Act, U.S., 146
Admiral, 15, 433
AEG-Telefunken, 38
Advanced Device Technology, Inc., 18
Advanced Micro Devices, 73, 74, 218
Alcoa NEC Communications, 188
All-Japan Telecommunications Workers' Union (Zendentsu), 370
Alps Electric, 407, 408
Amana Refrigeration, 165, 166
Amdahl, 229, 234, 236, 238, 239, 244, 245, 262, 263
Ando Electric, 215, 221
Anelva, 199, 217, 218, 219
anti-dumping action, U.S.: television components/sets, 157, 159, 160, 163, 168-170; microwave ovens, 157, 165-167, 170
anti-trust law, U.S., 182
anti-trust litigation, 141, 163-164, 169
Applied Materials, 218, 220
Aron, Paul, 320
array processors, 255, 256
Arthur D. Little (Japan) Inc., 371
artificial intelligence, 255
Arvin, 15
Asahi Chemical, 59, 298, 299
Asahi Research Center, 298
Asian electronics industries, 6-20, 411-440
Association of Home Appliance Manufacturers (AHAM), U.S., 157, 165
AST, 394
Atlas Electronics, 18, 422
ATT, 4, 364, 369, 393, 395, 420
audio makers, 180
auto industry, 27, 83, 130
automated warehouses, 305

Babbage, Charles, 256
Baker, Donald, 163
banking terminals, 350

BASF, 240, 245, 265
Beijing Computer Industry Corp., 439
Bell Laboratories, 5, 197, 357, 375, 420
Bell system, 368
Betamax group, 178
Boss, William, 152
Boston Consulting Group, 174, 175
Bridgestone, 287
Brookhaven National Institute, 290
Bucy, J.Fred, 197
BUNCH (Burroughs, Sperry Univac, NCR, Control Data and Honeywell), 245, 253
Burroughs, 362, 374

C. Itoh, 393, 397
cable television, 330, 351
CAC Taiwan, 285
CAD/CAM, 246, 253, 321, 322, 323; software, 322, 323, 324; systems, 305, 312, 313
CADAM (computer graphics augmented design and manufacturing), 321
CAE (computer-aided engineering), 322
calculator industry, 148-150
calculators, 7, 8, 29; imports, U.S. surcharges, 149; output, 97; prices, 97; revolution, 96
Calma, 321
Cambridge University, 269
Canadian Telidon System, 376
Canon, 30, 42, 45, 117, 199, 207, 214-216, 218, 223, 380, 383
Canon Sales, 218
capital costs, 65, 94, 111, 325; Japanese, 131-133, 136, 209; U.S., 132-133, 209, 211
capital markets, Japanese, 136; U.S., 131
capital supply, 94, 134-136
CAPTAIN Service Co., 375
CAPTAINS (Character and Pattern Telephone Access Information Network System), 334, 375, 376, 398
career advancement, 90
Carey, Frank, 233
Carter, Jimmy, 162

441

INDEX

National Research and Development Corporation, 268
National Semiconductor, 16, 102, 236, 263, 431
National Superspeed Computer Project, Japanese, 255, 266, 268
NC (numerical controls), 28; applications, 317; machine tools, 304-325, *passim*; technology, 7
NCR (National Cash Register), 8
NEC, 45, 59, 63, 73, 84; communications equipment, 69, 337, 342, 343, 356, 358, 360, 362, 369, 372, 373, 376; computers, 6, 42, 44, 228, 229, 231, 237, 238, 244, 246-248, 249-252, 260-261, 263, 265-268, 279, 280; education and training, 116, 125-126; facsimile, 380, 383, 384; factory automation, 312, 322; HDTV, 188; R&D, 33, 104, 108, 115-116; semiconductors, 80, 95, 96, 103, 202, 211, 431; semiconductor equipment, 199, 221; software, 266, 279, 284-285; SWQC information center, 282; VANs, 393, 395-397; VLSI technology, 8, 80, 206
NEC Information Services Ltd. (NEIS), 279, 393, 396
NEC-Toshiba, 80
Network Business Association (NBA), 393, 395
network information services industry, 391-399
Nevin, John, 151
New Japan Securities, 289
new media, 189
new product development, 79
NHK (Nippon Hoso Kyokai, Japan Broadcasting Corporation), 183, 184, 185, 187, 189
Nichimen, 376
NICs (newly industrializing countries), 414, 417, 426, 429, 430, 432, 433, 434, 435, 439
Nihon CDC, 297
Nihon Keizai Shimbun, 58, 60, 289, 291, 296, 297
Niigata Engineering, 305
Nikon, 215, 216, 223
Nippon Business Consultants, 280, 393
Nippon Columbia, 180

Nippon Denso, 30
Nippon Electric Company Ltd. *See* NEC Corp., *new name*
Nippon Express, 393
Nippon Hoso Kyokai (NHK), 357
Nippon Kogaku, 215
Nippon Sheet Glass, 357, 358
Nippon Steel, 59
Nippon Systems Development, 279
Nippon Telegraph and Telephone Public Corp. *See* NTT
Nippon Univac, 45, 250, 395, 396
Nippon Univac Information Systems Kaisha, 395
Nissan Motors, 59, 319
Nissei Sangyo, 215
Nisshin Electric, 215
Nisshin High Voltage, 220
Nitsuko, 372
Nitto Electric, 188
Nomura Computer System, 278
Nomura Research Institute, 43, 113, 191, 297, 307
North American Philips, 163
Northern Telecom, 364
Noyce, Robert, 132-133, 196
NTT, 228, 231, 353-365, 368-376; changing role, 337-346; competition, 341; database services, 291, 295, 296; data communications monopoly, 340-341; development program, 337; facsimile, 384, 385, 387; labor-management relations, 370; networks, 331, 333, 334; privatization, 353, 368-369; procurement, 338-342, 344-345; services, 339; strategies, 374; suppliers, 337, 340-342, 368-376; R&D, 50, 80, 81, 214, 221, 222, 268, 337, 375, 385; VANs, 288, 391, 392, 393, 394, 398; VLSI technology, 8
NTT family, 340, 368-376
Nyquist, 184

OEM supply, 362
office automation (OA), 298, 401-408; market, 45, 251, 408; sales, 407-408
offshore assembly, 17; investments, 103; production, 12, 13, 14, 15, 16, 412
Ohira, Masayoshi, 181
Ohkubo, Shigeru, 285

451

oil crises, 25, 37, 43, 83, 84, 98
Okamoto Seisakusho, 43
Okamura Corp., 311, 404, 405
Oki Electric, 211, 229, 42, 250, 251, 267, 337, 343, 362, 372, 373, 380, 383, 384, 395
Oki Electronics of America Inc., 343
Okuma Machinery Works, 324
Olivetti, 240, 241, 245, 263-9
Olympic Fishing Tackle Co., 288
Omi, Hanzo, 247
Omron, 117
optical fiber, 7, 347; applications, 356, 359; cable makers, 358; cable system, 342; communications equipment, 362, 363; links, 329, 334; optical fiber communications systems, 356-359; communications systems, applications, 356-357; industry, 357-358; production, 358; sales, 356-357; technology, applications, 353-354; technology, licensing, 358; transmission costs, 359; transmission systems, 334; —/satellite communications links, 359
optical lithographic equipment, 214
optical readers, 403
optical technologies, 331
orderly marketing agreements (OMA), 14-15; color TV, 143, 145, 162, 170, 172, 173, 174

PAL system, 183
Pacific Basin, 7, 20, 26, 79
Packard Bell, 15
packet networks, 348
packet radio, 348
packet switching, 376
Panafax Corporation, 382
parallel processing, 255, 256
Patent Office, Japanese, 56
patent pools, 79
patents, 46, 49, 76, 420, 421; France, 57; Japan, 56-60; semiconductor, 206; Switzerland, 57; U.K., 57, 58; U.S., 56, 57; West Germany, 56, 57
PCM (plug-compatible machines), 235-236, 243, 245, 246, 279; strategies, 246, 247, 248, 273; vendors, 243, 244, 245

Pentel, 311
People's Finance Corp., 309
Perkin-Elmer, 215
personal computers, 42, 43, 403; market share, 111; subcontracting, 245
personnel, mobility of, 73, 74
Pharma, 393, 398
Philco Corporation, 142, 433
Philco-Ford, 16
Philippines: electronics industry, 10, 18, 421, 423, 427, 437; semiconductors, 16, 17, 418, 431, 433
Philips, 4, 5, 15, 31, 32, 147, 152, 155, 173, 178, 236, 420, 431, 433
Philips Gloeilampenfabrieken, 163
Philips, North American, 147
Pilot, 311
Pioneer Electronic, 30, 117, 172, 178, 180, 354
Pitney-Bowes, 362
Planar transistor technology, 15
point of sale terminals, 350
Prestel, 334
privatization (of NTT), 353, 368-369
product diversification, 68, 252
product life cycles, 316, 417
production: costs, 417; efficiency in, 101; higher value-added in, 65; internationalization of, 135
production systems, 79
productivity, 39, 73, 99, 100, 129, 177, 201, 258, 282, 283, 316, 317, 404, 406
profit maximization, 74
protectionism, 13, 32, 37, 41, 42, 64-65, 141-155, 157, 159, 172, 196, 428
Public Electric Telecommunications Law, Japanese, 291
Public Facsimile Network, 361
pulse code modulation, 7, 349

quality: control, 76, 199-201; standards, 198, 200; testing 199, 200; workmanship, 199
Quasar Electronics, 175
Quotation Information Center, 290, 291, 296

R&D (Research and Development),

452

INDEX

Secom Co., 368, 393
Seiko, 30, 190
Seino Transportation Co., 393
Sekimoto, Tadahiro, 116
Semiconductor Equipment Association of Japan, 225
semiconductor equipment: export barriers, U.S., 212, 213; exports, 208, 213, 214; imports, 208; Japanese market share, 224; manufacturers, 208, 215-220, 224; specification modifications, 212; technology, 212, 214-220
semiconductor equipment industry, 99, 209-225; growth, U.S., 210; growth, Japanese, 210; investments, 224; newcomers, 214, 215; structural change, 223; structure, Japan, 208; structure, U.S., 208; ties to chipmakers, 209; world leadership, 223
Semiconductor Industry Association (SIA), U.S., 196
semiconductor industry: beginnings, 420; capital spending, North America, 212; capital spending, Japanese, 212; capital to revenue ratios, Japanese, 213; capital to revenue ratios, North American, 213; competition, 196, 200; finance for, 129, 130, 133; global strategies, 202; global structures, 209; growth rates, 224; market shares, 196; pricing policies, U.S., 201; structural change, 428-429; structural differences, U.S.-Japan, 201; U.S., global market share, 149
semiconductor technology, 6, 7, 8, 13, 15, 16, 44; advances, 217; and supercomputers, 255, 261, 265, 268; applications, 204-205; development, 203-207; East Asia, 15
semiconductor test equipment, 221; R&D, 222; leading makers, 221
semiconductors, 18, 195-202; assembly, 18; automation in production, 198, 200; companies, capital spending, 210; demand, 207; forecast, 203; global market shares, 203-204; innovation, 5, 103, 207; internationalization of production, 208; manufacturers, 29-30, 260, 431; manufacturing equipment, 199; market, 16; market share, 111; overseas

production, Japanese, 202; production, 203, 208; production/capital outlays, 210; R&D, 195, 213-214; R&D investment, 197; specialization in production, 97, 205; technology flow, 206; trade, 221; U.S.-Japan trade in, 195, 196, 197, 198; U.S. production in Japan, 201; world market share, 412
semiconductors/supercomputers integrated production, 268, 269
Senate, U.S., 98, 164
sensors, 305
Setchel-Carlson, 15
Sharp Corporation, 30, 42, 45, 103, 117, 146, 157, 166, 173, 175, 187, 211, 380, 425, 428
Sharp Electronics, 175, 191
Sherman Anti-Trust Act, 143, 158
Shibaura Seisakusho, 405
Shimizu, Sakae, 88, 278
Shin-Etsu Semiconductors, 207
Shinto, Hisashi, 374
Showa Denko, 218
Showa Electric Wire & Cable, 357, 358
Siemens, 31, 50, 234, 240, 245, 262, 263, 320, 362, 431
Siemens-AEG-Telefunken, 420
Signetics, 16, 431
Silicon Products, Inc., 18
Silicon Valley, 73, 79, 97, 130, 195, 201, 218, 414, 417
Singapore: electronics industry, 10, 20, 26, 32, 74, 110, 411, 423, 425, 427, 436, 437, 439; radios, 424; semiconductors, 15, 16, 17, 418, 419, 431, 434; television sets, 13, 14, 426, 428
Small Enterprise Loan Corp., 309
Society of Motion Picture and Television Engineers, 183
software: 273-288; communications, 353; computer industry, employment in, 278; development costs, 284; engineers and programers, 278, 283, 283, 284; IBM-compatibility, 253; Japanese language problems, 279, 283; quality control, 282; supercomputers, 261; training, 282, 283, 284
software applications, 276, 278, 279, 280, 283; aerospace U.S., 276; U.S. defense, 276

INDEX

software industry: development, 273; employment in, 278, 279, 281; future growth, 278-281, 283; international cooperation, 284-286; manpower problems, 278, 279; overseas marketing, 286; production, 283; revenues, 283; size, 273; structural shakeup, 286-8; structure, 274, 275, 276, 277, 278, 287
software services, expenditures for, 278
software subsidiaries, 278, 279
software systems: engineering, 274; innovation, 274
software technology: gap, 273-274; Japanese, 286
Sony, 9, 14, 66, 67, 81, 178, 207, 287, 295, 354, 393, 421; communications, 368; HDTV, 187, 188, 190; radios, 421, 422; R&D, 50, 75, 117; television, 103, 152, 172, 173, 175, 428; Trinitron, 70, 145, 187; VLSI technology, 207;
Sony America, 175
Sord Computer Systems, 285
Southern Pacific Communications, 362
Sperry Corp., 395
Sperry-Univac, 253
Sprague Electric, 16
sputtering equipment market, 219
Sri Lanka, 435
Stanford University, 122
steel industry, 84, investments in, 130
steppers, world market, 215, 216
Stevenson, Harold, 122
storage and conversion communications equipment (STOC), 385, 386
Strasbourg Astronomical Information Center, 290
Strauss, Robert, 142, 143, 144, 162
subcontracting, 11, 18, 102
Sugiura, Hideo, 92
Sumisho Computer Service, 278
Sumitomo Chemical Co., 59, 295
Sumitomo Chemical Information Center, 295
Sumitomo Corp., 278, 368, 376, 393, 397
Sumitomo Electric Industries, 333, 356, 357, 358
Sumitomo Group, 358
Sumitomo Metal Industries, 84

Sumitomo Metal Mining, 207
Sumitomo Special Metals, 207
supercomputer development trends, 264
supercomputer industry: leadership, 258; one-company industry, 257; production, 262-265; strategies, 259, 260, 261, 262-265
supercomputers, 130, 255-269; applications, 261, 266-267; architecture, 265, 268; cooling, 265; demand, 261; future competition, 269; genealogy, 263; global market, 262; IBM-compatibility, 260, 262; market, 260; market access, Japanese, 255; new semiconductor devices, 255; prices, 261; R&D commitment, Japanese, 269; software, 261, 262, 265-266, 268; techno-economic considerations, 266; U.S.government market, 259; world market domination, 255; worldwide competition, 259
supercomputers and semiconductor technology, 255, 261, 265, 268
supercomputers/semiconductor integrated production, 268, 269
Supreme Court, U.S., 143, 161, 169
Swedish Telecommunications Agency (STA), 362
Sylvania-Philco, 15, 142, 145
System Development Corporation of Japan (SDC-J), 291, 296
System Development Corporation (SDC), 291, 297
Systex Corp., 285

TASI. See time-assigned speech interpolation
TDK, 117
TEAC, 180
TEL-Varian, 219, 220
TI-Japan, 201
TRW, 239
TRW-Fujitsu, 239
Tabai Espec, 215
Tae Han, 173
Taihan, 426
Taihan Electric Wire, 17
Taiko Electric Works, 373
Taiwan: electronics industry, 10-14, 20, 32, 74, 110, 413, 419, 423, 413, 419,

455